2e

Student Solutions Manual for Kaseberg's

Intermediate Algebra A Just-in-Time Approach

Cindy Rubash
Lane Community College

Brooks/Cole
Thomson Learning

Australia • Canada • Mexico
Singapore • Spain • United Kingdom • United States

Assistant Editor: *Michelle Paolucci*
Senior Editorial Assistant: *Erin Wickersham*
Marketing Team: *Leah Thomson, Brooke Dill, Samantha Cabaluna*
Production Coordinator: *Dorothy Bell*
Cover Design: *Christine Garrigan*
Cover Illustration: *Harry Briggs*
Print Buyer: *Micky Lawler*
Printing and Binding: *Globus Printing*

For more information about this or any other Brooks/Cole product, contact:
BROOKS/COLE
511 Forest Lodge Road
Pacific Grove, CA 93950 USA
www.brookscole.com
1-800-423-0563 (Thomson Learning Academic Resource Center)

Printed in the United States of America

10 9 8 7 6 5 4 3 2 1

ISBN 0-534-37349-6

CONTENTS

Section 1.0

1. **a.** The set of numbers for counting is the set of *natural numbers.*

   b. Two numbers that add to zero are *opposites.*

   c. Two numbers that multiply to 1 are *reciprocals.*

   d. The set of natural numbers and zero is the set of *whole numbers.*

   e. The set of numbers that can be written $\frac{a}{b}$, a and b integers, b not equal to zero is the set of *rational numbers.*

3. **a.** is less than, <

   b. to perform the indicated operation, *simplify*

   c. is greater than or equal to, ≥

   d. a collection of objects or numbers, *set*

   e. is undefined in the real numbers, $\sqrt{-1}$

5. **a.** $a + b = b + a$ is an example of the *commutative property for addition.*

   b. A statement that one quantity is greater than or less than another quantity is an *inequality.*

5. **c.** $b \cdot c + b \cdot e = b(c + e)$ is an example of the *factored form of the distributive property.*

   d. $a(bc) = (ab)c$ is an example of the *associative property for multiplication.*

   e. Either a is true or b is true or both are true is an example of a *logic phrase "a or b".*

7. **a.** The answer when two negative numbers are multiplied is *positive.*

   b. The answer when two negative numbers are added is *negative.*

   c. The answer when two positive numbers are added is *positive.*

   d. The answer when zero and any number are multiplied is *zero.*

   e. The answer when two positive numbers are multiplied is *positive.*

9. $\frac{1}{2}$ is not an *integer.*

11. $\sqrt{9}$ is not *irrational.*

13. **a.** $-15 + (-3) = -18$

    $-15 - (-3) = -15 + 3 = -12$

    $-15 \cdot (-3) = 45$

    $-15 \div -3 = 5$

Section 1.0 (con't)

13. b. $-6 + 2 = -4$

$-6 - 2 = -6 + (-2) = -8$

$-6 \cdot 2 = -12$

$-6 \div 2 = -3$

c. $\frac{1}{4} + \left(-\frac{1}{2}\right) = \frac{1}{4} + \left(-\frac{2}{4}\right) = -\frac{1}{4}$

$\frac{1}{4} - \left(-\frac{1}{2}\right) = \frac{1}{4} + \frac{1}{2} = \frac{1}{4} + \frac{2}{4} = \frac{3}{4}$

$\frac{1}{4} \cdot \left(-\frac{1}{2}\right) = -\frac{1 \cdot 1}{4 \cdot 2} = -\frac{1}{8}$

$\frac{1}{4} \div \left(-\frac{1}{2}\right) = \frac{1}{4} \cdot \left(-\frac{2}{1}\right) = -\frac{2}{4} = -\frac{1}{2}$

d. $-\frac{5}{12} + \left(-\frac{1}{3}\right) = -\frac{5}{12} + \left(-\frac{4}{12}\right)$

$= -\frac{9}{12} = -\frac{3}{4}$

$-\frac{5}{12} - \left(-\frac{1}{3}\right) = -\frac{5}{12} + \frac{4}{12} = -\frac{1}{12}$

$-\frac{5}{12} \cdot \left(-\frac{1}{3}\right) = \frac{5 \cdot 1}{12 \cdot 3} = \frac{5}{36}$

$-\frac{5}{12} \div \left(-\frac{1}{3}\right) = -\frac{5}{12} \cdot \left(-\frac{3}{1}\right) = \frac{5 \cdot 3}{12}$

$= \frac{15}{12} = 1\frac{3}{12} = 1\frac{1}{4}$

e. $2.5 + (-0.25) = 2.5 - 0.25 = 2.25$

$2.5 - (-0.25) = 2.5 + 0.25 = 2.75$

$2.5 \cdot (-0.25) = -0.625$

$2.5 \div (-0.25) = -10$

f. $-\frac{3}{4} + \frac{5}{6} = -\frac{9}{12} + \frac{10}{12} = \frac{1}{12}$

$-\frac{3}{4} - \frac{5}{6} = -\frac{9}{12} + \left(-\frac{10}{12}\right) = -\frac{19}{12} = -1\frac{7}{12}$

$-\frac{3}{4} \cdot \frac{5}{6} = -\frac{3 \cdot 5}{4 \cdot 6} = -\frac{5}{8}$

$-\frac{3}{4} \div \frac{5}{6} = -\frac{3}{4} \cdot \frac{6}{5} = -\frac{3 \cdot 6}{4 \cdot 5} = -\frac{9}{10}$

15. a. $-4 - (-3) + (-8) = -4 + 3 + (-8)$

$= -1 + (-8) = -9$

b. $12 - (-4) + (-2) = 12 + 4 - 2$

$=16 - 2 = 14$

c. $4.5 + (-3.2) - (-2.8) = 4.5 - 3.2 + 2.8$

$= 1.3 + 2.8 = 4.1$

d. $6.2 - (-1.4) + (-3.5) = 6.2 + 1.4 - 3.5$

$= 7.6 - 3.5 = 4.1$

e. $2\tfrac{1}{4} - 3\tfrac{1}{2} + 4\tfrac{1}{4} = 2\tfrac{1}{4} + 4\tfrac{1}{4} - 3\tfrac{1}{2} = 6\tfrac{1}{2} - 3\tfrac{1}{2}$

$= 3$

f. $-3\tfrac{1}{2} + 2\tfrac{1}{4} - 5\tfrac{1}{4} = -3\tfrac{1}{2} - 3 = -6\tfrac{1}{2}$

17. a. $(-3)^2 = (-3)(-3) = 9$

b. $(-2)^3 = (-2)(-2)(-2) = 4(-2) = -8$

c. $-3^2 = -1(3)(3) = -9$

d. $(3x)^2 = (3x)(3x) = (3)(3)(x)(x) = 9x^2$

e. $3x^2$ is already simplified

f. $0 - 3x^2 = -3x^2$

Section 1.0 (con't)

19. a. $A = \frac{1}{2}(5.8)(3.6 + 6.0)\ m^2$

$A = \frac{1}{2}(5.8)(9.6)\ m^2$

$A = (2.9)(9.6)\ m^2$

$A \approx 27.8\ m^2$

b. $V = \frac{4}{3}\pi(1.875 \div 2)^3\ in^3$

$V = \frac{4}{3}\pi(0.9375)^3\ in^3$

$V \approx \frac{4}{3}\pi(0.824)\ in^3$

$V \approx 3.5\ in^3$

c. $d = (-6)^2 - 4(8)(-3)$

$d = 36 - 4(8)(-3)$

$d = 36 - (-96)$

$d = 36 + 96$

$d = 132$

d. $x = 12 - 8 \div 4 + 5 \cdot 6 - 3^3$

$x = 12 - \frac{8}{4} + 5 \cdot 6 - 27$

$x = 12 - 2 + 30 - 27$

$x = 10 + 3$

$x = 13$

21. a. $x = \dfrac{3-12}{3^2-6},\ x = \dfrac{3-12}{9-6},$

$x = \dfrac{-9}{3},\ x = -3$

21. b. $d = \sqrt{(6-(-2))^2 + (3-(-3))^2}$

$d = \sqrt{(8)^2 + (6)^2},\ d = \sqrt{64+36}$

$d = \sqrt{100},\ d = 10$

c. $m = \dfrac{3-(-3)}{6-(-2)},\ m = \dfrac{6}{8},\ m = \dfrac{3}{4}$

d. $x = \dfrac{-(-2) + \sqrt{(-2)^2 - 4(3)(-5)}}{2(3)}$

$x = \dfrac{-(-2) + \sqrt{4 - 4(3)(-5)}}{2(3)}$

$x = \dfrac{-(-2) + \sqrt{4 - (-60)}}{2(3)},\ x = \dfrac{-(-2) + \sqrt{64}}{2(3)},$

$x = \dfrac{-(-2) + 8}{2(3)},\ x = \dfrac{2+8}{6},\ x = \dfrac{10}{6},\ x = \dfrac{5}{3}$

e. $d =$

$\{|4.05 - 7.12| + |7.51 - 7.12| + |9.80 - 7.12|\} \div 3$

$d = (3.07 + 0.39 + 2.68) \div 3$

$d = 6.14 \div 3,\ d \approx 2.04\overline{6}$

23. a.

Multiply	x	-2
3x	$3x^2$	-6x

$3x(x - 2) = 3x^2 - 6x$

Section 1.0 (con't)

23. b.

Factor	a	-b
2a	$2a^2$	-2ab

$2a^2 - 2ab = 2a(a - b)$

25. a. $3(2x + 4) = 3(2x) + 3(4) = 6x + 12$

b. $15y + 10 = 5(3y) + 5(2) = 5(3y + 2)$

27. a. $x(x + 5) = x(x) + x(5) = x^2 + 5x$

b. $ab + bc = b(a + c)$

29. a. $-2(3 - 4x) = (-2)(3) + (-2)(-4x)$

$= -6 + 8x$

b. $-a(b - c) = (-a)(b) + (-a)(-c)$

$= -ab + ac$

c. $x(x^2 - 2x - 3) = x(x^2) + x(-2x) + x(-3)$

$= x^3 - 2x^2 - 3x$

31. a. gcf = 6, $6x - 54 = 6(x - 9)$

b. gcf = 15, $15x - 225 = 15(x - 15)$

c. gcf = 2a, $6a^2 - 8a = 2a(3a - 4)$

33. a. $4 + (-2) + (-8) + 6$

$= 4 + 6 + (-2) + (-8)$ (*commutative property for addition*)

$= 10 + (-10)$ (*associative property for addition*)

$= 0$

33. b. $-5(7)(-2)(3) = (-5)(-2)(7)(3)$

(*commutative property for multiplication*)

$= (10)(21)$ (*associative property for multiplication*)

$=210$

c. $5 + (-2) + 15 + 12$

$=5 + 15 + (-2) + 12$ (*commutative property for addition*)

$= 20 + 10$ (*associative property for addition*)

$= 30$

d. $\frac{1}{2}(19)(20) = \frac{1}{2}(20(19)$

(*commutative property for multiplication*)

$= (10)(19)$ (*associative property for multiplication*)

$= 190$

35. a. $\frac{1}{2} + \frac{1}{2} = 1, \frac{1}{2} \cdot \frac{1}{2} = \frac{1}{4}, 1 > \frac{1}{4}$

b. $\frac{1}{2} \div \frac{1}{2} = \frac{1}{2} \cdot \frac{2}{1} = 1, \frac{1}{2} + \frac{1}{2} = 1, 1 = 1$

c. $\frac{1}{4} + \frac{1}{4} = \frac{1}{2}, \frac{1}{2} \cdot \frac{1}{2} = \frac{1}{4}, \frac{1}{2} > \frac{1}{4}$

d. $\frac{1}{2} - \frac{1}{4} = \frac{1}{4}, \frac{7}{8} - \frac{3}{8} = \frac{4}{8} = \frac{1}{2}, \frac{1}{4} < \frac{1}{2}$

e. $0.3 \cdot 0.5 = 0.15$, $0.4 \cdot 0.4 = 0.16$,

$0.15 < 0.16$

f. $1.5 - 3.5 = -2$, $-2.5 - 0.5 = -3$, $-2 > -3$

Section 1.0 (con't)

37. $x \leq 2$

39. $-1 < x < 5$; x is between -1 and 5

41.

–1 0

x is greater than or equal to -1

43. $-2 < x < 4$, x is between -2 and 4

45. $x < 2$ or $x > 4$, x is less than 2 or x is greater than 4

47. $x < -3$ or $x > 2$, x is less than -3 or x is greater than 2

49. $x \leq -4$ or $x \geq 1$, x is less than or equal to -4 or x is greater than or equal to 1

51. Subtraction can be restated as addition of the opposite.

Section 1.1

1. a.

Input	Output
1	3
2	6
3	9
4	12
10	30

b. Change in outputs = 3

c. $y = 3x$

d.

Input	Output
50	$3(50) = 150$
100	$3(100) = 300$

3. a.

Input	Output
1	2
2	5
3	8
4	11
5	14
10	29

b. Change in outputs = 3

c. $y = 3x - 1$

3. d.

Input	Output
50	$3(50) - 1 = 149$
100	$3(100) - 1 = 299$

5. $10 - 2x = y$

7. $y = \frac{1}{2}x + 10$

9. $\frac{x}{15} + 2 = y$

11. $y = 7x - 14$

13. The output is five less than twice the input.

15. The input added to the output gives eleven.

17. The difference between twice the input and the output is five.

19. The quotient of the input and eight is the output.

21. $y = \$1.29x - \$0.45$

23. $y = \$1.25 - \$0.05x$

25. $y = \$125x + \$85$

27. a. When we substitute numbers into an expression or formula we *evaluate* the expression or formula.

b. Integers divisible by 2 are *even numbers.*

<u>**Section 1.1 (con't)**</u>

27. c. The numbers in the first column of an input-output table are the *input numbers.*

d. The answer to a division is the *quotient.*

e. A number, letter, or symbol whose value is fixed is a *constant.*

29. a. The answer to a multiplication problem is the *product.*

b. A letter that can take on any number from a given set is a *variable.*

c. Any combination of numbers and variables with operations is an *expression.*

d. The number x multiplied by itself is the *square of x.*

e. The answer to an addition problem is the *sum.*

31. a. In $2x + 3$ the variable is x, the numerical coefficient is 2, and the constant term is 3.

b. In πr^2 the variable is r, the numerical coefficient is π, there is no constant term.

c. In $4 - x$ the variable is x, the numerical coefficient is -1, and the constant term is 4.

31. d. In $x^2 + x - 1$ the variable is x, the numerical coefficients are 1 and 1, and the constant term is -1.

33. $3x^2 + 5x + 6$

35. $4a^2 + 3ab + 2b^2$

37. a. $3a - 2b + 4b - 2a = 3a - 2a - 2b + 4b$

$= 1a + 2b = a + 2b$

b. $6x^2 + 2x - 4x^2 - 3 = 6x^2 - 4x^2 + 2x - 3$

$= 2x^2 + 2x - 3$

39. a. $6ac - 3ab + 2ca + 7bc$

$= 6ac + 2ac - 3ab + 7bc$

$= 8ac - 3ab + 7bc$

(*Remember 2ca = 2ac; commutative property for multiplication*)

b. $x(4 + 2x) - x(x - 4)$

$= 4x + 2x^2 - x^2 + 4x$

$= 2x^2 - x^2 + 4x + 4x = 1x^2 + 8x$

$= x^2 + 8x$

41. a. gcf = 15, $\frac{15}{45} = \frac{1 \cdot 15}{3 \cdot 15} = \frac{1}{3}$

b. gcf = 16, $\frac{48}{160} = \frac{3 \cdot 16}{10 \cdot 16} = \frac{3}{10}$

c. gcf = 240, $\frac{5280}{3600} = \frac{22 \cdot 240}{15 \cdot 240} = \frac{22}{15}$

Section 1.1 (con't)

43. a. gcf = be, $\frac{abe}{ben} = \frac{a \cdot be}{n \cdot be} = \frac{a}{n}$

b. gcf = b, $\frac{ab + bc}{b} = \frac{b(a+c)}{b} = a + c$

c. gcf = 1, fraction is already simplified

45. a. gcf = x, $\frac{xy + xz}{xy} = \frac{x(y+z)}{xy} = \frac{y+z}{y}$

b. gcf = x, $\frac{x^2 + 2x}{2x} = \frac{x(x+2)}{x \cdot 2} = \frac{x+2}{2}$

c. gcf = xy, $\frac{xy^2 - 2x^2y}{2xy} = \frac{xy(y - 2x)}{xy \cdot 2}$

$= \frac{y - 2x}{2}$

47. a. $7^2 + 2(7) + 1 = 49 + 14 + 1 = 64$

b. $(7 + 1)^2 = 8^2 = 64$

49. a. $9^2 - 7^2 = 81 - 49 = 32$

b. $(9 - 7)^2 = 2^2 = 4$

51. a. $9 - 3(8 - 2) = 9 - 3(6) = 9 - 18 = -9$

b. $6(8 - 2) = 6(6) = 36$

53. a. $C = \frac{5}{9}(32 - 32), C = \frac{5}{9}(0), C = 0$

b. $C = \frac{5}{9}(-40 - 32), C = \frac{5}{9}(-72), C = -40$

c. $C = \frac{5}{9}(98.6 - 32), C = \frac{5}{9}(66.6), C = 37$

55. a. $L = 1 + (100 - 1)(2), L = 1 + 99(2),$

$L = 1 + 198, L = 199$

55. b. $L = 2 + (100 - 1)(2), L = 2 + 99(2),$

$L = 2 + 198, L = 200$

c. $L = 5 + (100 - 1)(5), L = 5 + 99(5),$

$L = 5 + 495, L = 500$

57. a. $V = \frac{4}{3}\pi(1.5 \text{ cm})^3,$

$V = \frac{4}{3}\pi(3.375) \text{ cm}^3, V \approx 14.1 \text{ cm}^3$

b. $V = \frac{4}{3}\pi(3 \text{ cm})^3, V = \frac{4}{3}\pi(27) \text{ cm}^3,$

$V \approx 113.1 \text{ cm}^3$

c. $V = \frac{4}{3}\pi(6 \text{ cm})^3, V = \frac{4}{3}\pi(216) \text{ cm}^3,$

$V \approx 904.8 \text{ cm}^3$

59. a. $D = \sqrt{(6-(-2))^2 + (3-(-3))^2},$

$D = \sqrt{(8)^2 + (6)^2}, D = \sqrt{64 + 36},$

$D = \sqrt{100}, D = 10$

b. $D = \sqrt{(14-2)^2 + (2-(-3))^2},$

$D = \sqrt{(12)^2 + (5)^2}, D = \sqrt{144 + 25},$

$D = \sqrt{169}, D = 13$

c. $D = \sqrt{(-4-11)^2 + (3-(-5))^2},$

$D = \sqrt{(-15)^2 + (8)^2}, D = \sqrt{225 + 64},$

$D = \sqrt{289}, D = 17$

Section 1.2

1.

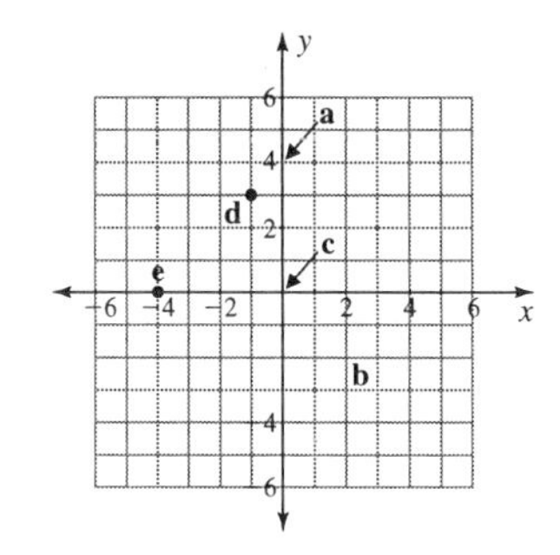

3. a. The vertical axis representing output is the *y-axis.*

b. The name of the point (0, 0) where the axes cross is the *origin.*

c. An ordered pair is *(x, y).*

d. The horizontal axis representing input is the *x-axis.*

e. Two lines forming a right angle are *perpendicular.*

5. a. The set with zero in it is *{0}.*

b. The empty set is *{ }.*

c. The distance between numbers labeled on the axes is the *scale.*

d. The direction in which the quadrants are numbered is *counterclockwise.*

e. The surface containing the horizontal and vertical axes is the *coordinate plane.*

7. a. (75, 167) is letter J

b. (10, 50) is letter E

c. (37, 98.6) is letter G

7. d. (100, 212) is letter K

9. a. Point B is ≈ (-18, 0)

b. Point A is (-40, -40)

11.

Input x	Output y = 2x + 3
-3	2(-3) + 3 = -6 + 3 = -3
-2	2(-2) + 3 = -4 + 3 = -1
-1	2(-1) + 3 = -2 + 3 = 1
0	2(0) + 3 = 3
1	2(1) + 3 = 2 + 3 = 5
2	2(2) + 3 = 4 + 3 = 7
3	2(3) + 3 = 6 + 3 = 9

13.

Input x	Output y = 3 - x
-3	3 - (-3) = 3 + 3 = 6
-2	3 - (-2) = 3 + 2 = 5
-1	3 - (-1) = 3 + 1 = 4
0	3 - 0 = 3
1	3 - 1 = 2
2	3 - 2 = 1
3	3 - 3 = 0

Section 1.2 (con't)

15.

Input x	Output $y = 3 - 2x$
-3	$3 - 2(-3) = 3 + 6 = 9$
-2	$3 - 2(-2) = 3 + 4 = 7$
-1	$3 - 2(-1) = 3 + 2 = 5$
0	$3 - 2(0) = 3$
1	$3 - 2(1) = 3 - 2 = 1$
2	$3 - 2(2) = 3 - 4 = -1$
3	$3 - 2(3) = 3 - 6 = -3$

17.

Input x	Output $y = x^2 - x$
-3	$(-3)^2 - (-3) = 9 + 3 = 12$
-2	$(-2)^2 - (-2) = 4 + 2 = 6$
-1	$(-1)^2 - (-1) = 1 + 1 = 2$
0	$0^2 - 0 = 0$
1	$1^2 - 1 = 1 - 1 = 0$
2	$2^2 - 2 = 4 - 2 = 2$
3	$3^2 - 3 = 9 - 3 = 6$

19.

Input x	Output $y = 2 - x^2$
-3	$2 - (-3)^2 = 2 - 9 = -7$
-2	$2 - (-2)^2 = 2 - 4 = -2$
-1	$2 - (-1)^2 = 2 - 1 = 1$
0	$2 - 0^2 = 2$
1	$2 - (1)^2 = 2 - 1 = 1$
2	$2 - (2)^2 = 2 - 4 = -2$
3	$2 - (3)^2 = 2 - 9 = -7$

21.

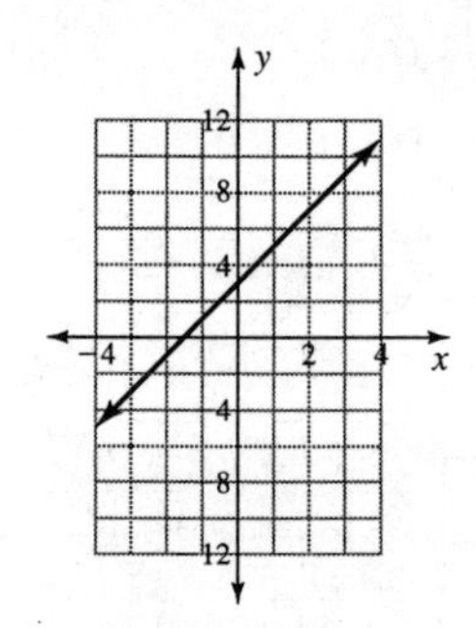

23.

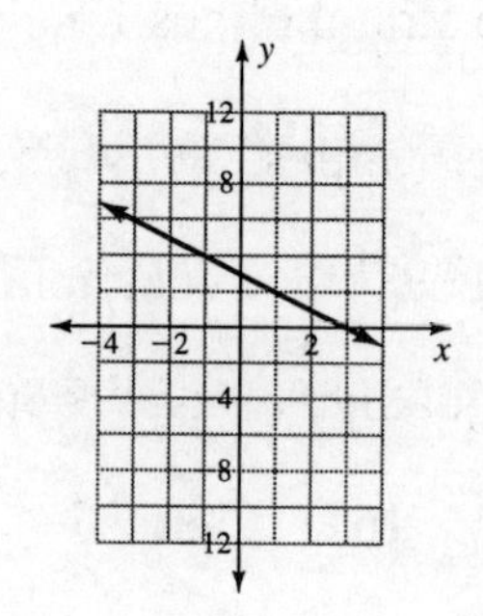

Section 1.2 (con't)

25.

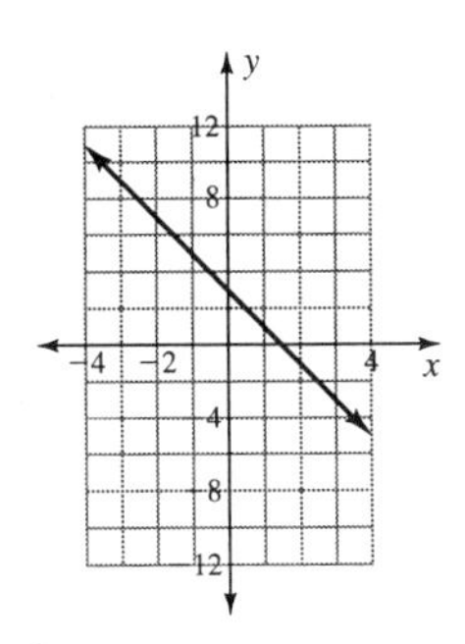

27.

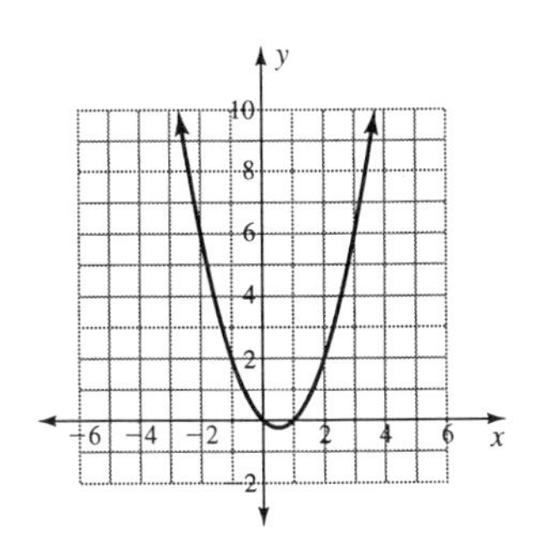

29.

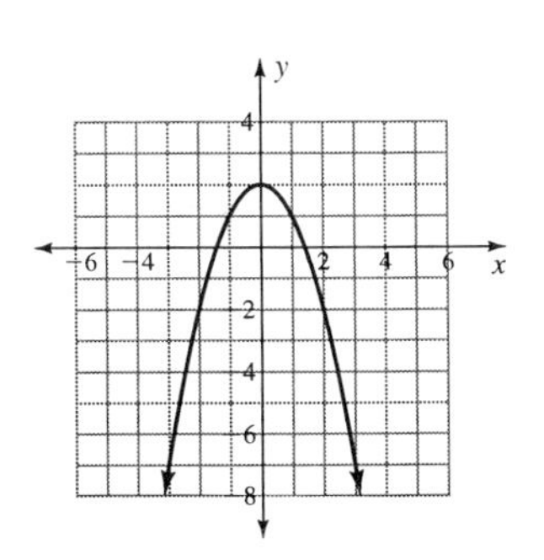

31. a. The graphs in exercises 21, 23 and 25 are straight lines.

b. The graphs in exercises 27 and 29 are parabolas.

c. Equation y = ax + b makes a straight line.

33. For height = 62 inches

Weight, lb	Index
100	$\frac{100(704.5)}{62^2} \approx 18.3$
110	$\frac{110(704.5)}{62^2} \approx 20.2$
120	$\frac{120(704.5)}{62^2} \approx 22.0$
130	$\frac{130(704.5)}{62^2} \approx 23.8$
140	$\frac{140(704.5)}{62^2} \approx 25.7$

35. For height = 70 inches

Weight, lb	Index
130	$\frac{130(704.5)}{70^2} \approx 18.7$
140	$\frac{140(704.5)}{70^2} \approx 20.1$
150	$\frac{150(704.5)}{70^2} \approx 21.6$
160	$\frac{160(704.5)}{70^2} \approx 23.0$
170	$\frac{170(704.5)}{70^2} \approx 24.4$

37. a. From the table W = 130 lb

b. From the table W = 110 lb

Section 1.2 (con't)

39. a. From the table $W = 160$ lb

b. From the table $W = 170$ lb

41. a. From the table in exercise 11, $x = -3$ when $y = -3$

b. From the table in exercise 11, $x = 2$ when $y = 7$

43. a. From the table in exercise 15, $x = -1$ when $y = 5$

b. From the table in exercise 15, $x = 3$ when $y = -3$

45. a. From the table in exercise 17, $x = -1$ or $x = 2$ when $y = 2$

b. From the graph in exercise 27, $x = -3$ or $x = 4$ when $y = 12$

c. From the graph in exercise 27, there is no real number solution when $y = 12$

d. From the graph in exercise 27, $x = 0.5$ when $y = -0.25$

47. a. From the table in exercise 19, $x = 0$ when $y = 2$

b. From the table in exercise 19, $x = -2$ or $x = 2$ when $y = -2$

c. From the graph in exercise 29, $x = -4$ or $x = 4$ when $y = -14$

d. From the graph in exercise 29, there is no real number solution when $y = 4$

49. On the table find the output for an input of 2.

51. On the graph find the x-coordinate of the intersection of $y = 3x + 4$ and $y = 7$

53. The independent variable is s, the dependent variable is A

55. The independent variable is s, the dependent variable is A

57. The independent variable is A, the dependent variable is E

59. The independent variable is the weight of the package, the dependent variable is the cost of shipping

61. Solving the equation $3x + 4 = -2$ is *finding the independent variable.*

63. Locating y, given the graph and x, is *finding the dependent variable.*

65. Locating x, given the graph and y, is *finding the independent variable.*

Mid-Chapter 1 Test

1. a. $-3 + (-5) - (-6) = -8 + 6 = -2$

b. $-\frac{1}{2}+\frac{1}{4}-\frac{3}{4}-(-\frac{1}{4}) = -\frac{1}{2}+\frac{1}{4}+\frac{1}{4}-\frac{3}{4}$

$= -\frac{1}{2}+\frac{1}{2}-\frac{3}{4} = -\frac{3}{4}$

c. $2.25 + 8.50 - 3.75 = 10.75 - 3.75 = 7$

d. $6.2 - 8.6 + 1.8 = 6.2 + 1.8 - 8.6$

$=8.0 - 8.6 = -0.6$

2. a. $2 \cdot 17 \cdot 50 = 2 \cdot 50 \cdot 17 = 100 \cdot 17 = 1700$

b. $\frac{1}{2}(-7)(-6) = \frac{1}{2}(42) = 21$

c. $-4(-8)(-5) = -4(40) = -160$

d. $\frac{3}{5}\left(\frac{11}{6}\right)\left(\frac{15}{121}\right) = \frac{3}{5} \cdot \frac{11}{2 \cdot 3} \cdot \frac{3 \cdot 5}{11^2}$

$= \frac{3 \cdot 3 \cdot 5 \cdot 11}{2 \cdot 3 \cdot 5 \cdot 11^2} = \frac{3}{22}$

3. a. $-4^2 = -1(4)(4) = -16$

b. $(-4)^2 = (-4)(-4) = 16$

c. $0 - 4^2 = 0 - 16 = -16$

d. $8 - 6 \div 3 + 3 \cdot 4 + \frac{1}{2}(8-3)$

$= 8 - \frac{6}{3} + 12 + \frac{1}{2}(5) = 8 - 2 + 12 + \frac{5}{2}$

$= 18 + 2\frac{1}{2} = 20\frac{1}{2}$

e. $\frac{3-6}{2-(-7)} = \frac{-3}{9} = \frac{-3}{3^2} = \frac{-1}{3}$

3. f. $\frac{-(-2)-\sqrt{(-2)^2-4(5)(-3)}}{2(5)}$

$= \frac{2-\sqrt{4-(-60)}}{10} = \frac{2-\sqrt{64}}{10} = \frac{2-8}{10}$

$= \frac{-6}{10} = -\frac{3}{5}$

g. $\sqrt{(2-(-7))^2+(3-6)^2} = \sqrt{9^2+(-3)^2}$

$= \sqrt{81+9} = \sqrt{90} = \sqrt{9(10)} = 3\sqrt{10} \approx 9.5$

h. $\frac{|1.25-2.83|+|2.56-2.83|+|4.68-2.83|}{3}$

$= \frac{1.58+0.27+1.85}{3} = \frac{3.7}{3} = \frac{37}{30} \approx 1.2$

4. a. $3(x + 2) = 3x + 6$

b. $-3(x - 2) = -3x + 6$

c. $7 - 4(x + 5) = 7 - 4x - 20$

$= -4x + 7 - 20 = -4x - 13$

d. $8 - 2(4 - x) = 8 - 8 + 2x = 2x$

5. a. gcf = 2, $4x - 18 = 2(2x - 9)$

b. gcf = x, $x^2 + 5x = x(x + 5)$

c. gcf = n, $mn^2 - np^2 = n(mn - p^2)$

d. gcf = 7xy, $63x^2y - 49xy^2$

$=7xy(9x - 7y)$

Mid-Chapter 1 Test (con't)

6. a. −4 0

b. 0 6

c. −3 0 2

d. 3 6

e. −2 0 3

f. 0

7. a. In -4 - x the variable is x, the constant term is -4, and the numerical coefficient is -1.

b. In $2x - x^2$ the variable is x, there is no constant term, and the numerical coefficients are 2 and -1.

8. a. $3(x^2 + 3x - 4) - x(x^2 + 3x - 4)$

$= 3x^2 + 9x - 12 - x^3 - 3x^2 + 4x$

$= -x^3 + 3x^2 - 3x^2 + 9x + 4x - 12$

$= -x^3 + 13x - 12$

b. $a(a^2 - 2ab + b^2) - b(a^2 + 2ab - b^2)$

$= a^3 - 2a^2b + ab^2 - a^2b - 2ab^2 + b^3$

$= a^3 - 2a^2b - a^2b + ab^2 - 2ab^2 + b^3$

$= a^3 - 3a^2b - ab^2 + b^3$

9. a. $\dfrac{x^2yz}{xy^2z} = \dfrac{xxyz}{xyyz} = \dfrac{x}{y}$

b. $\dfrac{x^2yz}{(xy)^2z} = \dfrac{x^2yz}{x^2y^2z} = \dfrac{1}{y}$

c. $\dfrac{xy - yz}{xy} = \dfrac{y(x-z)}{xy} = \dfrac{x-z}{x}$

10. a. Independent variable is s, dependent variable is h; $h = \dfrac{4\sqrt{3}}{2}$ yd, $h = 2\sqrt{3}$ yd

b. Independent variable is d, dependent variable is A; $A = \pi\left(\dfrac{5\text{ cm}}{2}\right)^2$,

$A = \pi\left(\dfrac{25\text{ cm}^2}{4}\right)$, $A = 6.25\pi\text{ cm}^2$,

11. Negative first number is a negative independent variable which occurs in quadrants 2 and 3.

12.

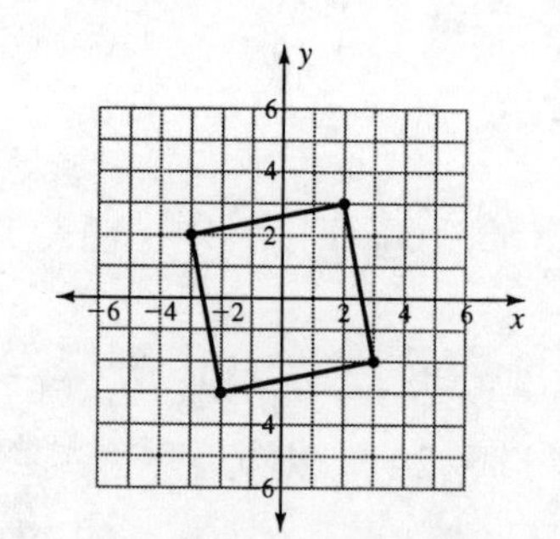

Shape is a square, opposite lines are parallel, adjacent lines are perpendicular.

Mid-Chapter 1 Test (con't)

13. a.

Input x	Output $y = 4 - 3x$
-2	$4 - 3(-2) = 4 + 6 = 10$
-1	$4 - 3(-1) = 4 + 3 = 7$
0	$4 - 3(0) = 4$
1	$4 - 3(1) = 4 - 3 = 1$
2	$4 - 3(2) = 4 - 6 = -2$

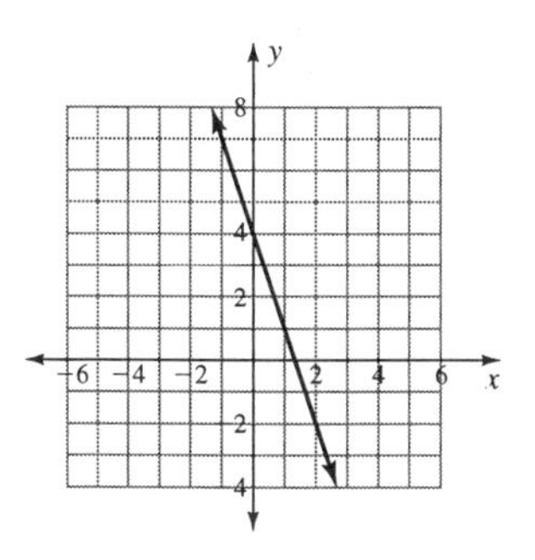

13. b.

Input x	Output $y = 3x - 2x^2$
-2	$3(-2) - 2(-2)^2 = 3(-2) - 2(4)$ $= -6 - 8 = -14$
-1	$3(-1) - 2(-1)^2 = 3(-1) - 2(1)$ $= -3 - 2 = -5$
0	$3(0) - 2(0)^2 = 0$
1	$3(1) - 2(1)^2 = 3 - 2 = 1$
2	$3(2) - 2(2)^2 = 3(2) - 2(4)$ $= 6 - 8 = -2$

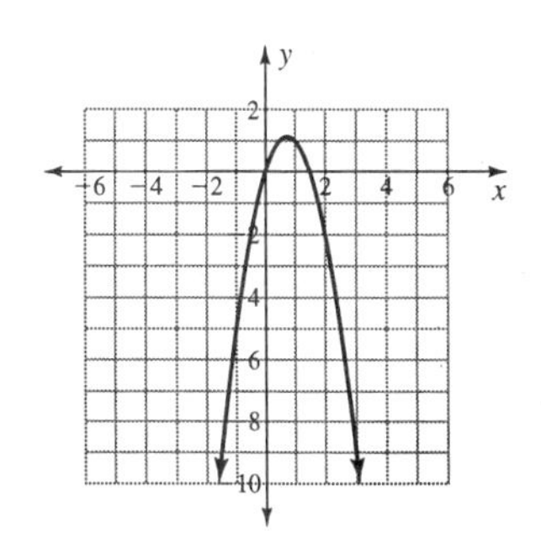

14. a. From the table, $x = -1$ or $x = 5$ when $y = 8$

b. From the table and the graph, $x = 1$ or $x = 3$ when $y = 0$

c. From the table, $x = 2$ when $y = -1$

Mid-Chapter 1 Test (con't)

15. a. Positive outputs, $y > 0$

b. Negative inputs, $x < 0$

c. Negative outputs, $y < 0$

d. Positive or zero inputs, $x \geq 0$

e. x-coordinate in 1st or 4th quadrant but not on y-axis, $x > 0$

f. y-coordinate in 3rd or 4th quadrant or on x-axis, $y \leq 0$

16. a.

Number of Pairs of Pens	Total Number of Panels
1	7
2	12
3	17
4	22
10	52
50	252
100	502

b. Each addition of a pair of pens requires 5 panels. The first pair required 5 + 2 panels. The rule is $y = 5x + 2$, where x is the number of pairs of pens and y is the total number of panels.

17. j, $4 + 2x$

18. a, $5x + 3$

19. g, $3x - 5$

20. h, $0.5x + 3$

21. i, $2x - 4$

22. b, $4x - 2$

23. e, $5 - 3x$

24. c, $\frac{1}{2} - 4x$

Section 1.3

1. **a.** $-4 - (6) = -4 + (-6) = -10$

b. $-5 - (-3) = -5 + 3 = -2$

c. $-\frac{1}{2} - \left(-\frac{3}{2}\right) = -\frac{1}{2} + \frac{3}{2} = 1$

d. $100 \cdot \frac{1}{4} = 100 \div 4 = 25$

e. $\frac{5}{8} \div \frac{5}{8} = \frac{5}{8} \cdot \frac{8}{5} = 1$

f. $-6 \div \frac{1}{3} = -6 \cdot 3 = -18$

3. **a.** $-3 + 3 = 0$

b. $5 \cdot \frac{1}{5} = 1$

c. $-2 + 3 = 1$

d. $8 \cdot 0 = 0$

e. $\frac{7}{8} \cdot \frac{8}{7} = 1$

f. $\frac{3}{8} + \left(-\frac{3}{8}\right) = 0$

5. **a.** $x - 8 = -4$, $x - 8 + 8 = -4 + 8$,

$x = 4$

b. $x + 3 = -6$, $x + 3 - 3 = -6 - 3$,

$x = -9$

c. $x - 6 = -8$, $x - 6 + 6 = -8 + 6$,

$x = -2$

d. $\frac{1}{2}x = -6$, $(2)\frac{1}{2}x = (2)(-6)$,

$x = -12$

e. $\frac{3}{4}x = 21$, $\left(\frac{4}{3}\right)\frac{3}{4}x = \left(\frac{4}{3}\right)(21)$,

$x = 28$

5. **f.** $\frac{3}{8}x = 12$, $\left(\frac{8}{3}\right)\frac{3}{8}x = \left(\frac{8}{3}\right)(12)$,

$x = 32$

g. $-\frac{2}{3}x = 24$, $\left(-\frac{3}{2}\right)\left(-\frac{2}{3}x\right) = \left(-\frac{3}{2}\right)(24)$,

$x = -36$

h. $-\frac{1}{4}x = 8$, $(-4)(-\frac{1}{4}x) = (-4)(8)$,

$x = -32$

i. $\frac{7}{8}x = 35$, $\left(\frac{8}{7}\right)\frac{7}{8}x = \left(\frac{8}{7}\right)(35)$,

$x = 40$

7. Inverse: Take off jacket, take off vest, take off shirt; activity: dressing and undressing

9. Inverse is not meaningful; activity is take a picture

11. Order is not important (unless you plan to dig the hole near the sprinkler)

13. **a.** $2x - 4 = 8$, $2x - 4 + 4 = 8 + 4$,

$2x = 12$, $2x \div 2 = 12 \div 2$, $x = 6$

b. $2x - 4 = 6$, $2x - 4 + 4 = 6 + 4$,

$2x = 10$, $2x \div 2 = 10 \div 2$, $x = 5$

c. $2x - 4 = -6$, $2x - 4 + 4 = -6 + 4$,

$2x = -2$, $2x \div 2 = -2 \div 2$, $x = -1$

d. $2x - 4 = 0$, $2x - 4 + 4 = 0 + 4$,

$2x = 4$, $2x \div 2 = 4 \div 2$, $x = 2$

Section 1.3 (con't)

15. a. $\frac{5}{9}(x - 32) = -25$, $\left(\frac{9}{5}\right)\frac{5}{9}(x - 32) = \left(\frac{9}{5}\right)(-25)$

$x - 32 = -45$, $x - 32 + 32 = -45 + 32$,

$x = -13$

b. $\frac{5}{9}(x - 32) = 10$, $\left(\frac{9}{5}\right)\frac{5}{9}(x - 32) = \left(\frac{9}{5}\right)(10)$

$x - 32 = 18$, $x - 32 + 32 = 18 + 32$,

$x = 50$

c. $\frac{5}{9}(x - 32) = 5$, $\left(\frac{9}{5}\right)\frac{5}{9}(x - 32) = \left(\frac{9}{5}\right)(5)$

$x - 32 = 9$, $x - 32 + 32 = 9 + 32$,

$x = 41$

d. $\frac{5}{9}(x - 32) = -30$, $\left(\frac{9}{5}\right)\frac{5}{9}(x - 32) = \left(\frac{9}{5}\right)(-30)$

$x - 32 = -54$, $x - 32 + 32 = -54 + 32$,

$x = -22$

17. $3x + 3 = 0$, $3x + 3 - 3 = 0 - 3$, $3x = -3$,

$3x \div 3 = -3 \div 3$, $x = -1$

19. $\frac{1}{3}x - 1 = 4$, $\frac{1}{3}x - 1 + 1 = 4 + 1$, $\frac{1}{3}x = 5$,

$(3)\frac{1}{3}x = (3)(5)$, $x = 15$

21. $\frac{3}{4}x + 5 = 23$, $\frac{3}{4}x + 5 - 5 = 23 - 5$,

$\frac{3}{4}x = 12$, $\left(\frac{4}{3}\right)\frac{3}{4}x = \left(\frac{4}{3}\right)(12)$, $x = 24$

23. $\frac{3}{8}x - 4 = 8$, $\frac{3}{8}x - 4 + 4 = 8 + 4$, $\frac{3}{8}x = 12$

$\left(\frac{8}{3}\right)\frac{3}{8}x = \left(\frac{8}{3}\right)(12)$, $x = 32$

25. $x - 2(4 - x) = -17$, $x - 8 + 2x = -17$,

$3x - 8 = -17$, $3x - 8 + 8 = -17 + 8$,

$3x = -9$, $3x \div 3 = -9 \div 3$, $x = -3$

27. $x - 3(4 + x) = 6$, $x - 12 - 3x = 6$,

$-2x - 12 = 6$, $-2x = 18$, $x = -9$

29. $x + 3(5 - x) = -9$, $x + 15 - 3x = -9$,

$-2x + 15 = -9$, $-2x = -24$, $x = 12$

31. $8 - 3(9 - x) = -14.5$, $8 - 27 + 3x = -14.5$,

$-19 + 3x = -14.5$, $3x = 4.5$, $x = 1.5$

33. $14 - 9(x + 2) = 45.5$, $14 - 9x - 18 = 45.5$,

$-9x - 4 = 45.5$, $-9x = 49.5$, $x = -5.5$

35. $37.45 = x + 0.07x$, $37.45 = 1.07x$,

$35 = x$

37. $46.17 = x + 0.15x + 0.065x$,

$46.17 = 1.215x$, $38 = x$

39. $D = rt$, $\frac{D}{r} = \frac{rt}{r}$, $\frac{D}{r} = t$

41. $y = mx + b$, $y - mx = b$

43. $V = \frac{4\pi r^3}{3}$, $V\left(\frac{3}{4\pi}\right) = \frac{4\pi r^3}{3}\left(\frac{3}{4\pi}\right)$,

$\frac{3V}{4\pi} = r^3$

45. $pN = 1$, $\frac{pN}{N} = \frac{1}{N}$, $p = \frac{1}{N}$

Section 1.3 (con't)

47. $A = \frac{1}{2}h(a+b),\ 2A = h(a+b),$

$\frac{2A}{h} = a + b,\ \frac{2A}{h} - a = b$

49. $a_n = a_1 + (n-1)d,\ a_n - (n-1)d = a_1$

51. $A = \frac{a+b+c}{3},\ 3A = a+b+c,$

$3A - b - c = a$

53. $S = \frac{1}{2}n(a_1 + a_n),\ S\left(\frac{2}{n}\right) = a_1 + a_n,$

$\frac{2S}{n} - a_n = a_1$

55. $E = \frac{T_h - T_c}{T_h},\ ET_h = T_h - T_c,$

$ET_h + T_c = T_h,\ T_c = T_h - ET_h$

57. $V = 344 + 0.6(T - 20),$

$V = 344 + 0.6T - 12,\ V = 332 + 0.6T,$

$V - 332 = 0.6T,\ \frac{V - 332}{0.6} = T$

or

$V = 344 + 0.6(T - 20),$

$V - 344 = 0.6(T - 20),\ \frac{V - 344}{0.6} = T - 20,$

$T = \frac{V - 344}{0.6} + 20$

59. $D = \left(\frac{D}{t}\right)t,\quad D = D$

61. $A = \frac{1}{2}\left(\frac{2A}{a+b}\right)(a+b),\quad A = A$

63. $y = 3\left(\frac{1}{3}(y+3)\right) - 3,\ y = 3\left(\frac{1}{3}y + 1\right) - 3,$

$y = y + 3 - 3,\ y = y$

65. $y = 2\left(\frac{1}{2}(y-4)\right) + 4,\ y = 2\left(\frac{1}{2}y - 2\right) + 4,$

$y = y - 4 + 4,\ y = y$

67. $S = \frac{n}{2}(2a + (n-1)d),$

$\frac{2S}{n} = 2a + (n-1)d,\ \frac{2S}{n} - 2a = (n-1)d,$

$\frac{\frac{2S}{n} - 2a}{n-1} = d$

Section 1.4

1. $-2 < -2x + 3,\ 2x - 2 < 3,\ 2x < 5,\ x < 2.5$

3. $-2x + 3 > 0,\ -2x > -3,\ x < 1.5$

5. $x + 4 > -\frac{1}{2}x + 1,\ \frac{3}{2}x + 4 > 1,\ \frac{3}{2}x > -3,$

$x > -2$

7. $0 < x + 4,\ -4 < x,\ x > -4$

9. $-\frac{1}{2}x + 1 \le 0,\ 1 \le \frac{1}{2}x,\ 2 \le x,\ x \ge 2$

11. $-2x + 3 \ge 3,\ -2x \ge 0,\ x \le 0$

13. $x + 4 \le 4,\ x \le 0$

15.

$x > -1$

17.

$x < \frac{3}{2}$

19.

$x < \frac{1}{3}$

21.

$x \ge \frac{1}{3}$

23.

$x \le -\frac{1}{3}$

25. $y = x - 2$ is always less than $y = x$

27. Answers will vary, one example would be $x - 2 > x$

29. $1 - 2x < x,\ 1 < 3x,\ \frac{1}{3} < x,\ x > \frac{1}{3}$

31. $-\frac{1}{2}x + 1 < 4 - x,\ \frac{1}{2}x + 1 < 4,$

$\frac{1}{2}x < 3,\ x < 6$

33. Multiplication by a negative changes the relative value of the two sides of the inequality. One example would be:

$3 < 5,\ (-2)(3) < (-2)(5),\ -6 > -10$

35.

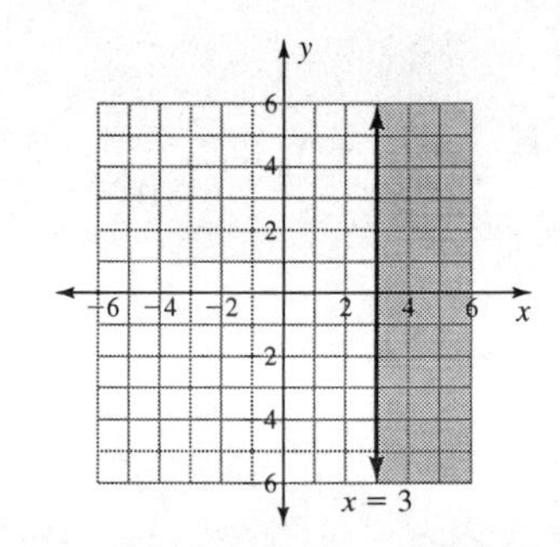

37.

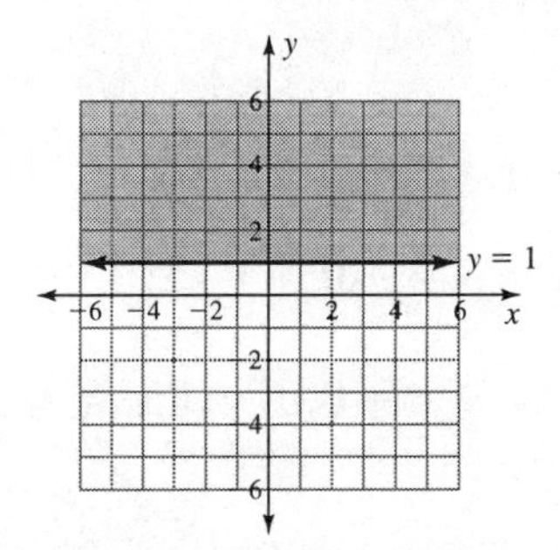

39.

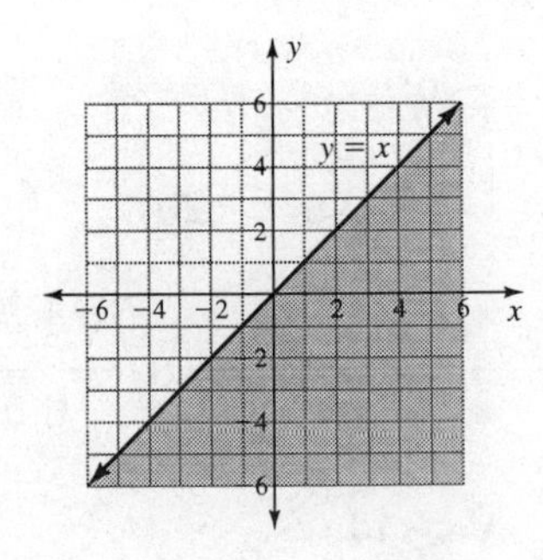

Section 1.4 (con't)

41.

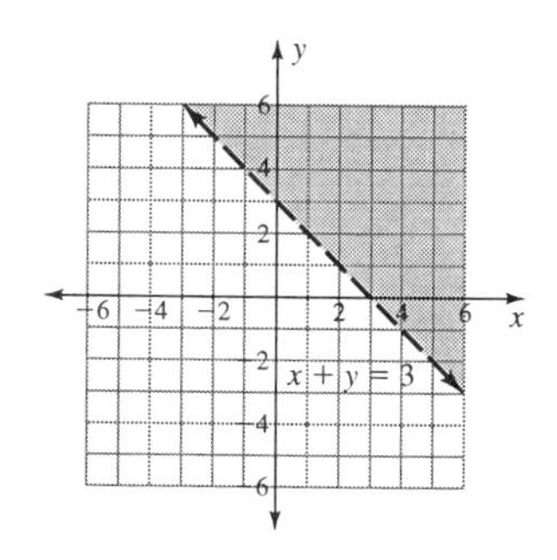

43.

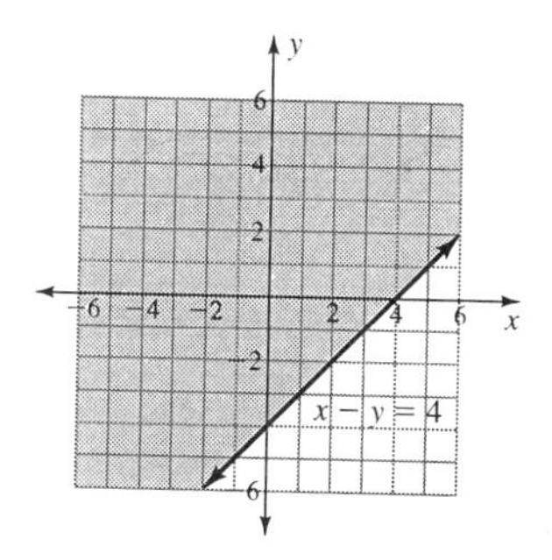

45.

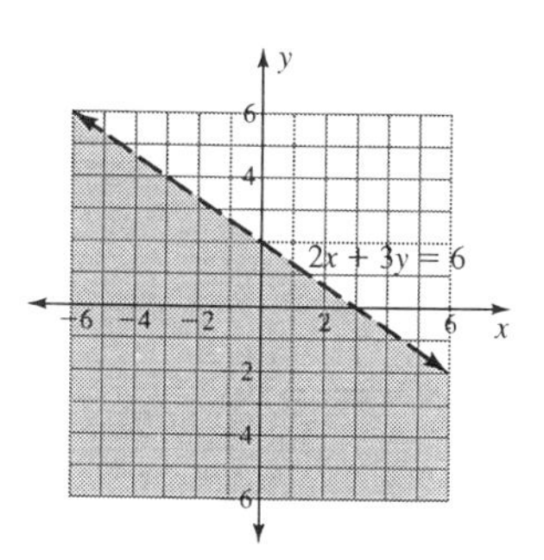

47.

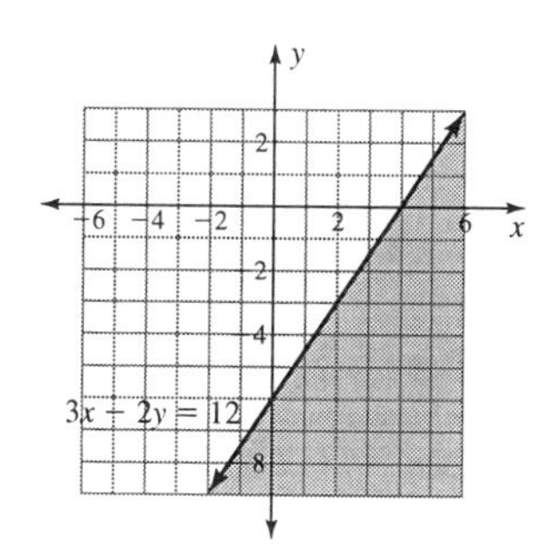

49. If x represents student tickets and y represents regular tickets the inequality is: $15x + 50y \geq 500{,}000$.

If only student tickets are sold, $y = 0$ and $x \geq 33{,}334$; if only regular tickets are sold, $x = 0$ and $y \geq 10{,}000$. The boundary line represents possible combinations of ticket sales to exactly meet the $500,000 goal.

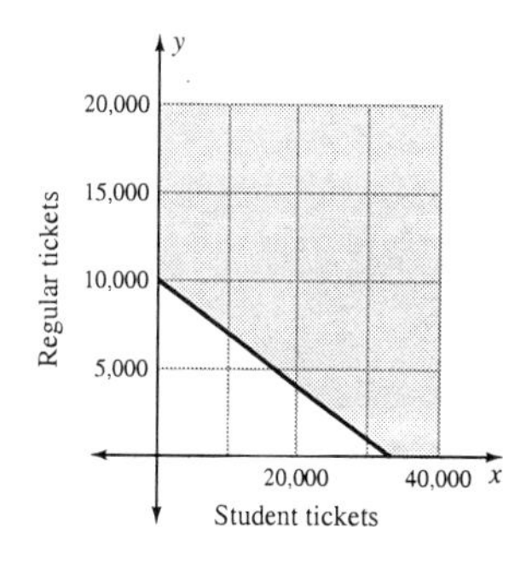

51. Let x = IV solution and

y = nutrition liquid

$500x + 250y \leq 2500$

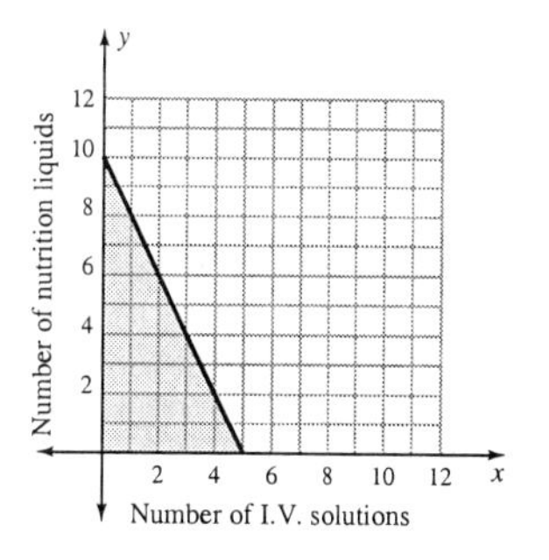

Section 1.4 (con't)

53. A dot indicates the value is a solution; a small circle indicates the value is excluded from the solution set.

55. Select a test point and use it to evaluate the inequality. If the test point makes the inequality true, shade the half-plane that contains the point. If the test point makes the inequality false, shade the other half-plane.

Chapter 1 Review

1. Associative properties for addition and multiplication, commutative properties for addition and multiplication, distributive property for multiplication over addition

3. Absolute value, braces, brackets, fraction bar, parentheses, square root

5. Boundary line, half-plane

7. Sum, difference, product, quotient

9. Input-output relationships, dependent variable, independent variable

11. Multiplicative inverses, reciprocals

13.

	Input x	Input y	Output x + y	Output x - y
a.	-7	3	-7 + 3 = -4	-7 - 3 = -10
b.	3	-7	3 + (-7) = -4	3 - (-7) = 10
c.	7	7 - 4 = 3	7 + 3 = 10	4

15.

	Input x	Input y	Output xy	Output x + y
a.	3	4	12	3 + 4 = 7
b.	-2	3	(-2)(3) = -6	-2 + 3 = 1
c.	-7	21 ÷ -7 = -3	21	-7 + -3 = -10
d.	-3	5	(-3)(5) = -15	-3 + 5 = 2

17. $\left(\frac{2}{3}+1\frac{1}{2}\right)+\frac{1}{2}=\frac{2}{3}+\left(1\frac{1}{2}+\frac{1}{2}\right)=\frac{2}{3}+2=2\frac{2}{3}$

Associative property for addition

19. $25 \cdot 13 \cdot 4 = 25 \cdot 4 \cdot 13 = 100 \cdot 13$

$= 1300$ Commutative and associative properties for multiplication

21. $-2^2 + 5 - 3(4 - 5) = -4 + 5 - 3(-1)$

$= -4 + 5 + 3 = 4$

23. $2\pi(1.5 \text{ in})^2 + 2\pi(1.5 \text{ in})(4 \text{ in})$

$= 2\pi(2.25 \text{ in}^2) + 2\pi(6 \text{ in}^2)$

$= 2\pi(2.25 + 6) \text{ in}^2 = 2\pi(8.25) \text{ in}^2$

$= 16.5\pi \text{ in}^2$

25. \$20.00 + (8 - 1)(-\$1.75)

= \$20 + 7(-\$1.75) = \$20 - \$12.25

= \$7.75

Chapter 1 Review (con't)

27. a. The difference between three and twice a number.

b. The product of three and the difference between a number and five.

29. $\frac{x}{15} + 8$

31. $9 - 3(2x - 5y) - 4(5x - 3y)$

$= 9 - 6x + 15y - 20x + 12y$

$= 9 - 6x - 20x + 15y + 12y$

$= -26x + 27y + 9$

33. $(-2)^2 - (-5)^2 = 4 - 25 = -21$

35. $A = \pi(3.5 \text{ in})^2$, $A = \pi(12.25) \text{ in}^2$,

$A = 12.25\pi \text{ in}^2$

37.

Input	Output
x	y = -x
-2	-(-2) = 2
-1	-(-1) = 1
0	-0 = 0
1	-1
2	-2

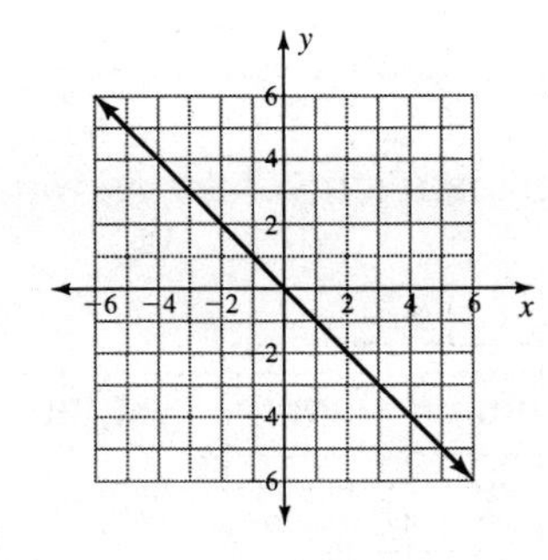

Chapter 1 Review (con't)

39.

Input x	Output $y = 30 - 3.5x$
-10	$30 - 3.5(-10) = 30 + 35$ $= 65$
-5	$30 - 3.5(-5) = 30 + 17.5$ $= 47.5$
0	$30 - 3.5(0) = 30$
5	$30 - 3.5(5) = 30 - 17.5$ $= 12.5$
10	$30 - 3.5(10) = 30 - 35$ $= -5$

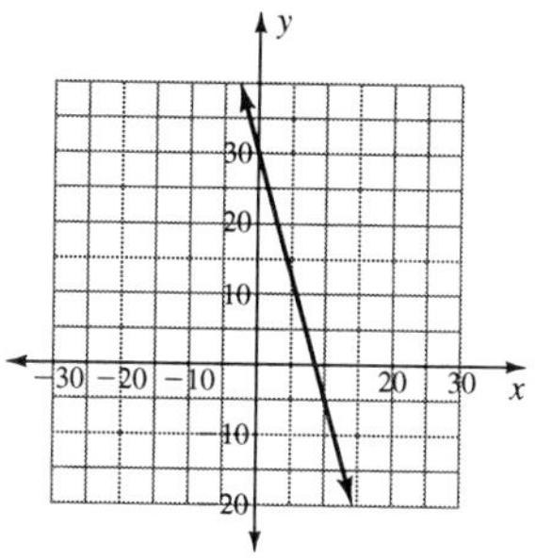

41. $2x + 2(1.5) = 11$, $2x + 3 = 11$,

$2x = 8$, $x = 4$

43. $40 = \frac{5}{9}(F - 32)$, $72 = F - 32$,

$F = 104$

45. $20 + (n - 1)(-1.75) = 0.75$,

$20 - 1.75n + 1.75 = 0.75$,

$21.75 - 1.75n = 0.75$, $-1.75n = -21$,

$n = 12$

47. $I = Prt$, $\frac{I}{Pt} = \frac{Prt}{Pt}$, $r = \frac{I}{Pt}$

49. $C = a + bY$, $C - a = bY$,

$\frac{C - a}{Y} = b$

51. gcf = 3x; $6x^2 + 15x = 3x(2x + 5)$

53. $x > 1$ 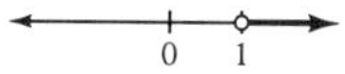

55. $x \geq 3$ 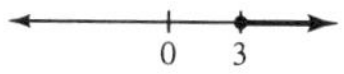

57. a. $x = 0$

b. $x > 0$

59. a. $x = -1$

b. $x \leq -1$

61. a. $\frac{1}{2}x - \frac{3}{2} = -2x + 1$, $\frac{5}{2}x - \frac{3}{2} = 1$,

$\frac{5}{2}x = \frac{5}{2}$, $x = 1$

b. $\frac{1}{2}x - \frac{3}{2} \leq -2x + 1$, $\frac{5}{2}x - \frac{3}{2} \leq 1$,

$\frac{5}{2}x \leq \frac{5}{2}$, $x \leq 1$

63.

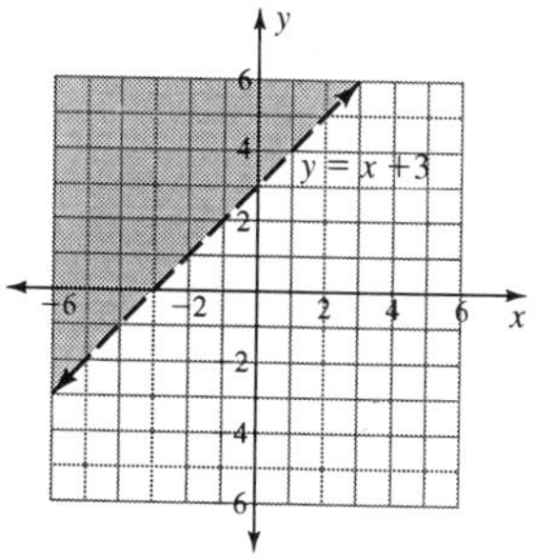

Chapter 1 Test

1. Another name for additive inverse is *opposite.*

2. An *ordered pair* is used to locate a point on the coordinate graph.

3. An *independent* variable is the input, x, or set of numbers on the horizontal axis.

5. a. $-27 + (-3) = -30$

$-27 - (-3) = -27 + 3 = -24$

$(-27)(-3) = 81$

$(-27) \div (-3) = 9$

b. $-1.5 + 0.5 = -1$

$-1.5 - 0.5 = -2$

$(-1.5)(0.5) = -0.75$

$(-1.5) \div (0.5) = -3$

c. $1\frac{1}{2} + \left(-\frac{1}{4}\right) = 1\frac{1}{4}$

$1\frac{1}{2} - \left(-\frac{1}{4}\right) = 1\frac{1}{2} + \frac{1}{4} = 1\frac{3}{4}$

$\left(1\frac{1}{2}\right)\left(-\frac{1}{4}\right) = \left(\frac{3}{2}\right)\left(-\frac{1}{4}\right) = -\frac{3}{8}$

$\left(1\frac{1}{2}\right) \div \left(-\frac{1}{4}\right) = \left(\frac{3}{2}\right)\left(-\frac{4}{1}\right) = -6$

6. a. $(348 + 295) + 105$

$= 348 + (295 + 105) = 348 + 400$

$= 748$, Associative property for addition

b. $230 + 689 + 70$

$= 230 + 70 + 689 = 300 + 689$

$= 989$, Commutative property for addition

7. a. $4\{2 - 3[4 + 5(2 - 5)] + 5\}$

$= 4\{2 - 3[4 + 5(-3)] + 5\}$

$= 4[2 - 3(4 - 15) + 5] = 4[2 - 3(-11) + 5]$

$= 4(2 + 33 + 5) = 4(40) = 160$

b. $-3^2 = -9$

c. $(-3)^2 = 9$

d. $\frac{1}{2} \cdot 25(17 + 23) = \frac{1}{2} \cdot 25(40)$

$= \frac{1}{2} \cdot 1000 = 500$

e. $3x - (8 - 3x) = 3x - 8 + 3x$

$= 6x - 8$

8. Four times the difference between x and 3.

9. $A = \dfrac{a + b + c}{3}$, $3A = a + b + c$,

$b = 3A - a - c$

10. $A = \dfrac{h}{2}(a + b)$, $\dfrac{2A}{h} = a + b$

$\dfrac{2A}{h} - a = b$

11. $15x^2y - 39xy^2 = (5)(3)xxy - (3)(13)xyy$

$= (3xy)(5x) - (3xy)(13y)$, gcf $= 3xy$

Chapter 1 Test (con't)

12. a. $x = -5$

b. $x = 2$

c. $x = -2$

d. $x > 2$

e. $x < -2$

f. $x < 1$

13. 12b. $-\frac{3}{4}x - \frac{1}{2} = -2,\ -\frac{3}{4}x = -\frac{3}{2},$

$x = (-\frac{3}{2})(-\frac{4}{3}),\ x = 2$

12f. $\frac{1}{3}x + \frac{5}{3} < 2, \frac{1}{3}x < \frac{1}{3}, x < 1$

14.

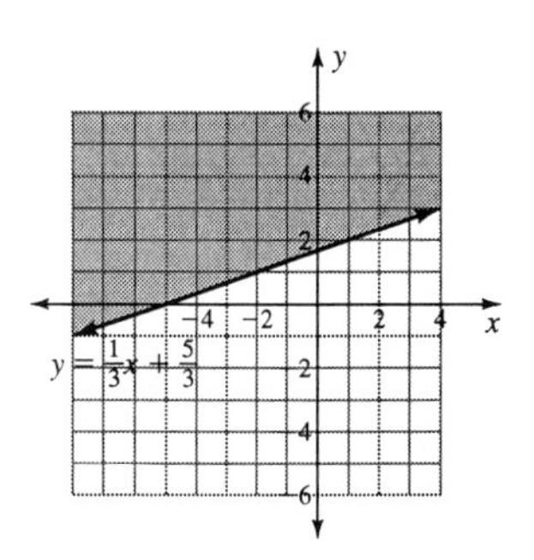

Section 2.0

1.

Small side	Side 2	Side 3	Total
2	6(2) + 3 = 15	7(2) - 3 = 11	2 + 15 + 11 = 28
5	6(5) + 3 = 33	7(5) - 3 = 32	5 + 33 + 32 = 70
x	6(x) + 3	7(x) - 3	91

$x + (6x + 3) + (7x - 3) = 91,$

$x + 6x + 7x + 3 - 3 = 91,$

$14x = 91, \; x = 6.5$

$6(6.5) + 3 = 42$

$7(6.5) - 3 = 42.5$

3.

Meal	Tip	Tax	Total
10	0.15(10) = 1.50	$\frac{1.50}{2.5}$ = 0.60	10 + 1.50 + 0.60 = 12.10
20	0.15(20) = 3.00	$\frac{3.00}{2.5}$ = 1.20	20 + 3 + 1.20 = 24.20
x	0.15(x)	$\frac{0.15x}{2.5}$	33.88

3. $x + 0.15x + \frac{0.15x}{2.5} = 33.88,$

$x + 0.15x + 0.06x = 33.88,$

$1.21x = 33.88, \; x = 28.00$

5.

Chocolates	Labels	Total $
(0.45)1000 = 450	8.50	450 + 8.50 = 458.50
(0.45)1200 = 540	17.00	540 + 17.00 = 557.00
0.45x	17.00	500

$0.45x + 17 = 500, \; 0.45x = 483,$

$x \approx 1073$ pieces

7. a. $y = \$4$ for $1 < x \leq 2$

$y = 4 + 0.50(x - 2)$ for $x > 2$

b. The only meaningful inputs are positive integers, a dot graph is appropriate.

c.

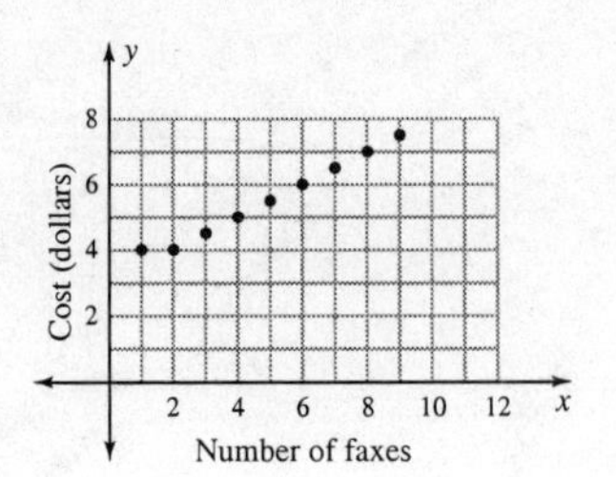

Section 2.0 (con't)

9. a. $y = \$20$ for $0 < x \leq 3$

$y = 20 + 5(x - 3)$ for $x > 3$

b. Inputs are rounded to the next largest integer, a step graph is appropriate.

c.

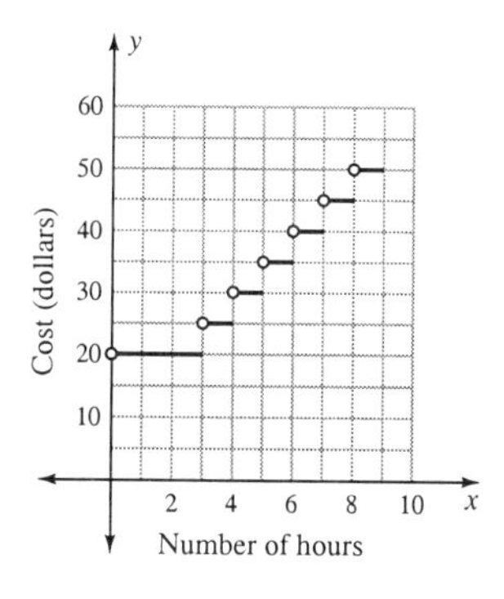

11. a. $y = \$65$ for $0 < x \leq 100$

$y = 65 + 0.15(x - 100)$ for $x > 100$

b. Inputs may be fractions of miles, a line graph is appropriate.

c.

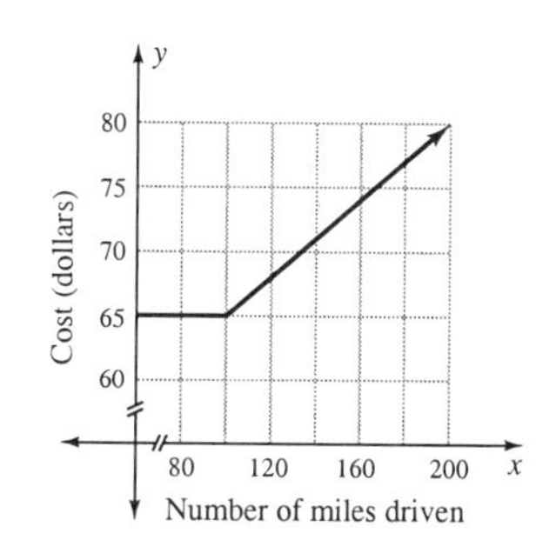

13. a. $y = \$85$ for $0 < x \leq 10$

$y = 85 + 4.75(x - 10)$ for $x > 10$

b. The only meaningful inputs are positive integers, a dot graph is appropriate.

c. The values would lie on the line shown in the graph below.

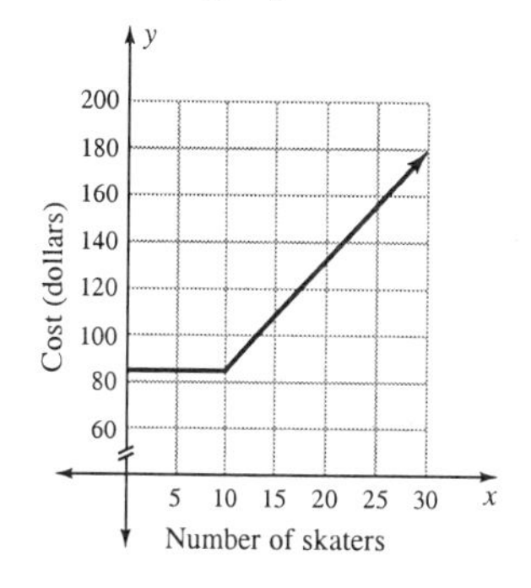

15. a. $y = \$19.95$ for $0 \leq x \leq 100$

$y = 19.95 + 0.25(x - 100)$ for $x > 100$

b. Inputs are rounded to the next largest integer, a step graph is appropriate.

c.

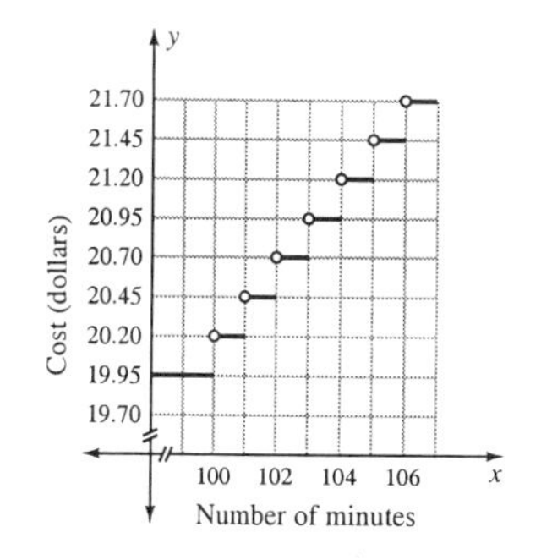

Section 2.0 (con't)

17. $4 + 0.50(x - 2) = 8.50$,

$4 + 0.50x - 1 = 8.50$, $0.50x + 3 = 8.50$,

$0.50x = 5.50$, $x = 11$ pages

19. $20 + 5(x - 3) = 55$, $20 + 5x - 15 = 55$,

$5x + 5 = 55$, $5x = 50$, $x = 10$ hours

21. $65 + 0.15(x - 100) = 245$

$65 + 0.15x - 15 = 245$, $0.15x + 50 = 245$

$0.15x = 195$, $x = 1300$ miles

23. $85 + 4.75(x - 10) = 100$

$85 + 4.75x - 47.50 = 100$,

$4.75x + 37.50 = 100$, $4.75x = 62.50$

$x \approx 13$ skaters

25. $19.95 + 0.25(x - 100) = 30$,

$19.95 + 0.25x - 25 = 30$,

$0.25x - 5.05 = 30$, $0.25x = 35.05$,

$x \approx 140$ minutes

Section 2.1

1. a.

Input x	Output $y = 15x - 4$
1	$15(1) - 4 = 15 - 4 = \$11$
2	$15(2) - 4 = 30 - 4 = \$26$
3	$15(3) - 4 = 45 - 4 = \$41$
4	$15(4) - 4 = 60 - 4 = \$56$

b. There is only one input for each output. The equation is a function.

$f(x) = 15x - 4$

3. a.

Input $x = y^2 + 2$	Output y
$(-4)^2 + 2 = 16 + 2 = 18$	-4
$(-2)^2 + 2 = 4 + 2 = 6$	-2
$(0)^2 + 2 = 2$	0
$(2)^2 + 2 = 4 + 2 = 6$	2
$(4)^2 + 2 = 16 + 2 = 18$	4

b. There are two outputs for each input. The equation is not a function.

5. a.

Input x	Output $y = 25 - x^2$
-5	$25 - (-5)^2 = 25 - 25 = 0$
-2	$25 - (-2)^2 = 25 - 4 = 21$
0	$25 - (0)^2 = 25$
2	$25 - (2)^2 = 25 - 4 = 21$
5	$25 - (5)^2 = 25 - 25 = 0$

b. There is one input for each output. The equation is a function.

$f(x) = 25 - x^2$

7. There will be one input for each output, this represents a function.

9. There could be more than one output for the same input, this is not a function.

11. There could be more than one output for the same input, this is not a function.

13. Does not pass the vertical-line test, this is not a function.

15. Passes the vertical-line test, this is a function.

17. Does not pass the vertical-line test, this is not a function.

19. Each output is two more than the corresponding input. Rule: $y = x + 2$

Section 2.1 (con't)

21. Each output is one third of the corresponding input. Rule: $y = \frac{1}{3}x$

23. Outputs are the first letter of the input. Rule: First letter in the name of a state.

25. a. On Thursday the alarm is set for 6:00 A.M.

b. No output is defined for an input of September.

c. The day after Sunday is Monday, the alarm is set for 6:00 A.M.

d. The day before Thursday is Wednesday, the alarm is set for 7:00 A.M.

27.

Input x	Output $f(x) = 5 + 2(x - 3)$
-2	$5 + 2(-2 - 3) = 5 - 10 = -5$
-1	$5 + 2(-1 - 3) = 5 - 8 = -3$
0	$5 + 2(0 - 3) = 5 - 6 = -1$
1	$5 + 2(1 - 3) = 5 - 4 = 1$
2	$5 + 2(2 - 3) = 5 - 2 = 3$
3	$5 + 2(3 - 3) = 5 + 0 = 5$
4	$5 + 2(4 - 3) = 5 + 2 = 7$

29.

Input x	Output $f(x) = x^2 - 2x - 3$
-2	$(-2)^2 - 2(-2) - 3 = 4 + 4 - 3 = 5$
-1	$(-1)^2 - 2(-1) - 3 = 1 + 2 - 3 = 0$
0	$(0)^2 - 2(0) - 3 = 0 + 0 - 3 = -3$
1	$(1)^2 - 2(1) - 3 = 1 - 2 - 3 = -4$
2	$(2)^2 - 2(2) - 3 = 4 - 4 - 3 = -3$
3	$(3)^2 - 2(3) - 3 = 9 - 6 - 3 = 0$
4	$(4)^2 - 2(4) - 3 = 16 - 8 - 3 = 5$

31.

Input x	Output $f(x) = 8 - x^2$
-2	$8 - (-2)^2 = 8 - 4 = 4$
-1	$8 - (-1)^2 = 8 - 1 = 7$
0	$8 - (0)^2 = 8 - 0 = 8$
1	$8 - (1)^2 = 8 - 1 = 7$
2	$8 - (2)^2 = 8 - 4 = 4$
3	$8 - (3)^2 = 8 - 9 = -1$
4	$8 - (4)^2 = 8 - 16 = -8$

Section 2.1 (con't)

33. a. $f(1) = 3 + 2(1 - 1)$, $f(1) = 3 + 2(0)$,

$f(1) = 3$

b. $f(5) = 3 + 2(5 - 1)$, $f(5) = 3 + 2(4)$,

$f(5) = 3 + 8$, $f(5) = 11$

c. $f(n) = 3 + 2(n - 1)$, $f(n) = 3 + 2n - 2$,

$f(n) = 2n + 1$

d. $f(n + m) = 3 + 2(n + m - 1)$,

$f(n + m) = 3 + 2n + 2m - 2$,

$f(n + m) = 2n + 2m + 1$

35. a. $f(3) = 3^3 + 3 - 2$, $f(3) = 9 + 3 - 2$,

$f(3) = 10$

b. $f(1) = 1^2 + 1 - 2$, $f(1) = 1 + 1 - 2$,

$f(1) = 0$

c. $f(\ ) = \ ^2 + \ - 2$

d. $f(n) = n^2 + n - 2$

e. $f(n - m) = (n - m)^2 + (n - m) -2$,

$f(n - m) = n^2 - 2nm + m^2 + n - m - 2$

37. A telephone number plus area code (in the United States) has 10 digits.

111 - 555 - 1234

39. A social security number has 9 digits

111 - 22 - 3333

41. The input - output sets match the graph in exercise 14.

43. The input - output sets match the graph in exercise 15.

45. In $f(x) = x^2 - 2$, x could be any real number; $f(x)$ will be ≥ -2. (x^2 is always greater than or equal to zero)

47. In $f(x) = 6 - x^2$, x could be any real number; $f(x)$ will be ≤ 6. (x^2 is always greater than or equal to zero)

49. a. The set is a domain. (inputs)

b. The inequality includes negative numbers plus zero.

51. a. The set is a range. (outputs)

b. The inequality includes positive numbers.

53. a. The set is a domain. (inputs)

b. The inequality includes positive numbers.

55. a. The set is a range. (outputs)

b. The inequality includes negative numbers plus zero.

57. a. $C(r) = 2\pi r$

b. $A(r) = \pi r^2$

59. a. $V(l, w, h) = lwh$

b. $V(r, h) = \pi r^2 h$

61. $I(P, r) = Pr$

Section 2.1 (con't)

63. a. $5 + 2(x - 3) = -3$, $5 + 2x - 6 = -3$

$2x - 1 = -3$, $2x = -2$, $x = -1$

b. $5 + 2(x - 3) = 5$, $5 + 2x - 6 = 5$

$2x - 1 = 5$, $2x = 6$, $x = 3$

c. $5 + 2(x - 3) = -11$, $5 + 2x - 6 = -11$

$2x - 1 = -11$, $2x = -10$, $x = -5$

d. $5 + 2(x - 3) = 13$, $5 + 2x - 6 = 13$

$2x - 1 = 13$, $2x = 14$, $x = 7$

65. a. From the table,

when $y = 0$; $x = 3$ or $x = -1$

b. Extending the table,

when $y = 21$; $x = -4$ or $x = 6$

c. From the table,

when $y = 5$; $x = -2$ or $x = 4$

d. Extending the table,

when $y = 60$; $x = -7$ or $x = 9$

67. a. From the table,

when $y = 4$; $x = -2$ or $x = 2$

b. From the table,

when $y = 7$; $x = -1$ or $x = 1$

c. Extending the table,

when $y = -8$; $x = -4$ or $x = 4$

d. Extending the table,

when $y = -28$; $x = -6$ or $x = 6$

Section 2.2

1. **a.** (-1, 0); (0, 2)

b. (3, 0); (0, -2)

3. (-1, 0), (1, 0); (0, -1)

5. The variable, x, has 1 as its exponent, this is a linear function.

7. The variable, x, has 1 as its exponent, this is a linear function.

9. The variable, x, has an exponent of 2, this is not a linear function.

11. The lines in exercise 2 both decrease, line 2b decreases less rapidly.

13. The lines in exercise 1 both increase, line 1a increases faster.

15. The function in exercise 4 increases for x < 0 and decreases for x > 0.

Answers for exercises 17 to 21 will vary.

23. $\frac{7-5}{-2-4} = \frac{2}{-6} = -\frac{1}{3}$

25. $\frac{-1-(-5)}{5-(-3)} = \frac{4}{8} = \frac{1}{2}$

27. $\frac{-3-0}{0-5} = \frac{-3}{-5} = \frac{3}{5}$

29. $\frac{-3-0}{0-(-2)} = \frac{-3}{2} = -\frac{3}{2}$

31. $\frac{0-(-2)}{4-4} = \frac{2}{0}$, slope is undefined

33. $\frac{-3-(-3)}{4-(-2)} = \frac{0}{6} = 0$

35. **a.** Between day 3 and day 4 the graph is a horizontal line, the slope is 0.

b. Between day 5 and day 6 the graph is rising, the slope is positive.

c. Between day 9 and day 10 the graph is a horizontal line, the slope is 0.

d. The slope between day 8 and day 9 is steeper than between day 2 and day 3, therefore the fastest drop is price is between day 8 and day 9.

e. There are only 2 segments where the graph is decreasing, between days 2 and 3, and between days 8 and 9.

37. The change in x is +10 between each entry, the change in y is +0.60 between each entry, the function is linear.

a. The slope is $\frac{\$0.60 \text{ tax}}{\$10 \text{ sales}} = \frac{\$0.06 \text{ tax}}{\$ \text{ sales}}$

b. A logical x-intercept point is (0, 0); for \$0 in sales the tax is \$0.

c. As in the previous answer the y-intercept point is (0, 0); there is \$0 tax for \$0 sales.

Section 2.2 (con't)

39. The change is x is +1 between each entry, the change in y is - 0.75 between each entry, the function is linear.

a. The slope is $\frac{-\$0.75}{1\text{ trip}}$

b. $20 - 0.75x = 0$, $20 = 0.75x$,

$20 \div 0.75 = x$, $x \approx 26.67$, the x-intercept point is $(26\frac{2}{3}, 0)$; the maximum number of trips is 26.

c. The y-intercept point is (0, 20); the original value of the ticket is \$20.

41. The change in x is 0.5 between each entry, the change in y varies between entries, the function is not linear.

43. $\frac{4-3}{3-1} = \frac{1}{2}$, $\frac{2-4}{4-3} = -2$,

$\frac{1-2}{2-4} = \frac{-1}{-2} = \frac{1}{2}$, $\frac{3-1}{1-2} = \frac{2}{-1} = -2$

Slopes of opposite lines are the same so opposite lines are parallel.

Slopes of adjacent lines are negative reciprocals so adjacent lines are perpendicular.

45.

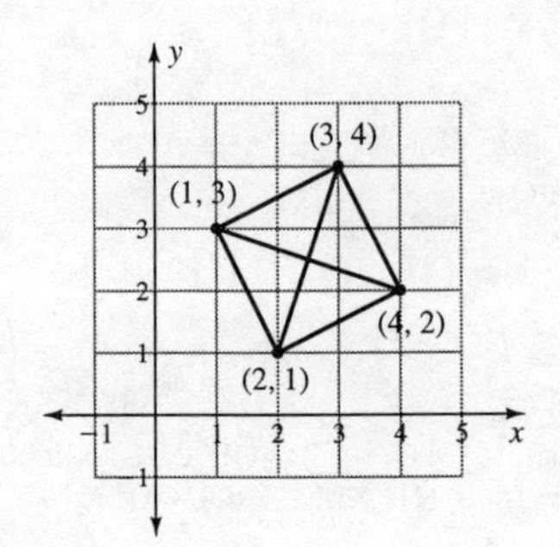

a. $\frac{4-1}{3-2} = \frac{3}{1} = 3$

b. $\frac{2-3}{4-1} = \frac{-1}{3} = -\frac{1}{3}$

Slopes are negative reciprocals, diagonals are perpendicular.

47. Using the intercepts to find the slope:

$\frac{b-0}{0-a} = -\frac{b}{a}$. The slope is the opposite of the y-intercept divided by the x-intercept.

Mid-Chapter 2 Test

1. True

2. False, the slope for a vertical line has zero in the denominator.

3. False, the change in output becomes smaller for a given change in input.

4. True

5. True

6. True

7. False, slopes of perpendicular lines multiply to -1.

8. False, $x = c$ is a vertical line.

9. The set of numbers $x \geq 0$ is called *non-negative.*

10. The set of inputs to a function is called the *domain.*

11. A vertical line has an *undefined* slope.

12. The ordered pair describing the intersection of a graph and the vertical axis is written *(0, y).*

13. a. $f(1) = 3(1) - 5,\ f(1) = -2$

b. $f(3) = 3(3) - 5,\ f(3) = 9 - 5,\ f(3) = 4$

c. $f(-5) = 3(-5) - 5,\ f(-5) = -15 - 5,$

$f(-5) = -20$

d. $f(a) = 3a - 5$

13. e. $f(a + b) = 3(a + b) - 5,$

$f(a + b) = 3a + 3b - 5$

14. $f(1) = 1^2 - 1,\ f(1) = 1 - 1,\ f(1) = 0$

$f(3) = 3^2 - 3,\ f(3) = 9 - 3,\ f(3) = 6$

$f(-5) = (-5)^2 - (-5),\ f(-5) = 25 + 5,$

$f(-5) = 30$

$f(a) = a^2 - a$

$f(a + b) = (a + b)^2 - (a + b),$

$f(a + b) = a^2 + 2ab + b^2 - a - b$

15. a. The domain is all real numbers

b. The range is $y \geq 0$

c. The graph describes a function.

16. a. The domain is all real numbers

b. The range is all real numbers

c. The graph describes a function

17. a. The domain is $-5 \leq x \leq 1$

b. The range is $-3 \leq y \leq 3$

c. The graph does not describe a function.

18. a. $\dfrac{-1-(-5)}{3-2} = \dfrac{4}{1} = 4$

b. $\dfrac{0-1}{2-(-2)} = \dfrac{-1}{4} = -\dfrac{1}{4}$

Mid-Chapter 2 Test (con't)

18. c. $\dfrac{1-2}{-1-(-4)} = \dfrac{-1}{2} = -\dfrac{1}{2}$

d. $\dfrac{-1-(-3)}{-2-(-3)} = \dfrac{2}{1} = 2$

e. $\dfrac{2-0}{-2-(-3)} = \dfrac{2}{1} = 2$

f. $\dfrac{-1-3}{-4-(-4)} = \dfrac{-4}{0}$, slope is undefined

d and e are parallel; a and b, c and d, c and e are perpendicular.

19. a.

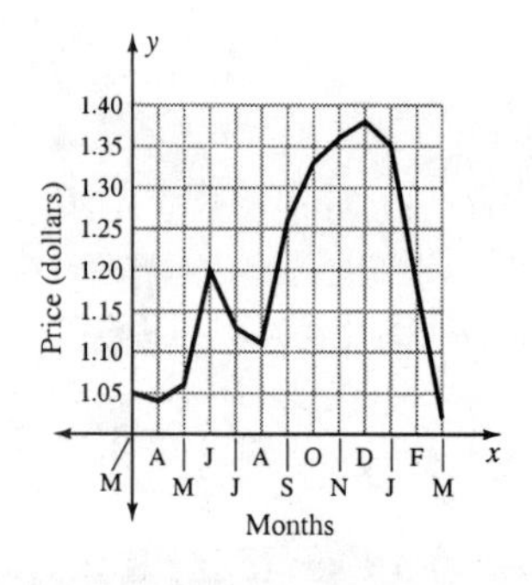

b. Steepest positive slope is between August and September;

$\dfrac{1.26-1.11}{1} = 0.15$

c. The flattest segment is between March and April; $\dfrac{1.04-1.05}{1} = -0.01$

d. Steepest negative slope is between January and February;

$\dfrac{1.18-1.35}{1} = -0.17$

19. e. The greatest change was between January and February.

f. The slope is the change in price per month.

g. Iraq's invasion of Kuwait and the Gulf war.

20. 3 days at $45.59 = 3(45.59) = $136.77,

tax, 136.77 (0.0725) = $9.92

license fee for 3 days = 0.97(3) = $2.91

159.03 - 136.77 - 9.92 - 2.91 = 9.43 for mileage plus tax.

Miles	0.149(x)+0.149(x)(0.0725) ≈ 0.1598x
20	0.1598(20) = 3.196
50	0.1598(50) = 7.99
60	0.1598(60) = 9.588
59	0.1598(59) = 9.428

The car was driven approximately 59 miles.

21. Let x equal minutes and y equal total cost.

$y = 16.45$ for $0 < x \le 30$

$y = 16.45 + 0.29(x - 30)$ for $x > 30$

Mid-Chapter 2 Test (con't)

22.

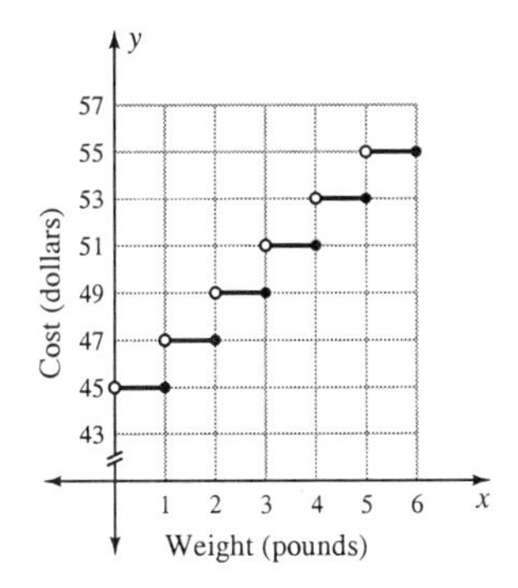

Find \$53 on the y-axis and move horizontally to graph, trace the segment vertically to the x-axis to find the number of pounds. A shipment costing \$53 has a weight of $4 < x \leq 5$ pounds.

Section 2.3

1. In $y = 0.055x$; the slope is \$0.055 per dollar and the y-intercept is zero.

3. In $y = 3.00x + 10$; the slope is \$3.00 per person and the y-intercept is \$10.

5. In $C = 2\pi r$; the slope is 2π and the y-intercept is zero.

7. In $F = \mu N$; the slope is μ and the y-intercept is zero.

9. In $C = a + bY$; the slope is b and the y-intercept is a.

For exercises 11 to 20 remember a linear equation is y = mx + b.

11. $m = 8,\ b = -4;\ \ y = 8x - 4$

13. $m = \frac{1}{2},\ b = -8;\ \ y = \frac{1}{2}x - 8$

15. $m = -2,\ b = 0;\ \ y = -2x$

17. $m = \dfrac{-2-6}{0-3},\ \ m = \dfrac{8}{3},\ b = -2;$

$y = \frac{8}{3}x - 2$

19. $m = \dfrac{-3-4}{5-(-2)},\ \ m = \dfrac{-7}{7},\ \ m = -1$

$b = y - mx,\ \ b = 4 - (-1)(-2),\ \ b = 2;$

$y = -x + 2$

21. a. Pulse rate is a function of age.

b. Answers will vary.

c. Let x = age and P = pulse rate

$P = 0.5(220 - x)$

d. $P = 0.7(220 - x)$

e. Lower limit $P = 0.5(220 - 50),\ P = 85$

Upper limit $P = 0.7(220 - 50),\ P = 119$

f. $95 = 0.5(220 - x),\ \ 190 = 220 - x,$

$x = 220 - 190,\ \ x = 30$

$133 = 0.7(220 - x),\ 190 = 220 - x,\ x = 30$

23. The fixed cost is the \$300 in fees; the variable cost per dollar is 2.5% or 0.025.

$C = 0.025x + 300$

25. $m = \dfrac{4-5}{5-2},\ \ m = \dfrac{-1}{3};$

$b = 5 - (-\frac{1}{3})(2),\ \ \ b = 5\frac{2}{3};$

$y = -\frac{1}{3}x + 5\frac{2}{3}$

27. $m = \dfrac{1-4}{4-5},\ \ m = \dfrac{-3}{-1},\ \ m = 3;$

$b = 4 - 3(5),\ \ b = 4 - 15,\ \ b = -11;$

$y = 3x - 11$

29. $m = \dfrac{5-1}{2-4},\ \ m = \dfrac{4}{-2},\ \ m = -2;$

$b = 1 - (-2)(4),\ \ b = 1 + 8,\ \ b = 9;$

$y = -2x + 9$

Section 2.3 (con't)

31. The two points are (250, \$11,000) and (300, \$13,185);

slope, V, is:

$$\frac{13{,}185 - 11{,}000}{300 - 250} = \frac{2185}{50} = 43.70$$

y-intercept, F, is:

$11{,}000 - 43.70(250) = 11{,}000 - 10{,}925$

$= 75$

$C = 43.70x + 75$; fixed cost is \$75; variable cost per pair is \$43.70

33. slope is: $\frac{212 - 32}{100 - 0} = \frac{180}{100} = \frac{9}{5}$;

y-intercept is 32; $F = \frac{9}{5}C + 32$

35. slope is: $\frac{4500 - 855}{250 - 60} = \frac{3645}{190} \approx 19.18$;

y-intercept is: $855 - 19.18(60) \approx -295.8$

or using the other point;

$4500 - 19.18(250) \approx -295$

The equation of the line is;

$y \approx 19.18x - 295$ which is nearly the same as example 9b.

37. First put the equation in standard form;

$2x + 3y = 6,\ 3y = -2x + 6,\ y = -\frac{2}{3}x + 2$;

parallel lines have the same slope, the y-intercept will be zero; $y = -\frac{2}{3}x$

39. Perpendicular lines have negative reciprocal slopes, the negative reciprocal of $-\frac{1}{2}$ is 2; the y-intercept is zero: $y = 2x$

41. Perpendicular lines have negative reciprocal slopes, the negative reciprocal of $\frac{5}{8}$ is $-\frac{8}{5}$; $b = 3 - (-\frac{8}{5})(2),\ b = 3 + \frac{16}{5}$,

$b = 6\frac{1}{5}$: $y = -\frac{8}{5}x + 6\frac{1}{5}$

43. First put equation into standard form,

$4x - 3y = 12,\ -3y = -4x + 12,\ y = \frac{4}{3}x - 4$

parallel lines have the same slope;

$b = 1 - \frac{4}{3}(-2),\ b = 3\frac{2}{3}$: $y = \frac{4}{3}x + 3\frac{2}{3}$

45. First put equation into standard form,

$5x - 2y = 8,\ -2y = -5x + 8,\ y = \frac{5}{2}x - 4$;

perpendicular lines have negative reciprocal slopes, negative reciprocal is

$-\frac{2}{5}$; $b = -1 - (-\frac{2}{5})(2),\ b = -\frac{1}{5}$:

$y = -\frac{2}{5}x - \frac{1}{5}$

Section 2.3 (con't)

47.

1st team ranking, x	2nd team ranking, y
1	16
2	15
3	14
4	13

Linear function, $\Delta x = 1$, $\Delta y = -1$;

m = -1; b = 16 - (-1)(1), b = 17;

y = -x + 17

49.

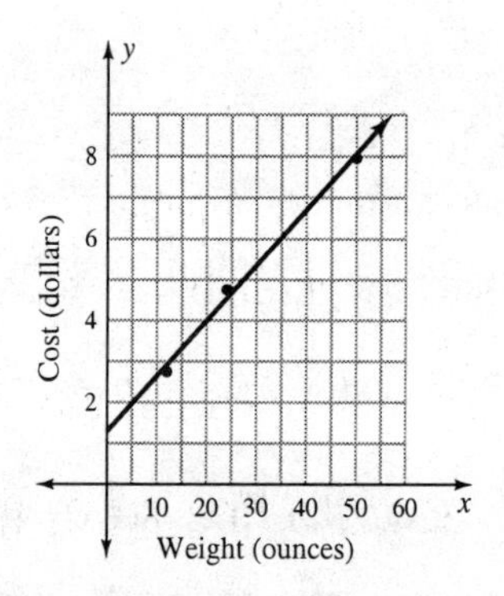

Using two points to fit a linear equation:

$$\text{slope} = \frac{7.95 - 2.75}{50 - 12} = \frac{5.20}{38} \approx 0.137;$$

y-intercept = 2.75 - 0.137(12) ≈1.11;

$y \approx 0.137x + 1.11$

Using linear regression on a calculator:

slope ≈ 0.135; y-intercept ≈ 1.29;

$y \approx 0.135x + 1.29$

51. Using linear regression on a calculator:

slope ≈ 1051.3; y-intercept ≈ -32.9;

$y \approx 1051.3x - 32.9$

Section 2.4

1. a. differences; -4, -2, -1

next term, $1 - \frac{1}{2} = \frac{1}{2}$

b. differences; -4, -4, -4

next term, -4 - 4 = -8

c. differences 8, 8, 8, 8

next term, 19 + 8 = 27

d. differences; -6, 10, -14, 18

next term, 10 - 22 = -12

3. b. $a_n = 8 + (n - 1)(-4)$, $a_n = 8 - 4n + 4$,

$a_n = -4n + 12$; matches $y = -4x + 12$

c. $a_n = -13 + (n - 1)8$, $a_n = -13 + 8n - 8$,

$a_n = 8n - 21$; matches $y = 8x - 21$

5. a. $a_1 = -1$, $d = 4$;

b. $a_{10} = 15 + 4(5) = 35$;

c. $a_n = -1 + (n - 1)4$, $a_n = -1 + 4n - 4$,

$a_n = 4n - 5$;

d. $y = 4x - 5$

7. a. $a_1 = 11$, $d = 6$;

b. $a_{10} = 35 + 6(5) = 65$;

c. $a_n = 11 + (n - 1)6$, $a_n = 11 + 6n - 6$,

$a_n = 6n + 5$;

d. $y = 6x + 5$

9. a. $a_1 = 9$, $d = 2$;

b. $a_{10} = 17 + 2(5) = 27$;

c. $a_n = 9 + (n - 1)2$, $a_n = 9 + 2n - 2$,

$a_n = 2n + 7$;

d. $y = 2x + 7$

11. a. $a_1 = 7$, $d = -3$;

b. $a_{10} = -5 + (-3)5 = -20$;

c. $a_n = 7 + (n - 1)(-3)$, $a_n = 7 - 3n + 3$,

$a_n = -3n + 10$;

d. $y = -3x + 10$

13. If the common difference in a sequence is constant the sequence can be modeled with a straight line.

15. Common difference is 80;

400, 320, 240, 160, 80, 0; it takes 5 months to repay the loan; total interest is 5($20) = $100; 100 ÷ 400 = 0.25 or 25%

17.

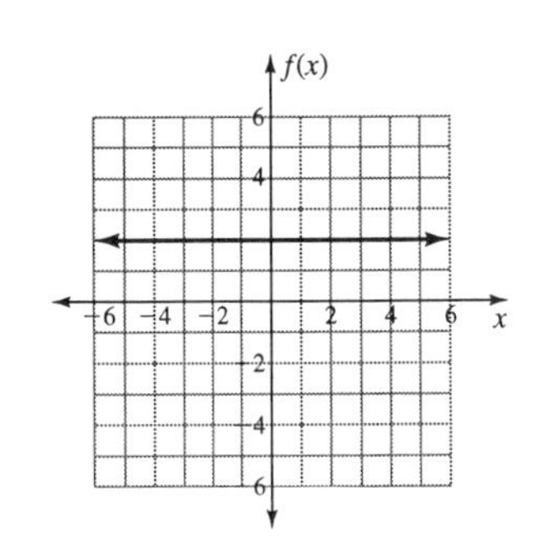

Domain: all real numbers; Range: y = 2

Section 2.4 (con't)

19.

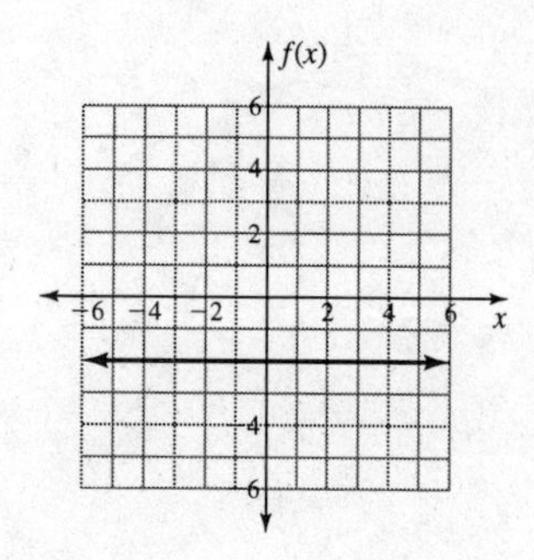

Domain: all real numbers,Range: y = -2

21. This is a constant function. Output is fixed.

23. This is a constant function, output is fixed.

25. This is a constant function, output is fixed.

27. This is an identity function, the output will always equal the input.

29.

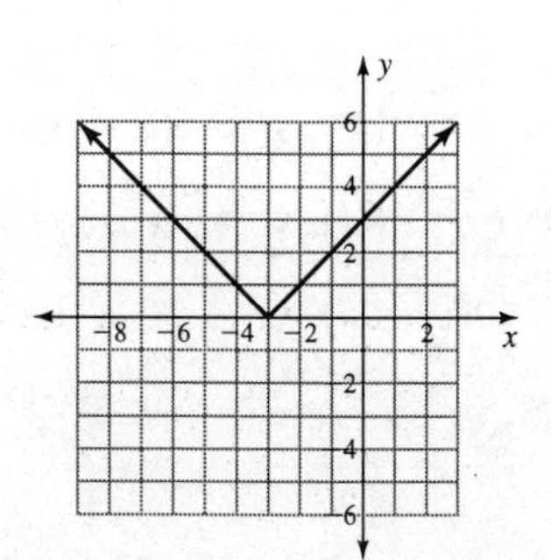

Domain is all real numbers.

Range is $y \geq 0$.

31.

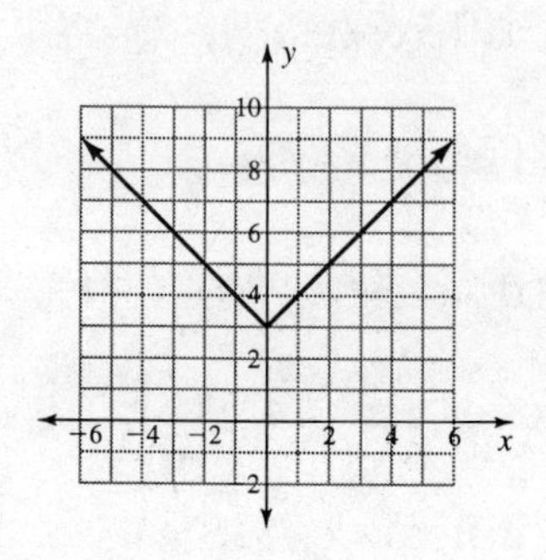

Domain is all real numbers.

Range is $y \geq 3$.

33.

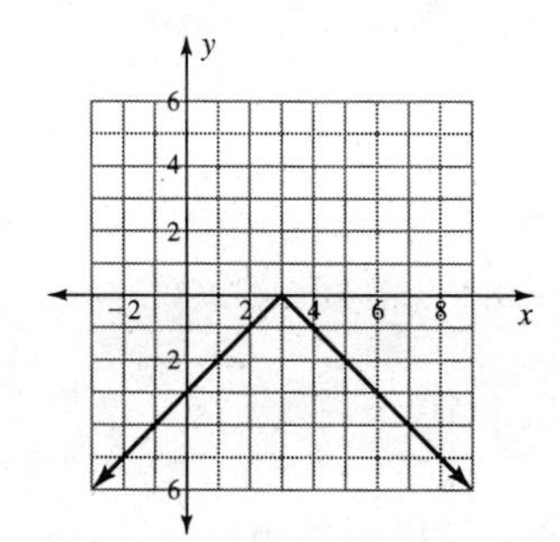

Domain is all real numbers.

Range is $y \leq 0$

35.

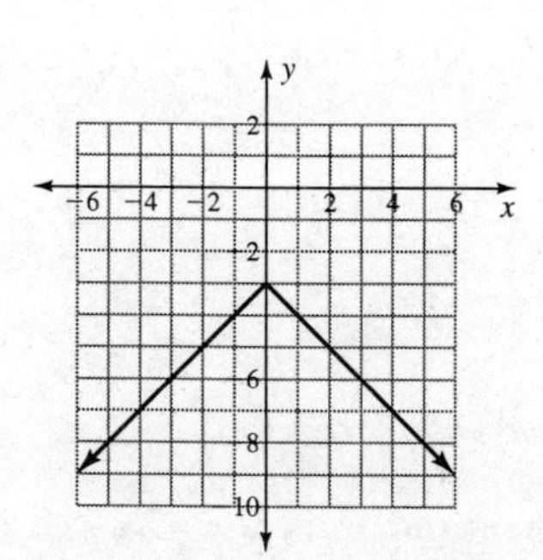

Domain is all real numbers.

Range is $y \leq -3$

Section 2.4 (con't)

37. a. From the graph, when $|2x - 3| = 5$, $x = -1$ or $x = 4$

b. From the graph, when $|2x - 3| = 3$, $x = 0$ or $x = 3$

c. From the graph, when $|2x - 3| = 0$, $x = 1.5$

d. From the graph, when $|2x - 3| = -2$, there is no solution

39. a. Picking 2 points on the left side from exercise 37, (0, 3) and (1.5, 0), the slope is $\frac{3}{-1.5} = -2$.

b. Picking 2 points on the right side from exercise 37, (1.5, 0) and (3, 3), the slope is $\frac{3-0}{3-1.5} = \frac{3}{1.5} = 2$.

c. The y-intercept is 3; an input of $x = 0$ gives the y-intercept.

d. The side of the graph containing the point (0, 3) has a slope of -2 (see **a**). The equation is $y = -2x + 3$ for $x \le 1.5$

e. $f(1.5) = -2(1.5) + 3$, $f(1.5) = 0$

f. The side of the graph containing the point (3, 3) has a slope of 2 (see **b**). The y-intercept is; $b = 3 - 2(3)$, $b = -3$. The equation is $y = 2x - 3$ for $x \ge 1.5$

39. g. $f(1.5) = 2(1.5) - 3$, $f(1.5) = 0$

41. a. From the graph, when $|0.5x + 2| = 3$, $x = -10$ or $x = 2$

b. When $|0.5x + 2| = -1$, there is no solution.

c. When $|0.5x + 2| = 1$, $x = -6$ or $x = -2$

d. Extending the graph, when $|0.5x + 2| = 4$, $x = -12$ or $x = 4$

43. a. When $|x + 2| = 0$, $x = -2$

b. $|x + 2| > 0$ is always true, the solution is all real numbers

c. $|x + 2| < 4$ when $-6 < x < 2$

d. $|x + 2| > -2$ is always true, the solution is all real numbers

e. $|x + 2| < -2$ is never true, there is no solution

f. $|x + 2| = 3$ when $x = -5$ or $x = 1$

g. $|x + 2| > 3$ when $x < -5$ or $x > 1$

Chapter 2 Review

1. The *vertical-line test* is used to find out if a graph is a function.

3. A graph for which the inputs are only integers is a *dot graph.*

5. Limits on inputs due to an application setting is the *relevant domain.*

7. The *common difference* tells if a sequence can be described by a linear equation.

9. A function where the output exactly matches the input is an *identity function.*

11. A function with a positive output for any real number input is the *absolute value function.*

13. *Subscripts* are small numbers placed below and to the right of letters or variables.

15. Ways to find a linear equation are *line of best fit, linear regression, point-slope equation, rule for the nth term of a sequence, and slope-intercept equation.*

17. Guess and Check Table:

Weight, lbs	Body mass index
110	$\frac{110(704.5)}{65^2} = 18.3$
115	$\frac{115(704.5)}{65^2} = 19.2$
114	$\frac{114(704.5)}{65^2} = 19.0$
140	$\frac{140(704.5)}{65^2} = 23.3$
145	$\frac{145(704.5)}{65^2} = 24.2$
144	$\frac{144(704.5)}{65^2} = 24.0$

19. Let x = amount of purchase, y = total cost; $y = x + 0.7x$, $y = 1.07x$; the graph will be a straight line

21. Let x = number of children, y = total cost; $y = 40$ for $0 < x \leq 10$; $y = 40 + 4.5(x - 10)$ for $x > 10$; inputs are integers only, graph is a dot graph

23. Let x = number of minutes, y = total cost; $y = 1.08$ for $0 < x \leq 2$, $y = 1.08 + 0.63(x - 2)$ for $x > 2$; inputs are rounded to next highest integer, graph is a step graph.

Chapter 2 Review (con't)

25. a. Domain is $-6.2 \le x \le 2.2$

b. Range is $-3 \le y \le 3$

c. The graph does not describe a function.

27. a. Domain is all real numbers

b. Range is $y \ge 0$

c. The graph describes a function.

29. a. From the graph, $f(0) = 1$

b. From the graph, $f(2) = 4$

c. From the graph, $f(-1) = 0.5$

d. From the graph, $f(3) = 8$

e. From the graph, $f(1) = 2$

f. From the graph, $f(x) \ge 2$ when $x \ge 1$

g. From the graph, $f(x) < 0$, never

h. From the graph, $f(x) \le 4$ when $x \le 2$

i. From the graph the y-intercept = 1

j. Domain is all real numbers

k. Range is $y > 0$

31. a. $f(x) = 0$, $\{ -1\}$

b. $f(x) = 9$, $\{-4, 2\}$

c. $f(x) = 16$, $\{-5, 3\}$

d. $f(x) = 4$, $\{-3, 1\}$

e. $f(x) = -2$, $\{ \}$

33. a. $f(5 \text{ km}) = -8.9° \text{ C}$

b. $f(3 \text{ km}) = 2.4° \text{ C}$

c. $f(x) = -3° \text{ C}$ when $x = 4$ km

d. $f(x) = 12° \text{ C}$ when $x = 1$ km

e. $f(0 \text{ km}) = 15.7° \text{ C}$

35. $f(1) = 2(1)^2 - 3(1) + 1$, $f(1) = 2 - 3 + 1$, $f(1) = 0$

37. $f(0.5) = 2(0.5)^2 - 3(0.5) + 1$, $f(0.5) = 0.5 - 1.5 + 1$, $f(0.5) = 0$

39. $f(-2) = 2(-2)^2 - 3(-2) + 1$, $f(-2) = 8 + 6 + 1$, $f(-2) = 15$

41. $f(\ \ ) = 2(\ \ )^2 - 3(\ \ ) + 1$

43. A perpendicular line will have a negative reciprocal slope, the negative reciprocal of 3 is $-\frac{1}{3}$.

a. The y-intercept is 0, $y = -\frac{1}{3}x$

b. $b = 4 - (-\frac{1}{3})(3)$, $b = 4 + 1$, $b = 5$, $y = -\frac{1}{3}x + 5$

45. Slope $= \dfrac{2-3}{2-1} = \dfrac{-1}{1} = -1$;

$b = 3 - (-1)(1) = 4$;

equation is $y = -x + 4$;

graph is neither horizontal or vertical;

x-intercept, $0 = -x + 4$, $x = 4$;

y-intercept is 4

Chapter 2 Review (con't)

47. Slope = $\frac{0-3}{0-(-3)} = \frac{-3}{3} = -1$;

equation is y = -x;

graph is neither horizontal or vertical;

x-intercept = 0;

y-intercept = 0

49. Slope = $\frac{3-(-1)}{1-(-1)} = \frac{4}{2} = 2$; b = 3 - 2(1),

b = 1;

equation is y = 2x + 1;

graph is neither horizontal or vertical

x-intercept, 0 = 2x + 1, $x = -\frac{1}{2}$;

y-intercept = 1

51. Slope = $\frac{3-3}{2-(-3)} = \frac{0}{5} = 0$;

equation is y = 3;

graph is horizontal;

there is no x-intercept;

y-intercept = 3

53. Exercises 45 and 47 have the same slope so they are parallel. There are no perpendicular lines in the specified exercises.

55. The graph of the function in exercise 28 is nearly linear, one data point, (0, 6), is not in line

57. a. y = 0.065x

b. Slope is 0.065, $ tax per $ purchased;

y-intercept is 0, no tax on zero purchases.

59. a. y = 500 + 45x

b. Slope is 45, $ per hour of repair;

y-intercept is 500, basic inspection cost

61. a. Using linear regression on a calculator:

$y \approx 6.40x + 0.19$;

Using the first and last data point:

Slope = $\frac{4.49-1.99}{0.625-0.25} = \frac{2.50}{0.375} \approx 6.67$

y-intercept = 1.99 - 6.67(.25) $\approx$ 0.33

$y \approx 6.67x + 0.33$

b. $y \approx 6.40(1) + 0.19$, $y \approx \$6.59$ or

$y \approx 6.67(1) + 0.33$, $y \approx \$7.00$

c. One possible answer is that a linear equation is not appropriate.

63. Using linear regression;

$y \approx -42x + 11{,}528$

65. Common difference is 7, next number is

36 + 7 = 43

Chapter 2 Review (con't)

67. Difference is half the previous number,

next number is 2 - 1 = 1

69. $a_1 = 4(1) - 3 = 1$; $a_2 = 4(2) - 3 = 5$,

$a_3 = 4(3) - 3 = 9$, $a_4 = 4(4) - 3 = 13$

71. Common difference is -2, $a_1 = 18$;

$a_n = 18 + (n - 1)(-2)$, $a_n = -2n + 20$

73. Common difference is 3, $a_1 = -5$;

$a_n = -5 + (n - 1)3$, $a_n = 3n - 8$

75. C = \$350, constant function

77. C = 5x + 350, increasing function

79. V = 35 - 1.50x, decreasing function

81.

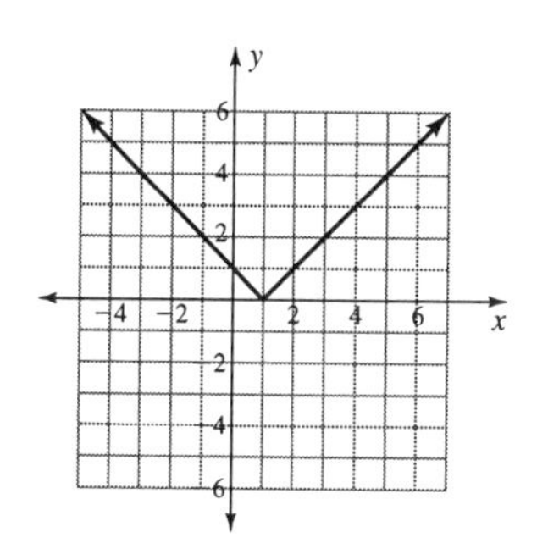

83. Domain is all real numbers,

Range is all real numbers

85. Domain is all real numbers

Range is $y \geq -1$

87. Domain is all real numbers

Range is $y \geq 0$

89. Domain is all real numbers

Range is $y = 1$

Chapter 2 Test

1.

# of Transcripts	Total Cost $
1	5
2	5
3	5 + 2 = 7
4	7 + 2 = 9
5	9 + 2 = 11
6	11 + 2 = 13

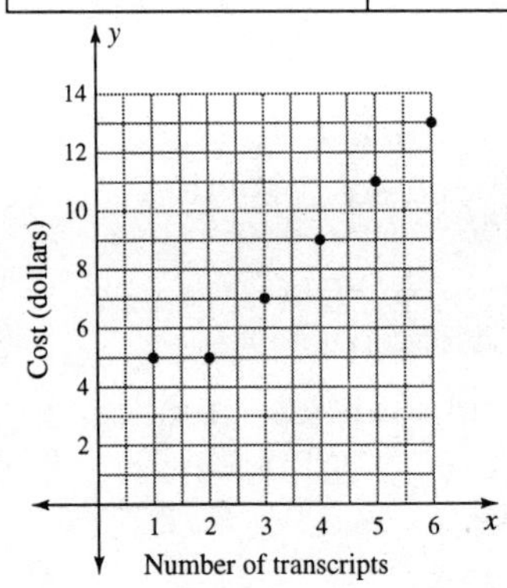

Points should not be connected because inputs are always whole numbers.

2. **a.** More than one output for one input, not a function.

b. One output for each input, function.

c. One output for each input, function.

d. More than one output for one input, not a function.

3. **a.** $f(-2) = 3(-2)^2 - 2(-2) - 4$,

$f(-2) = 3(4) + 4 - 4$, $f(-2) = 12$

b. $f(0) = 3(0)^2 - 2(0) - 4$,

$f(0) = -4$

c. $f(2) = 3(2)^2 - 2(2) - 4$,

$f(2) = 3(4) - 4 - 4$, $f(2) = 12 - 8$,

$f(2) = 4$

4. **a.** Slope $= \dfrac{4-2}{-2-5} = \dfrac{2}{-7} = -\dfrac{2}{7}$

b. $y = -\frac{2}{7}x + b$, $2 = -\frac{2}{7}(5) + b$,

$b = 2 + \frac{10}{7}$, $b = \frac{24}{7}$, $y = -\frac{2}{7}x + \frac{24}{7}$

c. A parallel line has the same slope as the original line, $-\dfrac{2}{7}$

d. A perpendicular line has a negative reciprocal slope, slope $= \dfrac{7}{2}$,

$b = -1 - \frac{7}{2}(2)$, $b = -1 - 7$, $b = -8$

$y = \frac{7}{2}x - 8$

5. **a.** The slope of a horizontal line is *zero.*

b. A line that falls from left to right has a *negative* slope and is said to be a *decreasing* function.

Chapter 2 Test (con't)

5. c. If the slope of a graph between all pairs of points is constant the graph is a *linear* function.

d. A horizontal linear graph is also called a *constant* function.

e. Linear or arithmetic sequences have a *constant* difference between terms.

5. f. The set of inputs to a sequence function is the *positive integers.*

6. a. $y = 7x + 2.50$

b. The equation is linear,

the slope is \$7 per mile

7. a. Input, time in minutes = x;

Output, cost in dollars = y

b. (1, 0.13), (19, 2.11)

c. Slope $= \dfrac{2.11 - 0.13}{19 - 1} = \dfrac{1.98}{18} = 0.11,$

$b = 0.13 - 0.11(1),\ b = 0.02,$

$y = 0.11x + 0.02$

8. Using calculator regression:

$y \approx 10.1x - 13.8$

9. a. Constant difference = 8, arithmetic sequence; next term is $42 + 8 = 50$

b. Differences are 6, 8, 10, 12; next term is $40 + 14 = 54$

c. Each term is the product of 3 and the previous term; next term is $81(3) = 243$

d. Constant difference = 7, arithmetic sequence; next term is $12 + 7 = 19$

10.

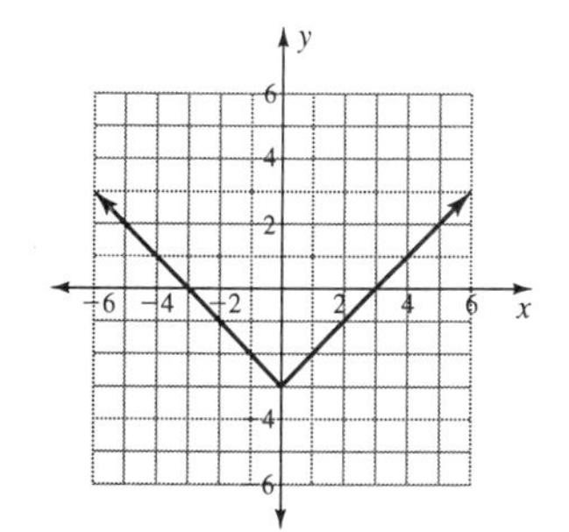

11.

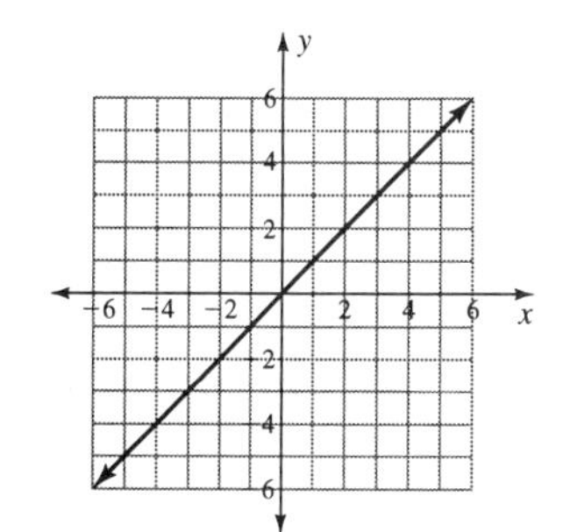

12.

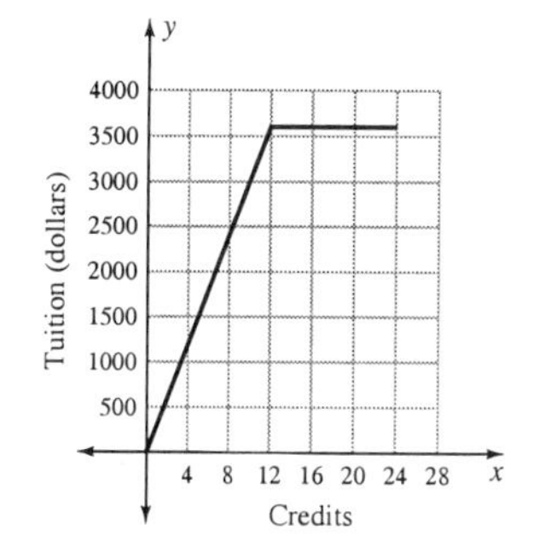

Chapter 2 Test (con't)

13.

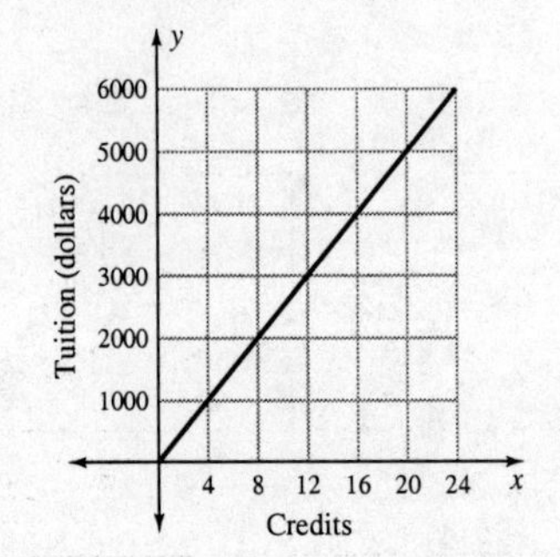

14.

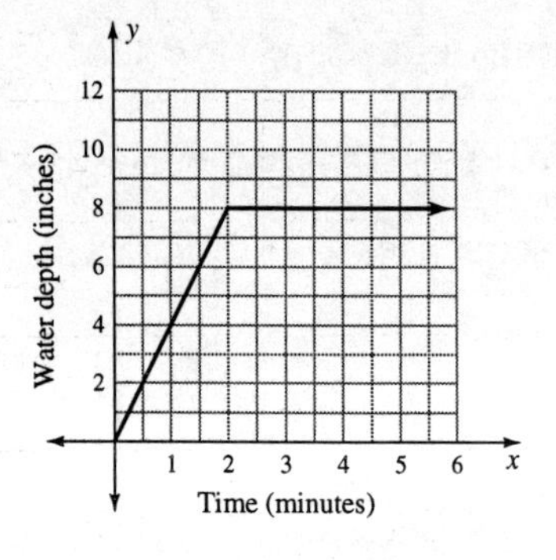

15.

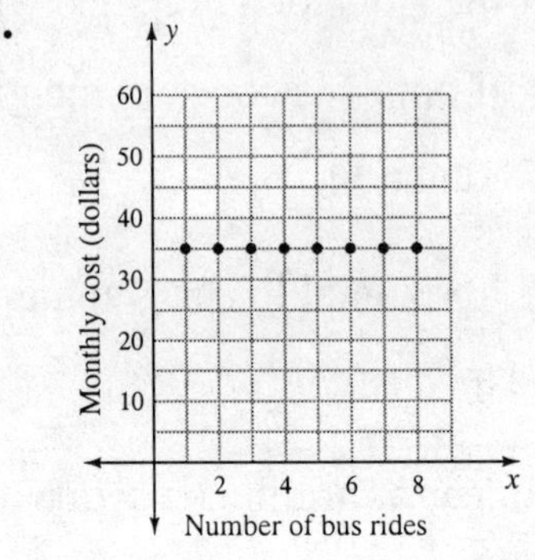

Cumulative Review

1.

Input x	Input y	Output xy	Output x + y	Output x - y
-2	4	(-2)(4) = -8	-2 + 4 = 2	-2 - 4 = -6
-3	7	(-3)(7) = -21	-3 + 7 = 4	-3 - 7 = -10
2	-6 ÷ 2 = -3	-6	2 + (-3) = -1	2 - (-3) = 5
-3	6 ÷ (-3) = -2	6	-3 + (-2) = -5	-3 - (-2) = -1
-1	-7 - (-1) = -6	(-1)(-6) = 6	-7	-1 - (-6) = 5
-7 - (-2) = -5	-2	(-5)(-2) = 10	-7	-5 - (-2) = -3
1 - (-2) = 3	-2	(3)(-2) = -6	1	3 - (-2) = 5
2	-7 - (2) = -9	(2)(-9) = -18	-7	2 - (-9) = 11

3. a. Two numbers, n and $-n$, that add to zero are *opposites.*

b. Two numbers or expressions, a and b, that are multiplied to obtain the product ab are *factors.*

c. Two numbers, n and $1/n$, that multiply to 1 are *reciprocals.*

d. Removing a common factor from two or more terms is *factoring.*

e. Collections of objects or numbers are *sets.*

5. $a(b + c) - b(a + c) + c(a - b)$

$= ab + ac - ab - bc + ac - bc = 2ac - 2bc$

7. $0.25\pi(15 \text{ ft})^2 = 0.25\pi(225 \text{ ft}^2)$

$= 56.25\pi \text{ ft}^2$

9. $\frac{16+21x}{6}$, is already simplified

11. $15 - 4x = 5(6 - x)$, $15 - 4x = 30 - 5x$,

$-4x + 5x = 30 - 15$, $x = 15$

13. $3x = x + 15$, $2x = 15$, $x = 7.5$

15. The multiplication $a(b + c)$ changes *a product to a sum.*

17. Inputs are zero or positive if we require $x \geq 0$.

<u>**Cumulative Review**</u> **(con't)**

19. To subtract real numbers, we may change *subtraction to the addition of the opposite number.*

21. $19 \le \frac{x(704.5)}{68^2}$, $19(68^2) \le x(704.5)$,

$\frac{19(68^2)}{704.5} \le x$, $124.7 \le x$;

$\frac{x(704.5)}{68^2} \le 24$, $x \le \frac{24(68^2)}{704.5}$, $x \le 157.5$

Range is $\approx 124.7 \text{ lb} \le x \le 157.5 \text{ lb}$

23. f(-1) = (-1) + 2, f(-1) = 1

f(0) = (0) + 2, f(0) = 2

f(1) = (1) + 2, f(1) = 3

f(2) = (2) + 2, f(2) = 4

25. $f(-1) = (-1)^2$, $f(-1) = 1$

$f(0) = (0)^2$, $f(0) = 0$

$f(1) = (1)^2$, $f(1) = 1$

$f(2) = (2)^2$, $f(2) = 4$

27. Slope: $\frac{1-(-3)}{-6-4} = \frac{4}{-10} = -\frac{2}{5} = -0.4$

b = 1 - (-0.4)(-6), b = 1 - 2.4, b = -1.4

y = -0.4x - 1.4

29. Slope is negative reciprocal of 3 or $-\frac{1}{3}$, the y-intercept is zero.

$y = -\frac{1}{3}x$

31. The output equals the input in each pair, the next pair is (4, 4), the rule is f(x) = x.

33. The output is 4, the next pair is (4, 4), the rule is f(x) = 4.

35. Let x be the number of workers and y the total cost. y = 65x + 500

37. The graph describes an identity function.

Section 3.0

1. a. $48 = 1 \cdot 48, 48 = 2 \cdot 24, 48 = 3 \cdot 16,$
$48 = 4 \cdot 12,\ 48 = 6 \cdot 8$

b. $36 = 1 \cdot 36, 36 = 2 \cdot 18, 36 = 3 \cdot 12,$
$36 = 4 \cdot 9, 36 = 6 \cdot 6$

c. $72 = 1 \cdot 72, 72 = 2 \cdot 36, 72 = 3 \cdot 24,$
$72 = 4 \cdot 18, 72 = 6 \cdot 12, 72 = 8 \cdot 9$

3. a. Square root is not a positive integer exponent, this is not a polynomial.

b. Two terms, positive integer exponents, this is a binomial.

5. a. One term, positive integer exponent, this is a monomial.

b. Three terms, positive integer exponents, this is a trinomial.

7.

m	n	m + n	m • n
-3	-5	-3 + (-5) = -8	(-3)(-5) = 15
3	4	7 = 3 + 4	12 = 3 · 4
2	6	2 + 6 = 8	2 · 6 = 12
3	5	8 = 3 + 5	15 = 3 · 5
-4	-6	-10 = -4 + (-6)	24 = (-4)(-6)
-2	-12	-14 = -2 + (-12)	24 = (-2)(-12)
-3	-8	-3 + (-8) = -11	(-3)(-8) = 24
2	-6	2 + (-6) = -4	(2)(-6) = -12

9. a. $(x^3 + 2x^2 + 4x) - (2x^2 + 4x + 8)$
$= x^3 + 2x^2 + 4x - 2x^2 - 4x - 8$
$= x^3 + 2x^2 - 2x^2 + 4x - 4x - 8 = x^3 - 8$

b. $(x^3 - 6x^2 + 9x) - (3x^2 - 18x + 27)$
$= x^3 - 6x^2 + 9x - 3x^2 + 18x - 27$
$= x^3 - 6x^2 - 3x^2 + 9x + 18x - 27$
$= x^3 - 9x^2 + 27x - 27$

11. a. $(x^3 + 2x^2 + x) + (x^2 + 2x + 1)$
$= x^3 + 2x^2 + x + x^2 + 2x + 1$
$= x^3 + 2x^2 + x^2 + x + 2x + 1$
$= x^3 + 3x^2 + 3x + 1$

b. $(a^3 - a^2b + ab^2) + (a^2b - ab^2 + b^3)$
$= a^3 - a^2b + ab^2 + a^2b - ab^2 + b^3$
$= a^3 - a^2b + a^2b + ab^2 - ab^2 + b^3 = a^3 + b^3$

13. $x(x^2 - 2xy + y^2) - y(x^2 - 2xy + y^2)$
$= x^3 - 2x^2y + xy^2 - x^2y + 2xy^2 - y^3$
$= x^3 - 2x^2y - x^2y + xy^2 + 2xy^2 - y^3$
$= x^3 - 3x^2y + 3xy^2 - y^3$

Section 3.0 (con't)

15.

Multiply	$2x$	$+3$
$3x$	$(2x)(3x) = 6x^2$	$(3)(3x) = 9x$
-1	$(2x)(-1) = -2x$	$(3)(-1) = -3$

$(2x + 3)(3x - 1) = 6x^2 + (9x - 2x) - 3$

$= 6x^2 + 7x - 3$

17.

Multiply	$3x$	$+1$
$3x$	$(3x)(3x) = 9x^2$	$(1)(3x) = 3x$
$+1$	$(3x)(1) = 3x$	$(1)(1) = 1$

$(3x + 1)(3x + 1) = 9x^2 + (3x + 3x) + 1$

$= 9x^2 + 6x + 1$

19.

Multiply	x^2	$-2x$	$+4$
x	$x \cdot x^2 = x^3$	$x(-2x) = -2x^2$	$x \cdot 4 = 4x$
$+2$	$2 \cdot x^2 = 2x^2$	$2(-2x) = -4x$	$2 \cdot 4 = 8$

$(x^2 - 2x + 4)(x + 2)$

$= x^3 + (-2x^2 + 2x^2) + (4x - 4x) + 8$

$= x^3 + 8$

21. $(x + 6)(x - 3) = x^2 + (6x - 3x) - 18$

$= x^2 + 3x - 18$

23. $(x - 6)(x + 3) = x^2 + (-6x + 3x) - 18$

$= x^2 - 3x - 18$

25. $(x - 9)(x - 2) = x^2 + (-9x - 2x) + 18$

$= x^2 - 11x + 18$

27. $(x + 4)(x - 4) = x^2 + (4x - 4x) - 16$

$= x^2 - 16$

29. $(x - 5)(x - 5) = x^2 + (-5x - 5x) + 25$

$= x^2 - 10x + 25$

31. $(2x - 3)(x + 4) = 2x^2 + (-3x + 8x) - 12$

$= 2x^2 + 5x - 12$

33. $(2x + 3)(x - 4) = 2x^2 + (3x - 8x) - 12$

$= 2x^2 - 5x - 12$

35. $(3x - 1)(3x + 2) = 9x^2 + (-3x + 6x) - 2$

$= 9x^2 + 3x - 2$

37. $(x + 2)^2 = x^2 + (2)(2x) + 4 = x^2 + 4x + 4$

39. $(2x - 1)(2x + 1) = 4x^2 + (2x - 2x) - 1$

$= 4x^2 - 1$

41.

Multiply	x^2	$+2x$	$+4$
x	$x \cdot x^2 = x^3$	$x(2x) = 2x^2$	$x \cdot 4 = 4x$
-2	$-2 \cdot x^2 = -2x^2$	$-2(2x) = -4x$	$-2 \cdot 4 = -8$

$(x - 2)(x^2 + 2x + 4)$

$= x^3 + (2x^2 - 2x^2) + (4x - 4x) - 8$

$= x^3 - 8$

Section 3.0 (con't)

43.

Multiply	x^2	$-3x$	-2
x^2	$x^2x^2 = x^4$	$x^2(-3x) = -3x^3$	$x^2(-2) = -2x^2$
$+3x$	$3xx^2 = 3x^3$	$3x(-3x) = -9x^2$	$3x(-2) = -6x$
-2	$-2x^2$	$-2(-3x) = 6x$	$-2(-2) = 4$

$(x^2 - 3x - 2)(x^2 + 3x - 2)$

$= x^4 + (3x^3 - 3x^3) + (-2x^2 - 9x^2 - 2x^2) +$

$(6x - 6x) + 4 = x^4 - 13x^2 + 4$

45.

Factor	**6x**	**+1**
2x	$12x^2$	$+2x$
-3	$-18x$	-3

$12x^2 - 16x - 3 = (2x - 3)(6x + 1)$

47.

Factor	**x**	**-3**
3x	$3x^2$	$-9x$
-4	$-4x$	$+12$

$3x^2 - 13x + 12 = (3x - 4)(x - 3)$

49. Diagonal sum is $7x$, product is $12x^2$;

Factor	**x**	**+3**
x	x^2	$+3x$
+4	$+4x$	$+12$

$x^2 + 7x + 12 = (x + 3)(x + 4)$

51. Diagonal sum is x, product is $-12x^2$

Factor	**x**	**+4**
x	x^2	$+4x$
-3	$-3x$	-12

$x^2 + x - 12 = (x + 4)(x - 3)$

53. Diagonal sum is $-7x$, product is $12x^2$;

Factor	**x**	**-3**
x	x^2	$-3x$
-4	$-4x$	12

$x^2 - 7x + 12 = (x - 3)(x - 4)$

55. Diagonal sum is $-11x$, product is $-12x^2$;

Factor	**x**	**-12**
x	x^2	$-12x$
+1	x	-12

$x^2 - 11x - 12 = (x - 12)(x + 1)$

Section 3.0 (con't)

57. Diagonal sum is $3x$, product is $-28x^2$;

Factor	**x**	**+7**
x	x^2	$7x$
-4	$-4x$	-28

$x^2 + 3x - 28 = (x + 7)(x - 4)$

59. Diagonal sum is $19x$, product is $60x^2$;

Factor	**2x**	**+5**
3x	$6x^2$	$+15x$
+2	$+4x$	$+10$

$6x^2 + 19x + 10 = (2x + 5)(3x + 2)$

61. Diagonal sum is $11x$, product is $-60x^2$;

Factor	**2x**	**5**
3x	$6x^2$	$15x$
-2	$-4x$	-10

$6x^2 + 11x - 10 = (2x + 5)(3x - 2)$

63. Diagonal sum is $-17x$, product is $60x^2$;

Factor	**6x**	**-5**
x	$6x^2$	$-5x$
-2	$-12x$	$+10$

$6x^2 - 17x + 10 = (6x - 5)(x - 2)$

65. Diagonal sum is $-x$, product is $-90x^2$;

Factor	**3x**	**-2**
5x	$15x^2$	$-10x$
+3	$+9x$	-6

$15x^2 - x - 6 = (3x - 2)(5x + 3)$

67. Diagonal sum is $13x$, product is $30x^2$;

Factor	**2x**	**+1**
3x	$6x^2$	$+3x$
+5	$+10x$	$+5$

$6x^2 + 13x + 5 = (2x + 1)(3x + 5)$

69. Diagonal sum is $61x$, product is $60x^2$;

Factor	**x**	**+6**
10x	$10x^2$	$+60x$
+1	$+x$	$+6$

$10x^2 + 61x + 6 = (x + 6)(10x + 1)$

71. First remove greatest common factor 2;

$2(3x^2 - 16x + 5)$;

Diagonal sum is $-16x$, product is $15x^2$;

Factor	**x**	**-5**
3x	$3x^2$	$-15x$
-1	$-x$	$+5$

$6x^2 - 32x + 10 = 2(x - 5)(3x - 1)$

Section 3.0 (con't)

73. First remove greatest common factor x;

$x(x^2 + 2x + 4)$; Diagonal sum is 2x, product is $4x^2$; There are no values that satisfy both conditions. Factored form is as shown above.

75. First remove greatest common factor x; $x(x^2 - 6x + 9)$; polynomial in parenthesis is a perfect square trinomial which factors to $(x - 3)^2$;

$x^2 - 6x^2 + 9x = x(x - 3)^2$

77. $a^3 - a^2b + ab^2 = a(a^2 - ab + b^2)$

79. $a^3 + a^2b + ab^2 = a(a^2 + ab + b^2)$

81. All are squares of binomials; exercises 28 and 29 are also squares of binomials.

83. b could be replaced by the sum of two numbers which multiply to +12. Factors of 12 are ± 1 and ± 12, ± 2 and ± 6, ± 3 and ± 4; giving values for b of ± 13, ± 8 and ± 7.

85. b could be replaced by the sum of two numbers which multiply to -20. Factors of -20 are ± 1 and ± 20, ± 2 and ± 10, ± 4 and ± 5. Keeping in mind that when one factor is positive the other must be negative, the values for b are ± 19, ± 8 and ± 1

87. Remove gcf 3; $3(x^2 - 16)$; binomial in parenthesis is the difference of squares which factors to: $(x + 4)(x - 4)$.

$3x^2 - 48 = 3(x + 4)(x - 4)$

89. Remove gcf 3; $3(4x^2 - 9)$; factor difference of squares, $(2x + 3)(2x - 3)$

$12x^2 - 27 = 3(2x + 3)(2x - 3)$

91. Remove gcf 5; $5(4x^2 - 9)$; factor difference of squares, $(2x + 3)(2x - 3)$

$20x^2 - 45 = 5(2x + 3)(2x - 3)$

93. Remove gcf 7; $7(4 - 9x^2)$; factor difference of squares, $(2 + 3x)(2 - 3x)$

$28 - 63x^2 = 7(2 + 3x)(2 - 3x)$

95. *Examples will vary*

a. Two or more numbers or expressions being multiplied are *factors.*

b. A one-term polynomial is a *monomial.*

c. A two-term polynomial is a *binomial.*

d. A three-term polynomial is a *trinomial.*

e. The result of multiplying a number times itself is a *square.*

f. The result of multiplying three identical factors is a *perfect cube.*

Section 3.1

1. $x^2 = 225,\ x = \pm\sqrt{225},\ x = \pm 15$

3. $x^2 = 1.21,\ x = \pm\sqrt{1.21},\ x = \pm 1.1$

5. $x^2 = 10000,\ x\ \pm\sqrt{10000},\ x = \pm 100$

7. $x^2 = -36$, no real number solution

9. $15^2 = 225$

11. $\sqrt{49} = 7$

13. $11^2 = 121$

15. $\sqrt{10}^2 = 10$

17. $\sqrt{1.96} = 1.4$

19. $\sqrt{75} \approx 8.660$

21. $\sqrt{40} \approx 6.325$

23. $\sqrt{200} \approx 14.142$

25. $3\sqrt{2} \approx 3(1.414) \approx 4.242$

$3\sqrt{2} = \sqrt{9\cdot 2} = \sqrt{18}$ (*ex.* 20)

27. $10\sqrt{2} \approx 10(1.414) \approx 14.142$

$10\sqrt{2} = \sqrt{100\cdot 2} = \sqrt{200}$ (*ex.* 23)

29. $3\sqrt{10} \approx 3(3.162) \approx 9.486$

$3\sqrt{10} = \sqrt{9\cdot 10} = \sqrt{90}$ (*ex.* 24)

31. $\sqrt{150} = \sqrt{25\cdot 6} = \sqrt{25}\cdot\sqrt{6} = 5\sqrt{6}$

33. $\sqrt{80} = \sqrt{16\cdot 5} = \sqrt{16}\cdot\sqrt{5} = 4\sqrt{5}$

35. $\sqrt{32} = \sqrt{16\cdot 2} = \sqrt{16}\cdot\sqrt{2} = 4\sqrt{2}$

37. $1+\sqrt{2} \approx 1+1.414 \approx 2.414$

39. $\dfrac{1-\sqrt{3}}{2} \approx \dfrac{1-1.732}{2} \approx -0.366$

41. $\dfrac{2+\sqrt{8}}{2} = \dfrac{2+\sqrt{4\cdot 2}}{2} = \dfrac{2+2\sqrt{2}}{2}$

$= \dfrac{2(1+\sqrt{2})}{2} = 1+\sqrt{2} \approx 2.414$

43. $\dfrac{3-\sqrt{27}}{6} = \dfrac{3-\sqrt{9\cdot 3}}{6} = \dfrac{3-3\sqrt{3}}{3\cdot 2}$

$= \dfrac{3(1-\sqrt{3})}{3\cdot 2} = \dfrac{1-\sqrt{3}}{2} \approx -0.366$

45. $\dfrac{4+\sqrt{8}}{4} = \dfrac{4+2\sqrt{2}}{4} = \dfrac{2(2+\sqrt{2})}{2\cdot 2} = \dfrac{2+\sqrt{2}}{2}$

$\dfrac{4+\sqrt{8}}{4} \approx 1.707,\ \dfrac{2+\sqrt{2}}{2} \approx 1.707$

47. $\dfrac{2+\sqrt{2}}{2} \approx 1.707$, already simplified

49. $\dfrac{4\sqrt{8}}{4} = \sqrt{8} = 2\sqrt{2},\ \dfrac{4\sqrt{8}}{4} \approx 2.828,$

$2\sqrt{2} \approx 2.828$

Section 3.1 (con't)

51. a. Graphs are identical for $y \geq 0$;

$y = \sqrt{x}$ is a function, $x = y^2$ is not.

b. *Examples will vary.* All examples should have the form $a = b^2, b = \sqrt{a}$; as in $9 = 3^2,\ 3 = \sqrt{9}$

c. $y=\sqrt{x}$ has the restriction $y \geq 0$; $x = y^2$ does not have this restriction.

d. x cannot be a negative number if y is a real number.

In the following exercises we are looking for a length, therefore all negative solutions have been discarded.

53. a. $x^2 = 4^2 + 8^2,\ x^2 = 16 + 64,\ x^2 = 80$

$x = \sqrt{80},\ x = \sqrt{16 \cdot 5},\ x = 4\sqrt{5}$

b. $x^2 = \sqrt{8}^2 + \sqrt{6}^2,\ x^2 = 8 + 6,\ x^2 = 14$

$x = \sqrt{14}$

c. $9^2 = 5^2 + x^2,\ x^2 = 81 - 25,\ x^2 = 56$

$x = \sqrt{56},\ x = \sqrt{4 \cdot 14},\ x = 2\sqrt{14}$

55. a. $x^2 = 8^2 + 8^2, x^2 = 2(64),\ x^2 = 128,$

$x = \sqrt{128},\ x = \sqrt{64 \cdot 2},\ x = 8\sqrt{2}$

b. $x^2 = 12^2 + 12^2,\ x^2 = 2(144),$

$x = \sqrt{2(144)},\ x = 12\sqrt{2}$

55. c. $x^2 + x^2 = 8^2,\ 2x^2 = 64,\ x^2 = 32,$

$x = \sqrt{32},\ x = \sqrt{2(16)},\ x = 4\sqrt{2}$

57. $12^2 + 16^2 = 20^2$?

$144 + 256 = 400$, Right triangle

59. $4.5^2 + 6^2 = 7.5^2$?

$20.25 + 36 = 56.25$, Right triangle

61. $8^2 + 9^2 = 12^2$?

$64 + 81 \neq 144$, Not a right triangle

63. $16^2 + 30^2 = 34^2$?

$256 + 900 = 1156$, Right triangle

65. $\sqrt{5}^2 + \sqrt{11}^2 = 4^2$?

$5 + 11 = 16$, Right triangle

67. The diagonal will be the hypotenuse;

$x^2 = 12^2 + 12^2$, $x = 12\sqrt{2}$ ft, $x \approx 16.97$ ft,

$x \approx 16$ ft 11.6 in

69. $x^2 = 13^2 + 19^2$, $x = \sqrt{169 + 361}$ ft,

$x = \sqrt{530}$ ft, $x \approx 23.02$ ft,

$x \approx 23$ ft 0.3 in

71. The antenna, wire and ground make a right triangle with the wire as the hypotenuse. $x^2 = 25^2 + 16^2$, $x^2 = 881$,

$x = \sqrt{881}$ ft, $x \approx 29.7$ ft.

Section 3.1 (con't)

73. First find the safe distance from the base of the wall; $\frac{4}{1}=\frac{16}{b}$, $b=\frac{16}{4}$, $b=4$

Length of ladder, x, is $x=\sqrt{16^2+4^2}$,

$x=\sqrt{272}$ ft, $x=4\sqrt{17}$ ft., $x\approx 16.5$ ft

For the following exercises, remember an isosceles right triangle has 2 equal sides

75. $x^2=2(20\text{ m})^2$, $x=\sqrt{2(20\text{ m})^2}$,

$x=20\sqrt{2}$ m

77. $x^2=2(\sqrt{2}\text{ ft})^2$, $x^2=4\text{ ft}^2$, $x=2$ ft

79. $x^2=2(\sqrt{18}\text{ in})^2$, $x^2=36\text{ in}^2$, $x=6$ in

81. $\frac{\sqrt{50}}{5}=\frac{5\sqrt{2}}{5}=\sqrt{2}$, $\frac{\sqrt{200}}{10}=\frac{10\sqrt{2}}{10}=\sqrt{2}$

Both ratios are $\sqrt{2}$.

In exercise 83 and 85 recall that the height will be one side of a right triangle and bisects the base of the equilateral triangle.

83. $8^2=h^2+4^2$, $h^2=64-16$, $h^2=48$,

$h=\sqrt{48}$, $h=\sqrt{16\cdot 3}$, $h=4\sqrt{3}$ in.

85. $5^2=h^2+2.5^2$, $h^2=25-6.25$, $h^2=18.75$,

$h=\sqrt{18.75}$, $h=\sqrt{\frac{75}{4}}$, $h=\frac{5\sqrt{3}}{2}$ in.

87. From example 15, $h=\frac{x\sqrt{3}}{2}$, the base of the equilateral triangle is x;

$A=\frac{1}{2}x\frac{x\sqrt{3}}{2}$, $A=\frac{x^2\sqrt{3}}{4}$

Section 3.2

1. a. Next number in the sequence is the previous number multiplied by 3, $243(3) = 729$

1st differences are: 2, 6, 18, 54, 162;

2nd differences are: 4, 12, 36, 108

Sequence is neither linear nor quadratic.

b. Next number is $39 + 7 = 46$,

1st difference is a constant , 7, sequence is linear.

c. Next number is $20 - 9 = 11$,

1st difference is a constant, -9, sequence is linear.

d. Next number is $31 + 18 = 49$,

1st differences are: 2, 6, 10, 14

2nd difference is a constant, 4, sequence is quadratic

3. a. $y = 1 - x^3$, has degree 3

b. $y = x^5 + 1$, has degree 5

c. $y = x^2 + x + 3$, has degree 2

d. $y = x - 1$, has degree 1

5. a. In $f(r) = \pi r^2 + 2\pi r(3)$ meters, the input variable is r, $a = \pi$, $b = 2\pi(3)$, $c = 0$

b. In $f(T) = \frac{g}{4\pi^2}T^2$, the input variable is T, $a = \frac{g}{4\pi^2}$, $b = 0$, $c = 0$

5. c. $y = \frac{1}{2}x(x+1)$, $y = \frac{1}{2}x^2 + \frac{1}{2}x$, the input variable is x, $a = \frac{1}{2}$, $b = \frac{1}{2}$, $c = 0$

d. $y = (x - 1)^2$, $y = x^2 - 2x + 1$, the input variable is x, $a = 1$, $b = -2$, $c = 1$

e. $y = 2(x + 1)^2 + 2$,

$y = 2(x^2 + 2x + 1) + 2$,

$y = 2x^2 + 4x + 2 + 2$, $y = 2x^2 + 4x + 4$,

the input variable is x, $a = 2$, $b = 4$, $c = 4$

7. x-intercept is where $y = 0$, from the table the x-intercepts are (1, 0) and (3, 0); y-intercept is where $x = 0$, from the table the y-intercept is (0, 3); the vertex is half-way between the x-intercepts, from the table the vertex is at (2, -1)

9.

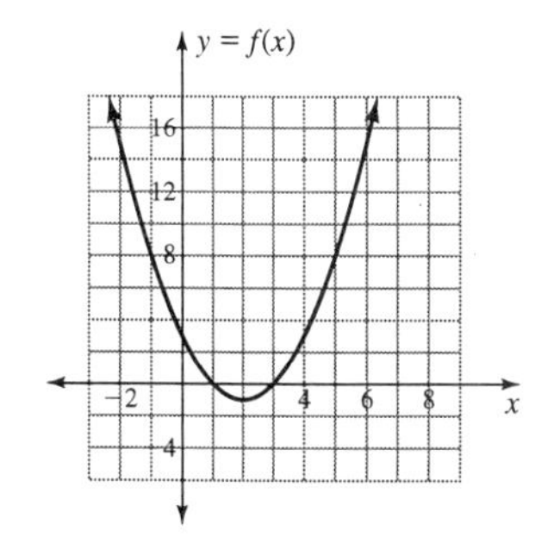

11. $x^2 - 4x + 3 = (x - 1)(x - 3)$; if we set the factors equal to zero and solve for x we get the x-intercepts.

13. Using the symmetry of the table to extend it we find $f(6) = 15$.

Section 3.2 (con't)

15. a. $x^2 - 4x + 3 = 8$, {-1, 5}

b. $x^2 - 4x + 3 = 0$, {1, 3}

c. $x^2 - 4x + 3 = 15$, {-2, 6}

d. $x^2 - 4x + 3 = -5$, { }

17. a. From the graph the x-intercept points are (-2, 0) and (4, 0).

b. From the graph the y-intercept point is (0, -8).

c. The axis of symmetry is half-way between the x-intercepts at x = 1.

d. $(1)^2 - 2(1) - 8 = -9$, the vertex is at (1, -9)

e. $x^2 - 2x - 8 = -9$, using the graph {1}

f. $x^2 - 2x - 8 = -5$, from the graph, {-1, 3}

g. $x^2 - 2x - 8 = 0$, from the graph, {-2, 4}

h. The range is $y \geq -9$

i. $x^2 - 2x - 8 = (x - 4)(x + 2)$, if we set the factors equal to zero and solve for x we get the x-intercepts.

19.

x	$f(x) = -x^2 - 2x + 8$
-5	$-(-5)^2 - 2(-5) + 8 = -7$
-4	$-(-4)^2 - 2(-4) + 8 = 0$
-3	$-(-3)^2 - 2(-3) + 8 = 5$
-2	$-(-2)^2 - 2(-2) + 8 = 8$
-1	$-(-1)^2 - 2(-1) + 8 = 9$
0	$-(0)^2 - 2(0) + 8 = 8$
1	$-(1)^2 - 2(1) + 8 = 5$
2	$-(2)^2 - 2(2) + 8 = 0$
3	$-(3)^2 - 2(3) + 8 = -7$

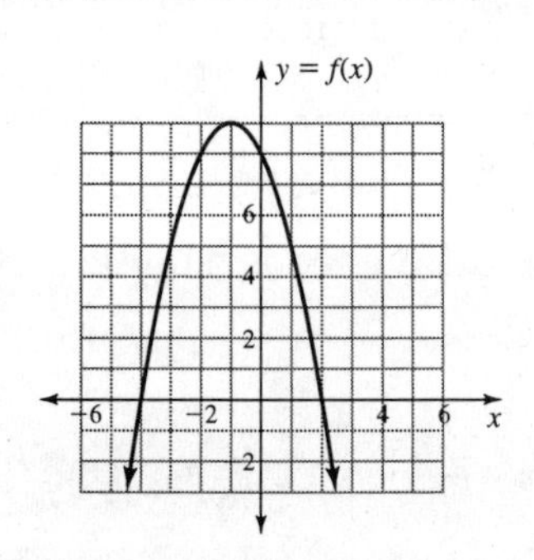

a. x-intercept points are (-4, 0) and (2, 0)

b. y-intercept point is (0, 8)

c. axis of symmetry is x = -1

d. vertex is at (-1, 9)

e. f(x) = 0 at {-4, 2}

f. f(x) = 8 at {-2, 0}

Section 3.2 (con't)

19. g. $f(x) = 10$, { }

h. $f(x) = 5$, {-3, 1}

i. Range is $y \leq 9$

j. $-x^2 - 2x + 8 = -(x + 4)(x - 2)$; if we set the factors equal to zero and solve for x, we get the x-intercepts.

21.

r	$A = 2\pi r^2 + 2\pi r(6)$
0	$2\pi(0)^2 + 2\pi(0)(6) = 0$
1	$2\pi(1)^2 + 2\pi(1)(6) \approx 44$
2	$2\pi(2)^2 + 2\pi(2)(6) \approx 100.5$
3	$2\pi(3)^2 + 2\pi(3)(6) \approx 169.6$
4	$2\pi(4)^2 + 2\pi(4)(6) \approx 251.3$
5	$2\pi(5)^2 + 2\pi(5)(6) \approx 345.6$
6	$2\pi(6)^2 + 2\pi(6)(6) \approx 452.4$

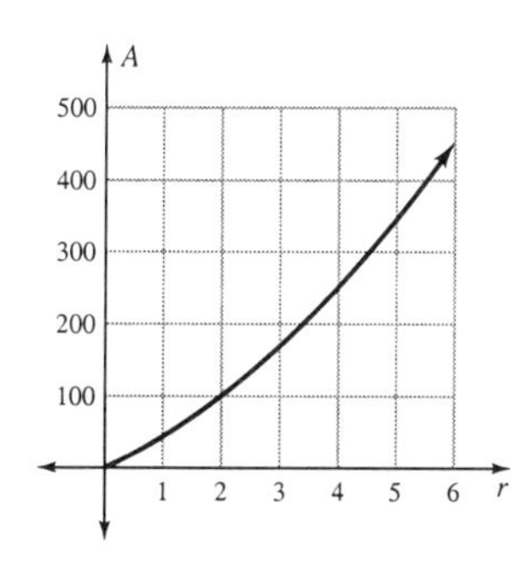

a. From the graph the radius will be approximately 3.4 in.

b. From the graph the radius will be approximately 5.5 in.

c. The top and bottom surface areas vary with the square of the radius.

23. a. The y-intercept is the initial height of 32.8 ft.

b. Values left of the y-axis have no meaning.

c. An estimate of -0.6 sec is reasonable but has no meaning in the problem setting.

d. The diver is at 32.8 ft at 0 sec. and at approximately 0.4 sec.

25. The new equation is:

$h = -16.1t^2 + 4t + 32.8$, using a graphing calculator to plot the new graph and then locating $h = 0$, we find $t \approx 1.5$ sec.

27. $h = 0$ at the water, $0 = -0.5(9.81)t^2 + 35$,

$0 = -4.905t^2 + 35$, $t \approx 2.7$ sec

29. $0 = -4.905t^2 + 1.5t + 35$, $t \approx 2.8$ sec

31. Answers will vary

33. Set the factors equal to zero and solve for x.

35. The other x-intercept will be an equal distance on the other side of the axis of symmetry, in this case 4 units to the left at $x = -4$.

Mid Chapter 3 Test

1. a. $x^3 - 6x^2 + 9x - x^2 + 18x - 27$

$= x^3 - 6x^2 - x^2 + 9x + 18x - 27$

$= x^3 - 7x^2 + 27x - 27$

b. $x^3 + 3x^2 + 9x + (-3x^2 - 9x - 27)$

$= x^3 + 3x^2 + 9x - 3x^2 - 9x - 27$

$= x^3 + 3x^2 - 3x^2 + 9x - 9x - 27$

$= x^3 - 27$

c. $16a + 4b + c - (9a + 3b + c)$

$= 16a + 4b + c - 9a - 3b - c$

$= 16a - 9a + 4b - 3b + c - c$

$= 7a + b$

d. $9a + 3b + c - (4a + 2b + c)$

$9a + 3b + c - 4a - 2b - c$

$9a - 4a + 3b - 2b + c - c$

$= 5a + b$

2. a. $(x - 5)(x + 5) = x^2 - 25$

b. $(1 - x)(1 - x) = 1 - 2x + x^2$

$= x^2 - 2x + 1$

c. $(1 - x)(x + 3) = x + 3 - x^2 - 3x$

$= -x^2 - 2x + 3$

d. $(2x + 3)(3 - 2x) = 6x - 4x^2 + 9 - 6x$

$= -4x^2 + 9$

3. a. Diagonal sum is 5x, product is $-36x^2$

Factor	x	+3
3x	$3x^2$	9x
-4	-4x	-12

$3x^2 + 5x - 12 = (x + 3)(3x - 4)$

b. $x^2 - x = x(x - 1)$

c. Diagonal sum is 5x, product is $-24x^2$

Factor	3x	+4
2x	$6x^2$	8x
-1	-3x	-4

$6x^2 + 5x - 4 = (3x + 4)(2x - 1)$

d. First remove gcf, 3x; $3x(x^2 + 2x + 1)$

$x^2 + 2x + 1 = (x + 1)^2$

$3x^3 + 6x^2 + 3x = 3x(x + 1)^2$

4. $2\sqrt{2} = \sqrt{4}\sqrt{2} = \sqrt{8}$

5. $5\sqrt{2} = \sqrt{25}\sqrt{2} = \sqrt{50}$

6. $4\sqrt{3} = \sqrt{16}\sqrt{3} = \sqrt{48}$

7. $6\sqrt{2} = \sqrt{36}\sqrt{2} = \sqrt{72}$

8. $3\sqrt{3} = \sqrt{9}\sqrt{3} = \sqrt{27}$

9. $2\sqrt{3} = \sqrt{4}\sqrt{3} = \sqrt{12}$

Mid Chapter 3 Test (con't)

10. $x^2 = 16,\ x = \pm\sqrt{16},\ x = \pm 4$

$x = \sqrt{16},\ x = 4$

$x^2 = 16$ includes both positive and negative solutions, $x = \sqrt{16}$ implies only the positive solution.

11. a. $\dfrac{3+\sqrt{6}}{3} \approx 1.816$

b. $\dfrac{\sqrt{30}-5}{10} \approx 0.048$

12. a. $7.5^2 + 10^2 = 12.5^2$?

$56.25 + 100 = 156.25$, right triangle

b. $8^2 + 15^2 = 17^2$?

$63 + 225 = 289$, right triangle

c. $6^2 + 8^2 = 12^2$?

$36 + 64 \neq 144$, not a right triangle

13. a. $13^2 = x^2 + 5^2,\ x^2 = 169 - 25,$

$x^2 = 144,\ x = 12$ (negative value discarded)

b. $20^2 = x^2 + 4^2,\ x^2 = 400 - 16,$

$x^2 = 384,\ x = \sqrt{384},\ x = \sqrt{64 \cdot 6},$

$x = 8\sqrt{6}$, (negative value discarded)

c. $x^2 = 2.5^2 + 6^2,\ x^2 = 6.25 + 36,\ x^2 = 42.25$

$x = 6.5$, (negative value discarded)

14. a. Next number is $14 + 10 = 24$,

1st differences are 2, 4, 6, 8

2nd difference is constant, 2, sequence is quadratic

b. Next number is $15 + 13 = 28$,

1st differences are -3, 1, 5, 9

2nd difference is constant, 4, sequence is quadratic

c. Next number is $20 + 3 = 23$,

1st difference is constant, 3, sequence is linear

d. Next number is $11 + 18 = 29$,

1st differences are 1, 3, 4, 7

2nd differences are 2, 1, 3, sequence is other

15. $y = \frac{1}{2}n(n+1)+1,\ y = \frac{1}{2}n^2 + \frac{1}{2}n + 1$

$a = \frac{1}{2},\ b = \frac{1}{2},\ c = 1$

16. $y = 1000(1 + x)^2,\ y = 1000(1 + 2x + x^2),$

$y = 1000x^2 + 2000x + 1000$

$a = 1000,\ b = 2000,\ c = 1000$

Mid Chapter 3 Test (con't)

17. a.

x	$A = 6x^2$
0	$6(0)^2 = 0$
1	$6(1)^2 = 6$
2	$6(2)^2 = 6(4) = 24$
3	$6(3)^2 = 6(9) = 54$
4	$6(4)^2 = 6(16) = 96$
5	$6(5)^2 = 6(25) = 150$
6	$6(6)^2 = 6(36) = 216$
7	$6(7)^2 = 6(49) = 294$
8	$6(8)^2 = 6(64) = 384$
9	$6(9)^2 = 6(81) = 486$
10	$6(10)^2 = 6(100) = 600$

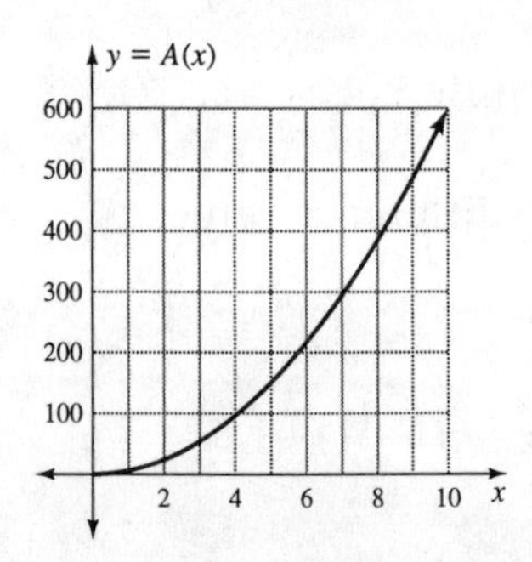

b. When we double the side the surface area is 4 times larger.

18. a. $h = -\frac{1}{2}gt^2 + v_0t + h_0$, $g = 32$ ft/sec^2

$h = -16t^2 + 60t$

18. b.

t	$h = -16t^2 + 60t$
0	$-16(0)^2 + 60(0) = 0$
0.5	$-16(0.5)^2 + 60(0.5) = 26$
1.0	$-16(1.0)^2 + 60(1.0) = 44$
1.5	$-16(1.5)^2 + 60(1.5) = 54$
2.0	$-16(2.0)^2 + 60(2.0) = 56$
2.5	$-16(2.5)^2 + 60(2.5) = 50$
3.0	$-16(3.0)^2 + 60(3.0) = 36$
3.5	$-16(3.5)^2 + 60(3.5) = 14$
4.0	$-16(4.0)^2 + 60(4.0) = -16$

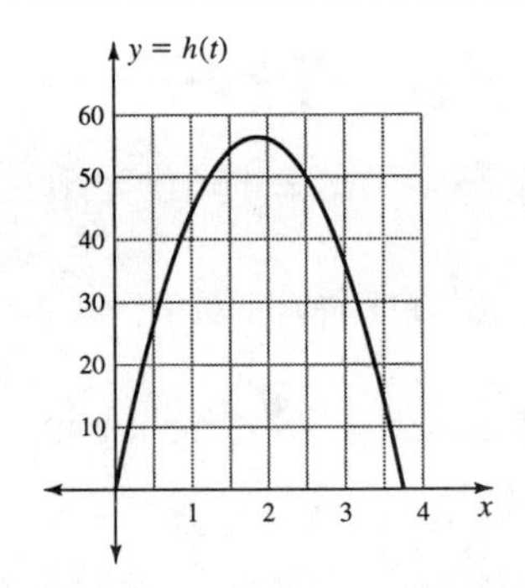

c. Using the table and the graph, the x-intercepts are at (0, 0) and (3.75, 0); the first is at the instant the ball is thrown, the second when it comes back to the ground.

d. From the graph, h = 56 at {1.75, 2}, on the way up and again on the way down.

e. Tracing the graph the highest point is approximately 56.25 ft.

Section 3.3

1. $x^2 - x - 6 = 0,\ (x - 3)(x + 2) = 0,$

$x - 3 = 0,\ x = 3,$ or

$x + 2 = 0,\ x = -2$

$\{-2, 3\}$

3. $2x^2 + x - 1 = 0,\ (x + 1)(2x - 1) = 0$

$x + 1 = 0,\ x = -1,$ or

$2x - 1 = 0,\ 2x = 1,\ x = \frac{1}{2}$

$\{-1, \frac{1}{2}\}$

5. $3x^2 - 48 = 0, 3(x^2 - 16) = 0,$

$3(x + 4)(x - 4) = 0,$

$x + 4 = 0, x = -4,$ or

$x - 4 = 0, x = 4$

$\{\pm 4\}$

Alternate method:

$3x^2 - 48 = 0,\ 3x^2 = 48, x^2 = 16,$

$x = \pm 4$

7. $x^2 + 4x + 4 = 9,\ x^2 + 4x - 5 = 0,$

$(x + 5)(x - 1) = 0,$

$x + 5 = 0,\ x = -5,$ or

$x - 1 = 0,\ x = 1$

$\{-5, 1\}$

9. $x^2 - 10x + 25 = 36,\ x^2 -10x - 11 = 0,$

$(x - 11)(x + 1) = 0,$

$x - 11 = 0,\ x = 11,$ or

$x + 1 = 0,\ x = -1$

$\{-1, 11\}$

11. $x^2 + 8x + 16 = 9,\ x^2 + 8x + 7 = 0,$

$(x + 7)(x + 1) = 0,$

$x + 7 = 0,\ x = -7,$ or

$x + 1 = 0,\ x = -1$

$\{-7, -1\}$

13. $x^2 - 4x + 4 = 16,\ x^2 - 4x - 12 = 0,$

$(x - 6)(x + 2) = 0,$

$x - 6 = 0,\ x = 6,$ or

$x + 2 = 0,\ x = -2$

$\{-2, 6\}$

15. $4x^2 - 1 = 0,\ (2x + 1)(2x - 1) = 0$

$2x + 1 = 0,\ 2x = -1,\ x = -\frac{1}{2},$ or

$2x - 1 = 0,\ 2x = 1,\ x = \frac{1}{2}$

$\{\pm \frac{1}{2}\}$

Alternate method:

$4x^2 - 1 = 0,\ 4x^2 = 1, x^2 = \frac{1}{4},$

$x = \pm \frac{1}{2}$

Section 3.3 (con't)

17. $4x^2 + 4x + 1 = 0$, $(2x + 1)(2x + 1) = 0$

$2x + 1 = 0$, $2x = -1$, $x = -\frac{1}{2}$

$\{-\frac{1}{2}\}$

19. $4x^2 - 16 = 0$, $(2x + 4)(2x - 4) = 0$,

$2x + 4 = 0$, $2x = -4$, $x = -2$, or

$2x - 4 = 0$, $2x = 4$, $x = 2$

$\{\pm 2\}$

Alternate method:

$4x^2 - 16 = 0$, $4x^2 = 16$, $x^2 = 4$,

$x = \pm 2$

21. $x^2 = 225$, $x = \pm 15$

23. $x^2 - 121 = 0$, $(x + 11)(x - 11) = 0$,

$x + 11 = 0$, $x = -11$, or

$x - 11 = 0$, $x = 11$

$\{\pm 11\}$

Alternate method:

$x^2 - 121 = 0$, $x^2 = 121$, $x = \pm 11$

25. $3x^2 - 48 = 0$, $3x^2 = 48$, $x^2 = 16$

$x = \pm 4$

27. $5x^2 - 45 = 0$, $5(x^2 - 9) = 0$,

$5(x + 3)(x - 3) = 0$,

$x + 3 = 0$, $x = -3$, or

$x - 3 = 0$, $x = 3$

$\{\pm 3\}$

Alternate method:

$5x^2 - 45 = 0$, $5x^2 = 45$, $x^2 = 9$,

$x = \pm 3$

29. $16x^2 - 48 = 0$, $16x^2 = 48$, $x^2 = 3$,

$x = \pm\sqrt{3}$, $x \approx \pm 1.7$

31. $9x^2 - 45 = 0$, $9x^2 = 45$, $x^2 = 5$,

$x = \pm\sqrt{5}$, $x \approx \pm 2.2$

33. a.

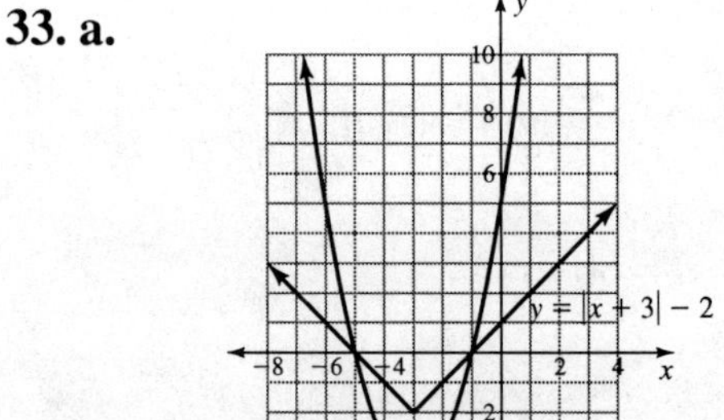
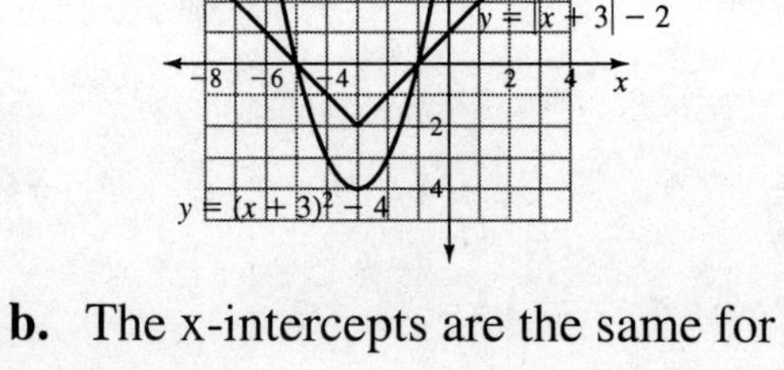

b. The x-intercepts are the same for both graphs.

c. The x-intercepts are the solutions.

35. $y = (x - 4)(x + 3)$, x-intercepts are 4 and -3.

Section 3.3 (con't)

37. $y = (2x - 5)(x + 3)$, x-intercepts are $\frac{5}{2}$ and -3.

39. $y = (3x - 4)(2x + 1)$, x-intercepts are $\frac{4}{3}$ and $-\frac{1}{2}$.

41. a. $A = \frac{\pi d^2}{4}$, $\frac{4A}{\pi} = d^2$, $d = \sqrt{\frac{4A}{\pi}}$,

$d = 2\sqrt{\frac{A}{\pi}}$, *negative solution was discarded.*

b. $\frac{\pi}{4} = 0.7854$, the keys form a square in the upper left portion of the keyboard and can be entered by starting in the upper left corner and going clockwise around the square.

43. $A = 5026 \text{ in}^2$, $d = 2\sqrt{\frac{5026}{\pi}}$,

$d \approx 80$ in or ≈ 6.7 ft.

45. $A = 4\pi r^2$, $A = 100 \text{ in}^2$,

$r = \sqrt{\frac{100}{4\pi}}$, $r \approx 2.8$ in.

47. a.

x	f(x) = |x - 2|
-3	|-3 - 2| = |-5| = 5
-1	|-1 - 2| = |-3| = 3
0	|0 - 2| = |-2| = 2
1	|1 - 2| = |-1| = 1
3	|3 - 2| = |1| = 1
5	|5 - 2| = |3| = 3

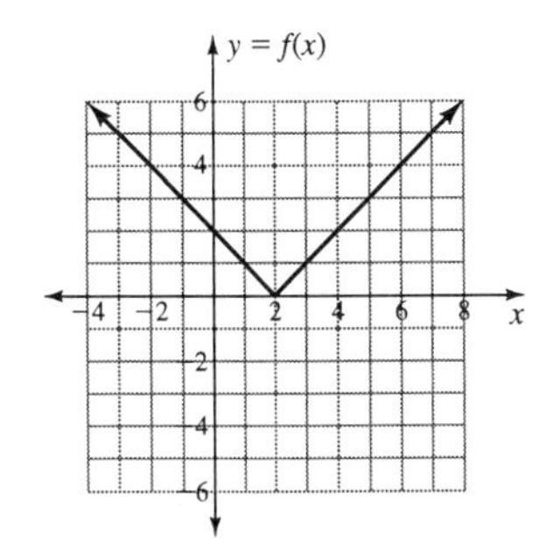

b. $f(0) = f(4) = 2$

c. The line of symmetry of $f(x) = |x - 2|$ is $x = 2$

49.

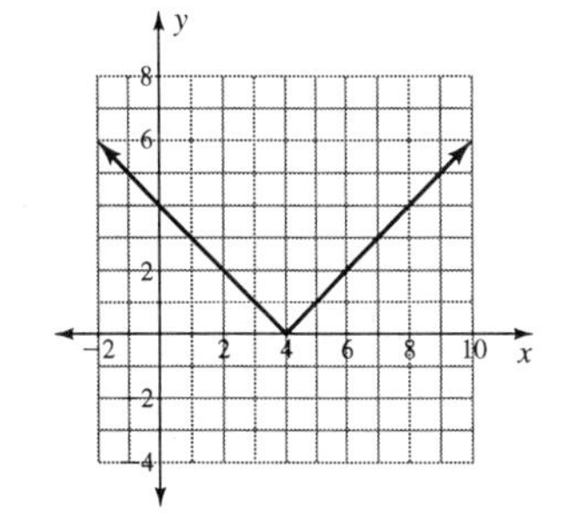

Section 3.3 (con't)

51.

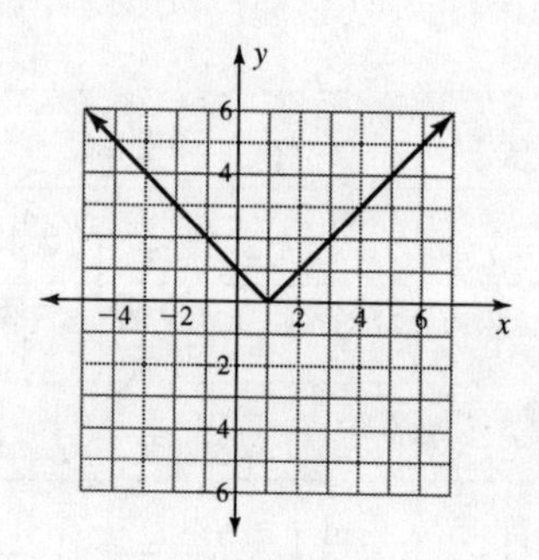

53. $|x| = 4$, $x = 4$ or $x = -4$; $\{\pm 4\}$

55. $|x+2| = 3$, $x + 2 = 3$, $x = 1$, or

$x + 2 = -3$, $x = -5$; $\{-5, 1\}$

57. $|x - 5| = 2$, $x - 5 = 2$, $x = 7$, or

$x - 5 = -2$, $x = 3$; $\{3, 7\}$

59. $|x - 4| = 2$, $x - 4 = 2$, $x = 6$, or

$x - 4 = -2$, $x = 2$; $\{2, 6\}$

61. The vertex of f(x) = abs x - 2 is at (0, -2). The vertex of f(x) = abs (x - 2) is at (2, 0), $f(x) = |x - 2|$ is f(x) = abs (x - 2).

63. $\sqrt{b^2a} = b\sqrt{a}$

65. $\sqrt{a^2b} = a\sqrt{b}$

67. $\sqrt{a^4b} = a^2\sqrt{b}$

69. When $v_0 = 0$ the equation is:

$h = -\frac{1}{2}gt^2 + h_0$,

when it hits the ground $h = 0$,

$0 = -\frac{1}{2}gt^2 + h_0$, $\frac{1}{2}gt^2 = h_0$

$t^2 = \frac{2h_0}{g}$, $t = \sqrt{\frac{2h_0}{g}}$

71. $t = \sqrt{\frac{2 \cdot 1368}{32.2}}$, $t \approx 9.2$ sec.

73. a. $r = \sqrt{12L}$, $r^2 = 12L$, $L = \frac{r^2}{12}$

b. $L = \frac{55^2}{12}$, $L \approx 252$ ft

c. $r = \sqrt{12 \cdot 100}$, $r \approx 35$ mph

Section 3.4

1. $(x + 3)^2 = x^2 + 2(3x) + 3^3 = x^2 + 6x + 9$

3. $(x - 6)^2 = x^2 + 2(-6x) + (-6)^2$

$= x^2 - 12x + 36$

5. $(x - \frac{1}{2})^2 = x^2 + 2(-\frac{1}{2}x) + (-\frac{1}{2})^2$

$= x^2 - x + \frac{1}{4}$

7. $(2x - 3)^2 = (2x)^2 + 2(-3)(2x) + (-3)^2$

$= 4x^2 - 12x + 9$

9. $x^2 - 4x + 4 = x^2 + 2(-2x) + (-2)^2 = (x - 2)^2$

11. $x^2 + 8x + 16 = x^2 + 2(4x) + 4^2 = (x + 4)^2$

13. $x^2 + 18x + 81 = x^2 + 2(9x) + 9^2$

$= (x + 9)^2$

15. $x^2 - 16x + 64 = x^2 + 2(-8x) + (-8)^2$

$= (x - 8)^2$

17.

	a	$+7$
a	a^2	$7a$
$+7$	$7a$	$+49$

$(a + 7)^2 = a^2 + 14a + 49$

19.

	x	± 8
x	x^2	$\pm 8x$
± 8	$\pm 8x$	$+64$

$(x + 8)^2 = x^2 + 16x + 64$, or

$(x - 8)^2 = x^2 - 16x + 64$

21. $x^2 + 4x = x^2 + 2x + 2x$, we add 2^2 to complete the square; $x^2 + 4x + 4$

23. $x^2 - 18x = x^2 - 9x - 9x$, we add $(-9)^2$ to complete the square; $x^2 - 18x + 81$

25. $x^2 - 7x = x^2 - \frac{7}{2}x - \frac{7}{2}x$, we add $(-\frac{7}{2})^2$ to complete the square; $x^2 - 7x + \frac{49}{4}$

27. $x^2 + 12x = x^2 + (2)(6x)$, add 6^2;

$x^2 + 12x + 36 = 13 + 36$

$x^2 + 12x + 36 = 49$

$(x + 6)^2 = 49$

29. $x^2 - 11x = x^2 + 2(-\frac{11}{2}x)$, add $(-\frac{11}{2})^2$

$x^2 - 11x + \frac{121}{4} = -18 + \frac{121}{4}$

$x^2 - 11x + \frac{121}{4} = \frac{49}{4}$

$(x - \frac{11}{2})^2 = \frac{49}{4}$

31. $3x - 4x^2 = 5$, $4x^2 - 3x + 5 = 0$

$a_2 = 4$, $a_1 = -3$, $a_0 = 5$

33. $4 = 5 - x^2$, $x^2 - 1 = 0$

$a_2 = 1$, $a_1 = 0$, $a_0 = -1$

35. $5 - 3x^2 = 4$, $3x^2 - 1 = 0$

$a_2 = 3$, $a_1 = 0$, $a_0 = -1$

37. $A = \pi r(4 + r)$, $A = 4\pi r + \pi r^2$,

$A = \pi r^2 + 4\pi r$; input variable is r,

$a = \pi$, $b = 4\pi$, $c = 0$

Section 3.4 (con't)

39. $r = n(n + 1)$, $r = n^2 + n$;

input variable is n,

$a = 1, b = 1, c = 0$

41. $A = P(1 + r)^2$, $A = P(1 + 2r + r^2)$,

$A = P + 2Pr + Pr^2$, $A = Pr^2 + 2Pr + P$;

input variable is r, $a = P$, $b = 2P$, $c = P$

43. $x = \dfrac{-(-4) - \sqrt{(-4)^2 - 4(1)(-5)}}{2(1)}$,

$x = \dfrac{4 - \sqrt{16 + 20}}{2}$, $x = \dfrac{4 - \sqrt{36}}{2}$,

$x = \dfrac{4-6}{2}$, $x = \dfrac{-2}{2}$, $x = -1$;

$a = 1, b = -4, c = -5$;

$x^2 - 4x - 5 = 0$

45. $x = \dfrac{-5 + \sqrt{5^2 - 4(2)(-12)}}{2(2)}$,

$x = \dfrac{-5 + \sqrt{25 + 96}}{4}$, $x = \dfrac{-5 + \sqrt{121}}{4}$,

$x = \dfrac{-5 + 11}{4}$, $x = \dfrac{6}{4}$, $x = \dfrac{3}{2}$;

$a = 2, b = 5, c = -12$;

$2x^2 + 5x - 12 = 0$

47. $x = \dfrac{-1 + \sqrt{1 - 4(3)(-4)}}{2(3)}$,

$x = \dfrac{-1 + \sqrt{1 + 48}}{6}$, $x = \dfrac{-1 + \sqrt{49}}{6}$,

$x = \dfrac{-1 + 7}{6}$, $x = \dfrac{6}{6}$, $x = 1$;

$a = 3, b = 1, c = -4$;

$3x^2 + x - 4 = 0$

49. $x^2 + 6x - 8 = 0$; $a = 1, b = 6, c = -8$;

$x = \dfrac{-6 \pm \sqrt{6^2 - 4(1)(-8)}}{2(1)}$,

$x = \dfrac{-6 \pm \sqrt{36 + 32}}{2}$, $x = \dfrac{-6 \pm \sqrt{68}}{2}$,

$x = \dfrac{-6 \pm 2\sqrt{17}}{2}$, $x = -3 \pm \sqrt{17}$

$\{-3 \pm \sqrt{17}\}$, solutions are irrational

51. $x^2 + 6x + 8 = 0$, $a = 1, b = 6, c = 8$;

$x = \dfrac{-6 \pm \sqrt{6^2 - 4(1)(8)}}{2(1)}$,

$x = \dfrac{-6 \pm \sqrt{36 - 32}}{2}$, $x = \dfrac{-6 \pm \sqrt{4}}{2}$,

$x = \dfrac{-6 \pm 2}{2}$,

$x = \dfrac{-6 + 2}{2}$, $x = \dfrac{-4}{2}$, $x = -2$ or

$x = \dfrac{-6 - 2}{2}$, $x = \dfrac{-8}{2}$, $x = -4$

$\{-4, -2\}$ solutions are rational

Section 3.4 (con't)

53. $2x^2 + 3x + 1 = 0$, $a = 2$, $b = 3$, $c = 1$;

$$x = \frac{-3 \pm \sqrt{3^2 - 4(2)(1)}}{2(2)},$$

$$x = \frac{-3 \pm \sqrt{9-8}}{4}, x = \frac{-3 \pm 1}{4};$$

$$x = \frac{-3+1}{4}, x = \frac{-2}{4}, x = -\frac{1}{2} \text{ or}$$

$$x = \frac{-3-1}{4}, x = \frac{-4}{4}, x = -1;$$

$\{-1, -\frac{1}{2}\}$, solutions are rational

55. $2x^2 + 3x + 2 = 0$, $a = 2$, $b = 3$, $c = 2$;

$$x = \frac{-3 \pm \sqrt{3^2 - 4(2)(2)}}{2(2)},$$

$$x = \frac{-3 \pm \sqrt{9-16}}{4}, x = \frac{-3 \pm \sqrt{-7}}{4},$$

$\{\frac{-3 \pm \sqrt{-7}}{4}\}$, not a real number solution

57. $3x^2 + 5x - 3 = 0$, $a=3$, $b = 5$, $c = -3$

$$x = \frac{-5 \pm \sqrt{5^2 - 4(3)(-3)}}{2(3)},$$

$$x = \frac{-5 \pm \sqrt{25+36}}{6}, x = \frac{-5 \pm \sqrt{61}}{6}$$

$\{\frac{-5 \pm \sqrt{61}}{6}\}$, solutions are irrational

59. $-0.12x^2 + 2.4x - 8 = 0$,

$a = -0.12$, $b = 2.4$, $c = -8$;

$$x = \frac{-2.4 \pm \sqrt{2.4^2 - 4(-0.12)(-8)}}{2(-0.12)},$$

$$x = \frac{-2.4 \pm \sqrt{5.76 - 3.84}}{-0.24},$$

$$x = \frac{-2.4 \pm \sqrt{1.92}}{-0.24}, x = \frac{-2.4 + \sqrt{1.92}}{-0.24}, \text{ or}$$

$$x = \frac{-2.4 - \sqrt{1.92}}{-0.24}$$

$\approx \{4.226, 15.774\}$

61. $-0.12x^2 + 2.4x - 10 = 0$,

$a = -0.12$, $b = 2.4$, $c = -10$

$$x = \frac{-2.4 \pm \sqrt{2.4^2 - 4(-0.12)(-10)}}{2(-0.12)},$$

$$x = \frac{-2.4 \pm \sqrt{5.76 - 4.80}}{-0.24},$$

$$x = \frac{-2.4 \pm \sqrt{0.96}}{-0.24}, x = \frac{-2.4 + \sqrt{0.96}}{-0.24}, \text{ or}$$

$$x = \frac{-2.4 - \sqrt{0.96}}{-0.24}$$

$\approx \{5.918, 14.082\}$

Section 3.4 (con't)

63. a. $-0.5(32.2)t^2 + 6t + 32.8 = 0$,

$-16.1t^2 + 6t + 32.8 = 0$;

$a = -16.1, b = 6, c = 32.8$

$$t = \frac{-6 \pm \sqrt{6^2 - 4(-16.1)(32.8)}}{2(-16.1)},$$

$$t = \frac{-6 + \sqrt{2148.32}}{-32.2}, \text{ or}$$

$$t = \frac{-6 - \sqrt{2148.32}}{-32.2}$$

$t \approx -1.3$ (discard negative) or $t \approx 1.6$ sec.

b. $-16.1t^2 + 6t + 32.8 = 23$,

$-16.1t^2 + 6t + 9.8 = 0$;

$a = -16.1, b = 6, c = 9.8$

$$t = \frac{-6 \pm \sqrt{6^2 - 4(-16.1)(9.8)}}{2(-16.1)},$$

$$t = \frac{-6 + \sqrt{667.12}}{-32.2}, \text{ or}$$

$$t = \frac{-6 - \sqrt{667.12}}{-32.2}$$

$t \approx -0.6$ (discard negative) or $t \approx 1.0$ sec.

63. c. $-16.1t^2 + 6t + 32.8 = 33$,

$-16.1t^2 + 6t - 0.2 = 0$;

$a = -16.1, b = 6, c = -0.2$

$$t = \frac{-6 \pm \sqrt{6^2 - 4(-16.1)(-0.2)}}{2(-16.1)},$$

$$t = \frac{-6 + \sqrt{23.12}}{-32.2}, \text{ or}$$

$$t = \frac{-6 - \sqrt{23.12}}{-32.2}$$

$t \approx 0.04$ sec or $t \approx 0.3$ sec

d. $-16.1t^2 + 6t + 32.8 = 35$,

$-16.1t^2 + 6t - 2.2 = 0$;

$a = -16.1, b = 6, c = -2.2$

$$t = \frac{-6 \pm \sqrt{6^2 - 4(-16.1)(-2.2)}}{2(-16.1)},$$

$$t = \frac{-6 \pm \sqrt{-105.68}}{-32.2},$$

No real number solution

Section 3.4 (con't)

63. e. $-16.1t^2 + 4t + 32.8 = 0$,

$a = -16.1$, $b = 4$, $c = 32.8$

$$t = \frac{-4 \pm \sqrt{4^2 - 4(-16.1)(32.8)}}{2(-16.1)},$$

$$t = \frac{-4 \pm \sqrt{2128.32}}{-32.2},$$

$$t = \frac{-4 - \sqrt{2128.32}}{-32.2}, \text{ or}$$

$$t = \frac{-4 + \sqrt{2128.32}}{-32.2}$$

$t \approx -1.3$ (discard negative) or $t \approx 1.6$ sec.

f. $-0.5(9.81)t^2 + 3t + 10 = 0$,

$-4.905t^2 + 3t + 10 = 0$;

$a = -4.905$, $b = 3$, $c = 10$

$$t = \frac{-3 \pm \sqrt{3^2 - 4(-4.905)(10)}}{2(-4.905)},$$

$$t = \frac{-3 \pm \sqrt{205.2}}{-9.81}$$

$t \approx -1.2$ (discard negative) or $t \approx 1.8$ sec

63. g. $-4.905t^2 + 2t + 10 = 5$,

$-4.905t^2 + 2t + 5 = 0$;

$a = -4.905$, $b = 2$, $c = 5$

$$t = \frac{-2 \pm \sqrt{2^2 - 4(-4.905)(5)}}{2(-4.905)},$$

$$t = \frac{-2 \pm \sqrt{102.1}}{-9.81}$$

$t \approx -0.8$ (discard negative) or $t \approx 1.2$ sec.

65. $\sqrt{16 + x^2}$, can not be simplified, it does not contain a perfect square

67. $\sqrt{4x^2 - 4x + 1} = \sqrt{(2x-1)^2} = |2x - 1|$

Section 3.5

	Inequality	Interval	Words	Number Line
1.	$-3 < x < 5$	$(-3, 5)$	Set of numbers greater than -3 and less than 5.	−3 0 5
3.	$-4 < x \leq 2$	$(-4, 2]$	Set of numbers greater than -4 and less than 2, including 2.	−4 0 2
5.	$x > 5$	$(5, \infty)$	Set of numbers greater than 5.	
7.	$x < -2$	$(-\infty, -2)$	Set of numbers less than -2.	−2 0
9.	$x \leq -3$	$(-\infty, -3]$	Set of numbers less than -3, including -3.	−3 0
11.	$x \geq 4$	$[4, \infty)$	Set of numbers greater than 4, including 4.	0 4

13. The graph in Example 5 shows

$2x^2 - 3x - 5 > 0$ when $x < -1$ or $x > 2.5$

15. $x^2 - 5x + 6 = (x - 2)(x - 3)$,

x-intercepts are $x = 2$ and $x = 3$,

the graph opens upward, so:

$x^2 - 5x + 6 > 0$ when $x < 2$ or $x > 3$

17. $x^2 - 5x - 14 = (x - 7)(x + 2)$,

x-intercepts are $x = 7$ and $x = -2$,

the graph opens upward, so;

$x^2 - 5x - 14 \leq 0$ on the interval $-2 \leq x \leq 7$

19. $x^2 - 5x - 6 = (x - 6)(x + 1)$,

x-intercepts are $x = 6$ and $x = -1$,

the graph opens upward, so;

$x^2 - 5x - 6 \geq 0$ when $x \leq -1$ or $x \geq 6$

21. $x^2 + 6x + 9 = (x + 3)^2$,

x-intercept is $x = -3$,

the graph opens upward, so;

$x^2 + 6x + 9 > 0$ for all real numbers, $x \neq -3$

Section 3.5 (con't)

23. $4x^2 - 4x + 1 = (2x - 1)^2$,

x-intercept is $x = \frac{1}{2}$,

the graph opens upward, so:

$4x^2 - 4x + 1 < 0$, never, { }

25. $x^2 + 2x - 15 = (x + 5)(x - 3)$,

x-intercepts are $x = -5$ and $x = 3$,

the graph opens upward, so;

$x^2 + 2x - 15 < 0$ on the interval;

$-5 < x < 3$

27. $x^2 + 10x - 24 = (x + 12)(x - 2)$,

x-intercepts are $x = -12$ and $x = 2$,

the graph opens upward, so;

$x^2 + 10x - 24 \geq 0$ when $x \leq -12$ or $x \geq 2$

29. $x^2 - 3x - 18 = (x + 3)(x - 6)$,

x-intercepts are $x = -3$ and $x = 6$,

the graph opens upward, so;

$x^2 - 3x - 18 \leq 0$ on the interval $-3 \leq x \leq 6$

31. $x^2 + 17x - 18 = (x + 18)(x - 1)$,

x-intercepts are $x = -18$ and $x = 1$,

the graph opens upward, so;

$x^2 + 17x - 18 > 0$ when $x < -18$ or $x > 1$

33. $x^2 + 8x + 16 = (x + 4)^2$,

x-intercept is $x = -4$,

the graph opens upward, so;

$x^2 + 8x + 16 \geq 0$, for all real numbers,□

35. $x^2 - 14x + 49 = (x - 7)^2$,

x-intercept is $x = 7$,

the graph opens upward, so;

$x^2 - 14x + 49 < 0$, never, { }

37. Sketching a line at $y = 10$ we can estimate the height will be greater than 10 ft on the interval $12 < x < 100$

39. The highest point appears to be ≈ 28 ft.

41. The ball would be 50 ft high.

43. If we let the origin be at the center of the bridge, then the equation for the area the team can inspect is $\frac{4}{125}x^2 \leq 20$.

Solving for the end points; $\frac{4}{125}x^2 = 20$,

$x^2 = 625,\ x = \pm\sqrt{625},\ x = \pm 25$

The section the team can inspect will be between the endpoints on the interval $-25 \leq x \leq 25$.

45. Seasonal products such as winter sports equipment, heating units, etc. would have this type of graph.

Section 3.5 (con't)

47. The 182nd day from July 1 is Dec. 29.

49. a. For revenue; $b \approx 245 - 120$, $b \approx 125$

$h \approx 80{,}000$; $A \approx \frac{2}{3}(125)(80{,}000)$,

rounding to the nearest 100 thousand,

$A \approx \$6{,}700{,}000$

b. For cost; $b \approx 365$, $h \approx 17{,}500$;

$A \approx \frac{2}{3}(365)(17{,}500)$, rounding again,

$A \approx \$4{,}300{,}000$

c. Revenues are larger than costs, there will be a profit.

51. For both curves $b \approx 30$ and $h \approx 42{,}000$;

Total area is: $A \approx 2(\frac{2}{3})(30)(42{,}000)$,

$A \approx 1{,}680{,}000$ metric tons

53. $y \leq x^2$ is graph c

55. $y \geq 0$ and $y \leq -x^2 + 3$ is graph f,

note graph opens downward ($-x^2$), and y-intercept is 3

57. $y \geq 0$ and $y \leq x^2 - 3$ is graph a.

59. x^2 is always ≥ 0, if $y \geq x^2$ than y must also be ≥ 0.

Chapter 3 Review

1. **a.** Two terms, binomial

   b. Not a polynomial, exponents are not positive integers

   c. Exponent not a positive integer, not a polynomial

   d. Two terms, binomial

   e. Three terms, trinomial

   f. Not a polynomial, absolute value function.

3. **a.** $(x - 3)(x + 3) = x^2 - 9$

   b. $(2x - 5)(2x - 5)$

   $= (2x)^2 + (2)(2x)(-5) + (-5)^2$

   $= 4x^2 - 20x + 25$

   c.

Multiply	x^2	$+x$	$+1$
x	x^3	x^2	x
-1	$-x^2$	$-x$	-1

$(x - 1)(x^2 + x + 1) = x^3 - 1$

   d. $(n + 4)(n + 4) = n^2 + (2)(4n) + 4^2$

   $= n^2 + 8n + 16$

3. **e.**

Multiply	$2x$	-3
$3x$	$6x^2$	$-9x$
$+4$	$8x$	-12

$(2x - 3)(3x + 4) = 6x^2 - x - 12$

f.

Multiply	x^2	$+x$	-1
x^2	x^4	x^3	$-x^2$
$-x$	$-x^3$	$-x^2$	x
-1	$-x^2$	$-x$	$+1$

$(x^2 + x - 1)(x^2 - x - 1) = x^4 - 3x^2 + 1$

5. **a.** Diagonal product $= -4x^2$, sum $= 3x$

Factor	x	$+4$
x	x^2	$+4x$
-1	$-x$	-4

$x^2 + 3x - 4 = (x + 4)(x - 1)$

   b. $2x^2 - 3x = x(2x - 3)$

   c. Diagonal product $= -6x^2$, sum $= +x$

Factor	x	-1
$2x$	$2x^2$	$-2x$
$+3$	$3x$	-3

$2x^2 + x - 3 = (2x + 3)(x - 1)$

Chapter 3 Review (con't)

5. d. $9x^x + 12x + 4 = (3x + 2)^2$

e. $x^2 - x = x(x - 1)$

f. $3x^2 + 6x + 3 = 3(x^2 + 2x + 1)$

$= 3(x + 1)^2$

7. a. $\sqrt{75} = \sqrt{25 \cdot 3} = 5\sqrt{3}$

b. $\sqrt{8} = \sqrt{4 \cdot 2} = 2\sqrt{2}$

c. $\sqrt{32} = \sqrt{16 \cdot 2} = 4\sqrt{2}$

9. a. $\dfrac{3 - 3\sqrt{6}}{3} = \dfrac{3(1 - \sqrt{6})}{3} = 1 - \sqrt{6}$

b. $\dfrac{3 + 3\sqrt{6}}{3} = \dfrac{3(1 + \sqrt{6})}{3} = 1 + \sqrt{6}$

11. a. $7.5^2 + 10^2 = 12.5^2$?

56.25 + 100 = 156.25, right triangle

b. $18^2 + 24^2 = 30^2$?

324 + 576 = 900, right triangle

c. $(\sqrt{5})^2 + (\sqrt{8})^2 = (\sqrt{13})^2$?

5 + 8 = 13, right triangle

d. $4^2 + (\sqrt{20})^2 = 6^2$?

16 + 20 = 36, right triangle

13. $z^2 = x^2 + x^2$, $z^2 = 2x^2$, $z = \pm\sqrt{2x^2}$,

$z = x\sqrt{2}$; $\dfrac{z}{x} = \sqrt{2}$ (note negative value for z was discarded as z is a length)

15. False, $\sqrt{x}$ is always positive, on the graph $y \geq 0$.

17. First differences, 1, 5, 9, 13, ...

Second differences, 4, 4, 4, 4, ...

Next number = 25 + 17 = 42;

Quadratic sequence

19. First differences, 4, 4, 4, 4, ...

Next number = 10 + 4 = 14;

Linear sequence

21. a.

x	$x^2 - x - 6$
-2	$(-2)^2 - (-2) - 6 = 0$
-1	$(-1)^2 - (-1) - 6 = -4$
0	$0^2 + 0 - 6 = -6$
1	$1^2 - 1 - 6 = -6$
2	$2^2 - 2 - 6 = -4$

b. Extending the table, x-intercepts are (-2, 0) and (3, 0)

c. The axis of symmetry is between $x = 0$ and $x = 1$ at $x = 0.5$

d. $(0.5)^2 - (0.5) - 6 = -6.25$

vertex is at (0.5, -6.25)

Chapter 3 Review (con't)

21. e.

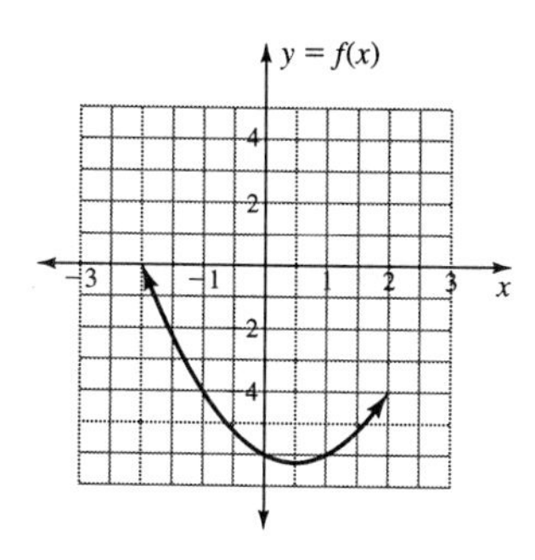

f. $x^2 - x - 6 = 6,\ x^2 - x - 12 = 0,$

$(x - 4)(x + 3) = 0;\ x - 4 = 0, x = 4$

$x + 3 = 0,\ x = -3;$

$\{-3, 4\}$

g. $x^2 - x - 6 = 0, (x - 3)(x + 2) = 0,$

$x - 3 = 0,\ x = 3; x + 2 = 0, x = -2$

$\{-2, 3\}$

h. $x^2 - x - 6 = -7$, from the graph, { },

never.

23.

r	$A = 20{,}000(1 - r)^2$
0	$20{,}000(1 - 0)^2 = 20{,}000$
1% = 0.01	$20{,}000(1 - 0.01)^2 = 19{,}602$
2% = 0.02	$20{,}000(1 - 0.02)^2 = 19{,}208$
3% = 0.03	$20{,}000(1 - 0.03)^2 = 18{,}818$
4% = 0.04	$20{,}000(1 - 0.04)^2 = 18{,}432$
5% = 0.05	$20{,}000(1 - 0.05)^2 = 18{,}050$
6% = 0.06	$20{,}000(1 - 0.06)^2 = 17{,}672$
7% = 0.07	$20{,}000(1 - 0.07)^2 = 17{,}298$
8% = 0.08	$20{,}000(1 - 0.08)^2 = 16{,}928$

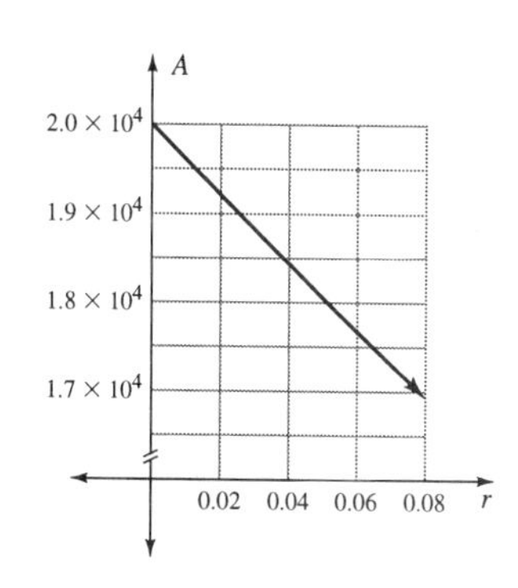

25. $C = 2\pi\sqrt{\dfrac{a^2 + b^2}{2}},\ \dfrac{C}{2\pi} = \sqrt{\dfrac{a^2 + b^2}{2}},$

$$\frac{C^2}{4\pi^2} = \frac{a^2 + b^2}{2},\ \frac{C^2}{2\pi^2} = a^2 + b^2,$$

$$\frac{C^2}{2\pi^2} - b^2 = a^2,\ a = \sqrt{\frac{C^2}{2\pi^2} - b^2}$$

Chapter 3 Review (con't)

27. $x^2 - 5x + 6 = 0$; $(x - 3)(x - 2) = 0$,

$(x - 3) = 0$, $x = 3$; or $(x - 2) = 0$, $x = 2$

$\{2, 3\}$;

$a = 1$, $b = -5$, $c = 6$

$$x = \frac{-(-5) \pm \sqrt{(-5)^2 - 4(1)(6)}}{2(1)},$$

$$x = \frac{5 \pm \sqrt{1}}{2},\ x = \frac{5+1}{2},\ x = 3, \text{ or}$$

$$x = \frac{5-1}{2},\ x = 2$$

$\{2, 3\}$

29. $x^2 = 6x - 9$, $x^2 - 6x + 9 = 0$, $(x - 3)^2 = 0$

$x - 3 = 0$, $x = 3$

$\{3\}$;

$a = 1$, $b = -6$, $c = 9$

$$x = \frac{-(-6) \pm \sqrt{(-6)^2 - 4(1)(9)}}{2(1)},$$

$$x = \frac{6 \pm \sqrt{0}}{2},\ x = 3$$

$\{3\}$

31. $3x^2 + x - 4 = 0$, $(x - 1)(3x + 4) = 0$,

$x - 1 = 0$, $x = 1$; or

$3x + 4 = 0$, $3x = -4$, $x = -\frac{4}{3}$

$\{-\frac{4}{3}, 1\}$;

$a = 3$, $b = 1$, $c = -4$

$$x = \frac{-1 \pm \sqrt{1^2 - 4(3)(-4)}}{2(3)},$$

$$x = \frac{-1 \pm \sqrt{49}}{6},\ x = \frac{-1+7}{6},\ x = 1, \text{ or}$$

$$x = \frac{-1-7}{6},\ x = \frac{-4}{3}$$

$\{-\frac{4}{3}, 1\}$;

33. $9x^2 = 4$, $9x^2 - 4 = 0$, $(3x + 2)(3x - 2) = 0$,

$3x + 2 = 0$, $3x = -2$, $x = -\frac{2}{3}$ or

$3x - 2 = 0$, $3x = 2$, $x = \frac{2}{3}$

$\{\pm \frac{2}{3}\}$;

$a = 9$, $b = 0$, $c = -4$

$$x = \frac{0 \pm \sqrt{0 - 4(9)(-4)}}{2(9)},$$

$$x = \frac{\pm\sqrt{144}}{18},\ x = \frac{\pm 12}{18},\ x = \pm\frac{2}{3}$$

$\{\pm \frac{2}{3}\}$;

Chapter 3 Review (con't)

35. a. $h = -\frac{32}{2}t^2 + 72t + 0$, $h = -16t^2 + 72t$

b. Factoring the equation with h = 0,

$t(-16t + 72) = 0$, $t = 0$ or

$-16t + 72 = 0$, $-16t = -72$, $t = 4.5$

t-intercepts are (0, 0) and (4.5, 0);

(0, 0) is when the ball is thrown, (4.5, 0) is when it comes back down.

c. $-16t^2 + 72t = 80$, $-16t^2 + 72t - 80 = 0$

$a = -16$, $b = 72$, $c = -80$;

$$t = \frac{-72 \pm \sqrt{72^2 - 4(-16)(-80)}}{2(-16)},$$

$$t = \frac{-72 \pm 8}{-32}, t = \frac{-72 + 8}{-32}, t = 2 \text{ or}$$

$$t = \frac{-72 - 8}{-32}, t = 2.5$$

{2, 2.5}

The ball passes through 80 ft on the way up and on the way down.

d. The axis of symmetry is at t = 2.25,

$h = -16(2.25)^2 + 72(2.25)$, $h = 81$ ft.

37. $\frac{x^2}{2.000 - x} = 12$, $x^2 = 24 - 12x$,

$$x^2 + 12x - 24 = 0$$

$$x = \frac{-12 \pm \sqrt{12^2 - 4(1)(-24)}}{2(1)},$$

$$x = \frac{-12 \pm 4\sqrt{15}}{2}, \; x = -6 \pm 2\sqrt{15}$$

$x \approx -13.746$ (discard this answer, it is not within the restriction on x)

$x \approx 1.746$

39. $\frac{(2x)^2}{(2.000 - x)(1.000 - x)} = 24$

$$\frac{4x^2}{2 - 3x + x^2} = 24, \; 4x^2 = 48 - 72x + 24x^2$$

$$20x^2 - 72x + 48 = 0$$

$$x = \frac{-(-72) \pm \sqrt{(-72)^2 - 4(20)(48)}}{2(20)},$$

$$x = \frac{72 \pm 8\sqrt{21}}{40}, \; x = \frac{9 \pm \sqrt{21}}{5}$$

$x \approx 2.717$ (discard this answer, it is not within the restriction on x)

$x \approx 0.883$

Chapter 3 Review (con't)

41. Standard form

Add 1 to both sides

Add $\frac{1}{4}$ to both sides

Factor the left side

Add terms on the right side

Take the square root of both sides

Use definition of principal square root

Use definition of absolute value

Add $\frac{1}{2}$ to both sides, evaluate and round

43. $x^2 + 6x - 7 \geq 0$, $(x + 7)(x - 1) \geq 0$;

x-intercepts are (-7, 0) and (1, 0);

graph opens upward, function is ≥ 0

when $x \leq -7$ or $x \geq 1$

45. $2x^2 - 5x - 3 > 0$, $(2x + 1)(x - 3) > 0$;

x-intercepts are $(-\frac{1}{2}, 0)$ and (3, 0);

graph opens upward, function is > 0

when $x < -\frac{1}{2}$ or $x > 3$

47. $-4x^2 + 5x + 2 < -7$, $-4x^2 + 5x + 9 < 0$,

$(-4x + 9)(x + 1) < 0$; end points are at

(-1, 0) and $(\frac{9}{4}, 0)$; graph opens

downward, function is < 0 when $x < -1$

or $x > \frac{9}{4}$

49. True, the vertex is the maximum or

minimum y-value.

51. True, the vertex is on the axis of

symmetry.

53. a. $L = 500\left|1 + \frac{8}{3}\left(\frac{30}{500}\right)^2\right|$, $L = 504.8$ ft

b. $L = 500\left|1 + \frac{8}{3}\left(\frac{30}{500}\right)^2 - \frac{32}{5}\left(\frac{30}{500}\right)^4\right|$,

$L \approx 594.7585$ ft

c. $L =$

$500\left|1 + \frac{8}{3}\left(\frac{30}{500}\right)^2 - \frac{32}{5}\left(\frac{30}{500}\right)^4 + \frac{256}{7}\left(\frac{30}{500}\right)^6\right|$,

$L \approx 504.7594$ ft

55. a. $y < 0$, $y < -x^2 + 4$; graph 3

b. $y > 0$, $y > -x^2 + 4$; graph 1

c. $y < 0$, $y > -x^2 + 4$; graph 4

d. $y > 0$, $y < -x^2 + 4$; graph 2

Chapter 3 Test

1. $x = \dfrac{-b \pm \sqrt{b^2 - 4ac}}{2a}$

2.

Multiply	3x	-4
4x	$12x^2$	-16x
-3	-9x	12

$(3x - 4)(4x - 3) = 12x^2 - 25x + 12$

3.

Multiply	x^2	+3x	+9
x	x^3	$3x^2$	9x
-3	$-3x^2$	-9x	-27

$(x - 3)(x^2 + 3x + 9) = x^3 - 27$

4. Diagonal product is $12x^2$, sum is -7x;

Factor	2x	-3
x	$2x^2$	-3x
-2	-4x	+6

$2x^2 - 7x + 6 = (2x - 3)(x - 2)$

5. $0.04x^2 - 169 = 0$,

$(0.2x + 13)(0.2x - 13) = 0$,

$0.2x + 13 = 0$, $0.2x = -13$, $x = -65$,

$0.2x - 13 = 0$, $0.2x = 13$, $x = 65$

$\{\pm 65\}$

6. $\sqrt{98} = \sqrt{49 \cdot 2} = 7\sqrt{2}$

7. $\dfrac{4 - 2\sqrt{6}}{2} = \dfrac{2(2 - \sqrt{6})}{2} = 2 - \sqrt{6}$

8. **a.** $9^2 + x^2 = (2x)^2$, $81 + x^2 = 4x^2$

$3x^2 = 81$, $x^2 = 27$, $x = \sqrt{27}$, $x = 3\sqrt{3}$

b. $2x^2 = 7^2$, $2x^2 = 49$, $x^2 = \dfrac{49}{2}$,

$x = \sqrt{\dfrac{49}{2}}$, $x = \dfrac{7}{\sqrt{2}}$, $x = \dfrac{7\sqrt{2}}{\sqrt{2}\sqrt{2}}$, $x = \dfrac{7\sqrt{2}}{2}$

9. **a.** First differences; 8, 10, 12, 14

Second differences; 2, 2, 2

Next number = 38 + 16 = 54

Quadratic sequence

b. First differences; -8, -8, -8, -8

Next number = -9 + -8 = -17

Linear sequence

c. First differences; 6, 6, 6, 6

Next number = 26 + 6 = 32

Linear sequence

d. First differences; -8, -7, -6, -5

Second differences; 1, 1, 1

Next number = -1 + -4 = -5

Quadratic sequence

Chapter 3 Test (con't)

10. $|x - 2| = 1$, $x - 2 = \pm 1$,

$x - 2 = 1$, $x = 3$ or $x - 2 = -1$, $x = 1$

$\{1, 3\}$

11. a. $x^2 + 5x - 6 = -10$, one solution appears on the table, using symmetry another solution should be between -2 and 0, adding -1 to table;

$y = (-1)^2 + 5(-1) - 6$, $y = -10$

solutions are $\{-4, -1\}$

b. One solution appears on the table, another solution should be between -8 and -6, adding -7 to the table;

$y = (-7)^2 + 5(-7) - 6$, $y = 8$

solutions are $\{-7, 2\}$

c. One solution appears on the table, another solution should be between 0 and 2, adding 1 to the table;

$y = 1^2 + 5(1) - 6$, $y = 0$

solutions are $\{-6, 1\}$

d. There are no x values that will give a y value of -20, { }

e. Using the symmetry of the table, the axis of symmetry is half-way between -4 and -1 at $x = -2.5$.

f. $y = (-2.5)^2 + 5(-2.5) - 6$, $y = -12.25$, coordinates of the vertex: $(-2.5, -12.25)$

12. a. $x^2 + 5x - 6 \geq 0$, $(x + 6)(x - 1) \geq 0$;

x-intercepts are at (-6, 0) and (1, 0)

graph opens upward

function is ≥ 0 when $x \leq -6$ or $x \geq 1$

b. $x^2 + 5x - 6 \leq 0$, from (a) above:

x-intercepts are (-6, 0) and (1, 0)

graph opens upward, function is ≤ 0 on the interval $-6 \leq x \leq 1$

c. $x^2 + 5x - 6 > 0$, from (a) above:

x-intercepts are (-6, 0) and (1, 0)

graph opens upward, function is > 0 when $x < -6$ or $x > 1$

13. $2x^2 - 7x + 6 = 0$, $(2x - 3)(x - 2) = 0$;

$2x - 3 = 0$, $2x = 3$, $x = \frac{3}{2}$ or

$x - 2 = 0$, $x = 2$ $\{\frac{3}{2}, 2\}$

14. $a = 2$, $b = -7$, $c = 6$

$$x = \frac{-(-7) \pm \sqrt{(-7)^2 - 4(2)(6)}}{2(2)},$$

$$x = \frac{7 \pm \sqrt{49 - 48}}{4},\ x = \frac{7 \pm \sqrt{1}}{4},$$

$$x = \frac{7 + 1}{4},\ x = \frac{8}{4},\ x = 2;\ or$$

$$x = \frac{7 - 1}{4},\ x = \frac{6}{4},\ x = \frac{3}{2}$$

$\{\frac{3}{2}, 2\}$

Chapter 3 Test (con't)

15. a.

r	$A=20{,}000(1+r)^2$
0	$20{,}000(1+0)^2 = 20{,}000$
0.01	$20{,}000(1+0.01)^2 = 20{,}402$
0.02	$20{,}000(1+0.02)^2 = 20{,}808$
0.03	$20{,}000(1+0.03)^2 = 21{,}218$
0.04	$20{,}000(1+0.04)^2 = 21{,}632$
0.05	$20{,}000(1+0.05)^2 = 22{,}050$
0.06	$20{,}000(1+0.06)^2 = 22{,}472$
0.07	$20{,}000(1+0.07)^2 = 22{,}898$
0.08	$20{,}000(1+0.08)^2 = 23{,}328$
0.09	$20{,}000(1+0.09)^2 = 23{,}762$
0.10	$20{,}000(1+0.10)^2 = 24{,}200$

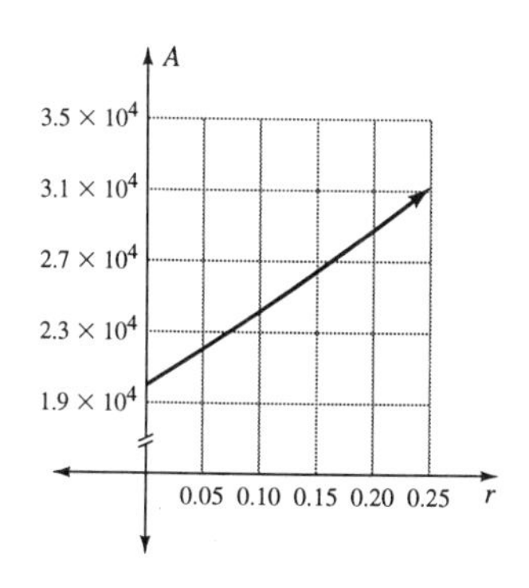

b. The salary will be \$22,898 after 2 years at 7%

c. Extending the graph a salary of \$30,000 requires a raise of ≈ 22.5%

16. $v = \sqrt{2gs}$, $v^2 = 2gs$, $s = \dfrac{v^2}{2g}$

17. If x = the length of one side of the block in steps, then $212^2 = 2x^2$; $212 = x\sqrt{2}$, $x \approx 150$, total steps ≈ 300

18. $x = \sqrt{4}$, $x = 2$;

$\sqrt{x^2} = 2$, $|x| = 2$, $x = \pm 2$

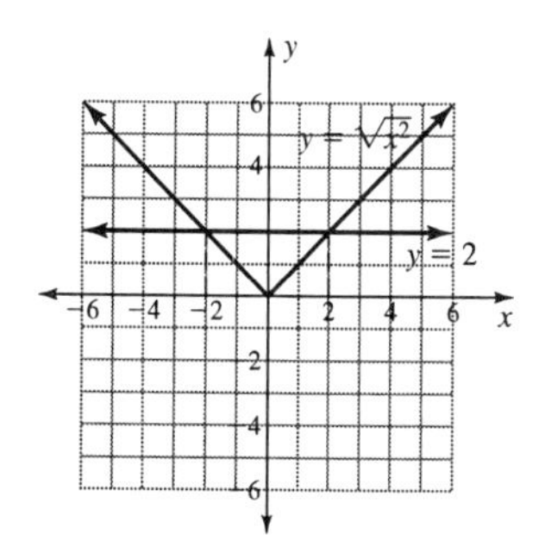

Section 4.0

1.

Multiply	x	-3
x	x^2	-3x
+4	4x	-12

$(x - 3)(x + 4) = x^2 + x - 12$

3.

Multiply	2x	+3
x	$2x^2$	3x
-4	-8x	-12

$(2x + 3)(x - 4) = 2x^2 - 5x - 12$

5. $(x + 3)(x + 3) = x^2 + 2(3x) + 3^2$
$= x^2 + 6x + 9$

7. $(x + 6)(x - 6) = x^2 - 6^2 = x^2 - 36$

9. $(2x - 5)^2 = (2x)^2 + 2(-5)(2x) + (-5)^2$
$= 4x^2 - 20x + 25$

11. Exercises 4, 5, 8, 9, and 10 contain binomial squares.

13. Shift the graph of $y = x^2$ to the right 4 units to get the graph of $y = (x - 4)^2$

15. $2(4) = 8$; $x^2 + \mathbf{8}x + 16 = (x + 4)^2$

17. $49 = 7^2$, $2(7) = 14$;

$x^2 - \mathbf{14}x + 49 = (x - \mathbf{7})^2$

19. $2.5^2 = 6.25$; $x^2 - 5x + \mathbf{6.25} = (x - 2.5)^2$

21. $7 = 2(3.5)$, $3.5^2 = 12.25$;

$x^2 + 7x + \mathbf{12.25} = (x + \mathbf{3.5})^2$

23. $24 = 2(12)$, $12^2 = 144$;

$x^2 - 24x + \mathbf{144} = (x - \mathbf{12})^2$

25. $(x - 4)(x + 4) = x^2 - 4^2 = x^2 - 16$

27. $(2x - 3)(2x + 3) = (2x)^2 - 3^2 = 4x^2 - 9$

29. The graph has been shifted down 16 units, the x-intercepts are at (-4, 0) and (4, 0)

31. $x^2 - 144 = x^2 - 12^2 = (x + 12)(x - 12)$

33. $0.25x^2 - 0.01 = (0.5x)^2 - 0.1^2$
$= (0.5x + 0.1)(0.5x - 0.1)$

35. a. To find the vertex let y = 0 and solve for x, or observe that if there is a minus sign between x and r the vertex is at r, if there is a plus sign the vertex is at -r.

b. From our observation above, the vertex is at x = -1 on the x-axis. Letting y = 0 and solving for x; x + 1 = 0, x = -1

c. The vertex is on the x-axis at x -= 4; check by letting y = 0, x - 4 = 0, x = 4

37. a. When r > 0 the graph shifts right $|r|$ units, when r < 0 the graph shifts left $|r|$ units.

b. $(x + 2)^2 = [x - (-2)]^2$, $r < 0$, graph shifts left 2 units

Section 4.0 (con't)

37. c. $(x - 1)^2$, $r > 0$, graph shifts right 1 unit.

39. $(x + 1)(x^2 - x + 1)$ is a sum of cubes, multiplies to $x^3 + 1$

41.

Multiply	x^2	$-6x$	$+9$
x	x^3	$-6x^2$	$9x$
-3	$-3x^2$	$18x$	-27

$(x - 3)(x^2 - 6x + 9) = x^3 - 9x^2 + 27x - 27$

43.

Multiply	x^2	$-2x$	$+1$
x	x^3	$-2x^2$	x
$+1$	x^2	$-2x$	1

$(x + 1)(x^2 - 2x + 1) = x^3 - x^2 - x + 1$

45. Exercises 39 and 44 contain a sum or difference of cubes

47. $x^3 - 125$ is a difference of cubes, it factors to $(x - 5)(x^2 + 5x + 25)$

49. a. $x^3 + 1$ is a sum of cubes, factors to $(x + 1)(x^2 - x + 1)$

b. $x^3 + 125$ is a sum of cubes, factors to $(x + 5)(x^2 - 5x + 25)$

c. $x^3 - 1000$ is a difference of cubes, factors to $(x - 10)(x^2 + 10x + 100)$

51. a. $x^2 + 2x + 1$ is a perfect square trinomial

b. $x^2 + 2x + 1 = (x + 1)^2$

c. $(x + 1)(x + 1)^2 = 0$, $x + 1 = 0$, $x = -1$

d. Graph crosses the x-axis once at $(-1, 0)$

53. a. $x^2 - 1$ is a difference of squares

b. $x^2 - 1 = (x + 1)(x - 1)$

c. $(x + 1)(x + 1)(x - 1) = 0$,

$x + 1 = 0$ or $x - 1 = 0$

$x = -1$ or $x = 1$; $\{-1, 1\}$

d. Graph crosses the x-axis once at $(1, 0)$ and touches at $(-1, 0)$

55. $y = (x + 1)^3$ and $y = x^3 + 1$ have the same general shape.

Compared to $y = x^3$; $y = (x + 1)^3$ is shifted left 1 unit, $y = x^3 + 1$ is shifted up 1 unit.

57. $f(x) = x^2 - 4x^5$ has degree 5, there will be 5 possible real number solutions.

59. $f(x) = 2x^3 - 2x^2$ has degree 3, there will be 3 possible real number solutions.

61. a. There is no output $y = n$ that will have no solutions.

b. One possible solution is $y = -6$

c. One possible solution is $y = -1$

Section 4.0 (con't)

61. d. One possible solution is $y = 0$

e. There is no output $y = n$ that will have 4 solutions.

63. a. One possible solution is $y = 6$

b. $y \approx 5.5$ is the only solution

c. One possible solution is $y = -6$

d. One possible solution is $y \approx -0.8$

e. One possible solution is $y = -1$

65. Tracing the graph $f(x) = 0$ when $x \approx \{0.382, 1, 2.618\}$

67. $y = x^3 + 1$ crosses the x-axis, $y = x^2 + 1$ does not.

69. Graph and trace to find solutions. Pick a value for y, graph the horizontal line and trace the points of interception.

Section 4.1

1. $y = a(x + 1)(x + 5)$, substitute $(-4, 3)$ for x and y then solve for a;

$3 = a(-4 + 1)(-4 + 5)$, $3 = a(-3)(1)$,

$3 = -3a$, $a = -1$

3. $y = a(x + 1)(x + 5)$, substitute $(0, 7.5)$ for x and y then solve for a;

$7.5 = a(0 + 1)(0 + 5)$, $7.5 = 5a$, $a = 1.5$

5. Using the x-intercepts given

$y = a(x - 3)(x + 2)$,

substituting $(-3, 12)$ and solving for a;

$12 = a(-3 - 3)(-3 + 2)$, $12 = 6a$, $a = 2$;

$y = 2(x - 3)(x + 2)$, $y = 2(x^2 - x - 6)$,

$y = 2x^2 - 2x - 12$

7. Using the x-intercepts given

$y = a(x - 3)(x + 2)$;

substituting $(0, 12)$ and solving for a,

$12 = a(0 - 3)(0 + 2)$, $12 = -6a$, $a = -2$;

$y = -2(x - 3)(x + 2)$, $y = -2(x^2 - x - 6)$,

$y = -2x^2 + 2x + 12$

9. $y = a(x + 3)(x - 5)$;

$6 = a(3 + 3)(3 - 5)$, $6 = -12a$, $a = -0.5$

$y = -0.5(x + 3)(x - 5)$,

$y = -0.5(x^2 - 2x - 15)$ $y = -0.5x^2 + x + 7.5$

11. $y = a(x + 2)(x - 5)$;

$-24 = a(2 + 2)(2 - 5)$, $-24 = -12a$, $a = 2$;

$y = 2(x + 2)(x - 5)$, $y = 2(x^2 - 3x - 10)$,

$y = 2x^2 - 6x - 20$

13. $y = a(x - 10)(x - 30)$;

$12 = a(20 - 10)(20 - 30)$, $12 = -100a$

$a = -0.12$;

$y = -0.12(x - 10)(x - 30)$,

$y = -0.12(x^2 - 40x + 300)$,

$y = -0.12x^2 + 4.8x - 36$

15. $y = ax^2$ *single x-intercept is at (0,0)*

$20 = a(25)^2$, $a = 0.032$;

$y = 0.032x^2$

17. $x^2 - 50x + 400 = (x - 10)(x - 40)$

$y = -\frac{3}{40}(x - 10)(x - 40)$; Average the x-intercept values to find the x coordinate of the vertex, substitute that value and solve for y.

Section 4.1 (con't)

19. First differences; 9, 11, 13, 15

Second differences; 2, 2, 2,

Quadratic function;

$2a = 2, a = 1;$

To find f(0), working backwards, first difference will be $9 - 2 = 7, 7 - 7 = 0$

$f(0) = 0,\ c = 0;$

$f(1) = 7, 7 = 1(1)^2 + b(1) = 0, 7 = 1 + b,$

$b = 6;$

$y = x^2 + 6x$

21. First differences; 4, 4, 4, 4, 4

Linear function;

$a_n = 4 + (n - 1)4,\ a_n = 4 + 4n - 4,$

$a_n = 4n,\ y = 4x$

23. First differences; 19, 23, 27, 31

Second differences; 4, 4, 4,

Quadratic function;

$2a = 4, a = 2;$

$f(0) = 8 - (19 - 4), f(0) = -7, c = -7$

$f(1) = 8, 8 = 2(1)^2 + b(1) - 7,$

$8 = b - 5,\ b = 13;$

$y = 2x^2 + 13x - 7$

25. First differences; 7, 9, 11, 13

Second differences; 2, 2, 2

Quadratic function;

$2a = 2, a = 1;$

$f(0) = 5 - (7 - 2), f(0) = 0, c = 0;$

$f(1) = 5,\ 5 = 1(1)^2 + b(1) + 0;$

$5 = 1 + b, b = 4;$

$y = x^2 + 4x$

27. First differences; 12, 12, 12, 12

Linear function;

$a_n = 13 + (n - 1)12,\ a_n = 13 + 12n - 12,$

$a_n = 12n + 1,\ y = 12x + 1$

29. First differences; 0, 1, 1, 2, 3

Second differences; 1, 0, 1, 1

Neither type of function

31. First differences; 7, 5, 3, 1

Second differences; -2, -2, -2

Quadratic function;

$2a = -2, a = -1$

$f(0) = 9 - [7 - (-2)],\ f(0) = 0, c = 0;$

$f(1) = 9,\ 9 = -1(1)^2 + b(1) + 0, 9 = b - 1,$

$b = 10;$

$y = -x^2 + 10x$

Section 4.1 (con't)

33. Use of quadratic regression makes working with non-integer data easier.

Quadratic regression gives the following formula; $y \approx 3.14x^2 - .007x + 0.002$,

If we consider the possibility of rounding errors in the data and therefore discount the last two terms in the equation we would have the common formula for the area of a circle with a radius of x; $y = \pi x^2$ or $y \approx 3.14x^2$

Section 4.2

1. In example 1, *a* is negative, indicating the parabola turns down.

3. Correct equation is $y = \frac{4}{125}x^2$, the coefficient of x^2 is positive indicating a parabola that turns up.

5.

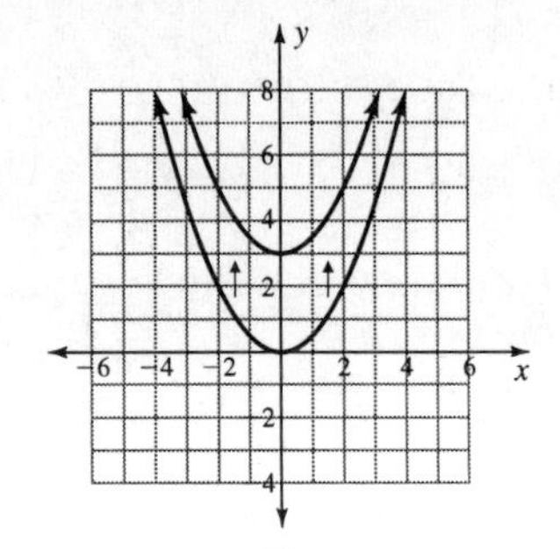

7.

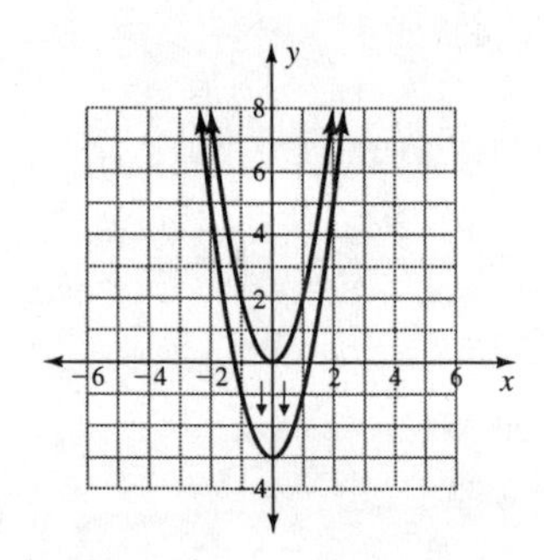

9. Coefficient on x^2 in g(x) is 0.5; h(x), m(x) and q(x) have the same coefficient.

11. For the parabola to turn up the coefficient on x^2 must be positive, f(x), g(x), h(x), j(x), m(x) and q(x) have positive coefficients.

13. The y-intercept is *c*, only j(x) has $c = 2$,
$j(x) = 2(x - 1)^2$, $j(x) = 2(x^2 - 2x + 1)$
$j(x) = 2x^2 - 4x + 2$

15. A larger coefficient on x^2 makes the parabola steeper, g(x) is steeper

17. j(x) has the larger coefficient on x^2

19. h(x) has the larger coefficient on x^2

21. p(x) has the larger coefficient on x^2

23. A coefficient on x^2 larger than 1 gives a graph *steeper* than the graph of $y = x^2$.

25. The slope, m, controls the steepness of a linear graph.

27.

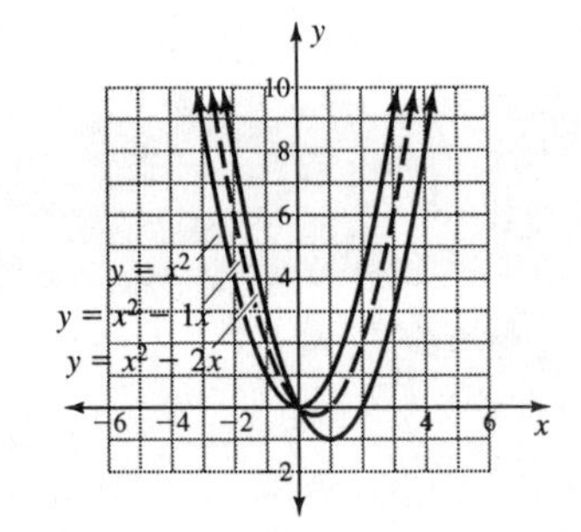

The vertex of $y = x^2$ is (0, 0),

The vertex of $y = x^2 - 1x$ is (0.5, -0.25)

The vertex of $y = x^2 - 2x$ is (1, -1)

The graphs shift down and to the right

29.

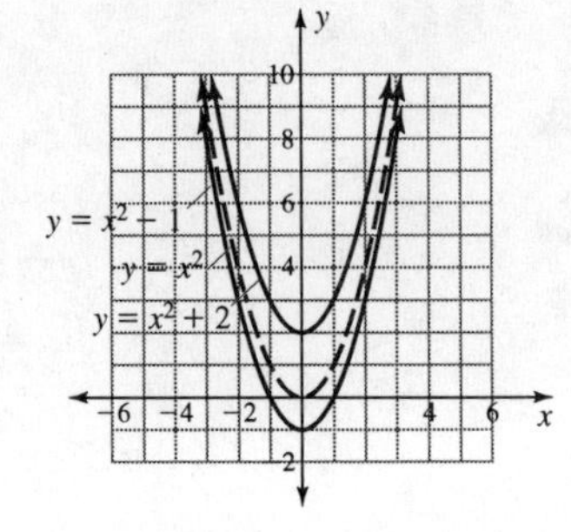

Section 4.2

31. Start with the equation $y = x^2$ and shift the graph down 3 units, the new equation is $y = x^2 - 3$. Another possibility is a graph turning down with the same shape, the starting equation is $y = -x^2$, shift down 3 units and the new equation is $y = -x^2 - 3$.

33. a. Moving the origin down 20 feet has the same effect as moving the graph up 20 feet. The new coordinates will have the same x value and the y value will be increased by 20. Cannon (0, 20); highest point (200, 170), fire (400, 20)

b. The shape of the graph has not changed but the y-intercept is now at (0, 20), the new equation is the same as the old with $c = 20$;

$y = -0.00375x^2 + 1.5x + 20$

Section 4.3

1. $4x^2 + 7x - 2$; $a = 4, b = 7, c = -2$;

$7^2 - 4(4)(-2) = 49 + 32 = 81$

There are 2 real solutions

3. $5x^2 + 4x + 1$; $a = 5, b = 4, c = 1$;

$4^2 - 4(5)(1) = 16 - 20 = -4$

There are 2 complex solutions

5. $4x^2 + 4x + 1$; $a = 4, b = 4, c = 1$;

$4^2 - 4(4)(1) = 16 - 16 = 0$

There is one real solution

7. $2x^2 + 6x + 5$; $a = 2, b = 6, c = 5$;

$6^2 - 4(2)(5) = 16 - 40 = -24$

There are 2 complex solutions

9. $\frac{1}{2}x^2 - 3$; $a = \frac{1}{2}, b = 0, c = -3$;

$0^2 - 4(\frac{1}{2})(-3) = 0 + 6 = 6$

There are 2 real solutions

11. a. $\sqrt{-20} = i\sqrt{20} = 2i\sqrt{5}$

b. $\sqrt{-40} = i\sqrt{40} = 2i\sqrt{10}$

c. $\sqrt{-72} = i\sqrt{72} = 6i\sqrt{2}$

13. a. $\sqrt{-3} = i\sqrt{3}$

b. $\sqrt{-54} = i\sqrt{54} = 3i\sqrt{6}$

c. $\sqrt{-27} = i\sqrt{27} = 3i\sqrt{3}$

15. a. $16 = 16 + 0i$

b. $\sqrt{-16} = i\sqrt{16} = 4i$;

$\sqrt{-16} = 0 + 4i$

c. $\sqrt{36} = 6; \sqrt{-6} = i\sqrt{6}$;

$\sqrt{36} + \sqrt{-6} = 6 + i\sqrt{6}$

17. a. $\dfrac{6 + \sqrt{-12}}{2} = \dfrac{6 + 2i\sqrt{3}}{2} = 3 + i\sqrt{3}$

b. $\dfrac{2 - \sqrt{-24}}{4} = \dfrac{2 - 2i\sqrt{6}}{4} = \dfrac{1 - i\sqrt{6}}{2}$

$= \dfrac{1}{2} - \dfrac{i\sqrt{6}}{2}$

19. $2 - 3i$

21. $i + 1 = 1 + i$; conjugate $= 1 - i$

23. $2i - 3 = -3 + 2i$; conjugate $= -3 - 2i$

25. $4 + 2i$

27. $2i + 3i - 4i - 5i + 6i + 7i - 8i - 9i = -8i$

29. $6 + 4i - 7i + 5 - 12i = 6 + 5 + 4i - 7i - 12i$

$= 11 - 15i$

31. $-a + bi - (a + bi) = -a + bi - a - bi$

$= -a - a + bi - bi = -2a$

33. $3 - 4(5 - i) = 3 - 20 + 4i = -17 + 4i$

35. $4i - 3(5 - i) = 4i - 15 + 3i = -15 + 7i$

Section 4.3 (con't)

37. a. This is a difference of squares

$(4 - 3i)(4 + 3i) = 4^2 - (3i)^2$

$= 16 - 9(-1) = 16 + 9 = 25$

b. $(3+4i)(3-4i) = 3^2 - (4i)^2$

$= 9 - 16(-1) = 9 + 16 = 25$

39. a. $(\sqrt{2}+i)(\sqrt{2}-1) = (\sqrt{2})^2 - i^2$

$= 2 - (-1) = 3$

b. $(1 - i\sqrt{2})(1 + i\sqrt{2}) = 1^2 - (i\sqrt{2})^2$

$= 1 - (-1)(2) = 3$

41. $(x+i)(x-i) = x^2 - i^2 = x^2 - (-1)$

$= x^2 + 1$

43. $(x-3i)(x+3i) = x^2 - (3i)^2 = x^2 - 9(-1)$

$= x^2 + 9$

45. $x^2 - 8 = 0, (x + \sqrt{-8})(x - \sqrt{-8}) = 0$

$x + \sqrt{-8} = 0, x = -\sqrt{-8}, x = -2i\sqrt{2}$
$x - \sqrt{-8} = 0, x = \sqrt{-8}, x = 2i\sqrt{2}$

$\{\pm 2i\sqrt{2}\}$

47. $x = \dfrac{-2 \pm \sqrt{2^2 - 4(1)(4)}}{2(1)}$,

$x = \dfrac{-2 \pm \sqrt{-12}}{2}, x = \dfrac{-2 \pm 2i\sqrt{3}}{2},$

$x = -1 \pm i\sqrt{3}$

49. $x = \dfrac{-(-4) \pm \sqrt{(-4)^2 - 4(1)(8)}}{2(1)}$,

$x = \dfrac{4 \pm \sqrt{-16}}{2}, x = \dfrac{4 \pm 4i}{2}, x = 2 \pm 2i$

51. $x^2 + 3x - 4 = 0$; $(x + 4)(x - 1) = 0$;

$x + 4 = 0$, $x = -4$; $x - 1 = 0$, $x = 1$

$\{-4, 1\}$

53. $x^2 - 4x + 4 = 0$; $(x - 2)^2 = 0$,

$x - 2 = 0$, $x = 2$

55. $(x + 1)^2 + 1 = 0$, $x^2 + 2x + 1 + 1 = 0$,

$x^2 + 2x + 2 = 0$; $x = \dfrac{-2 \pm \sqrt{2^2 - 4(1)(2)}}{2(1)}$,

$x = \dfrac{-2 \pm \sqrt{-4}}{2}, x = \dfrac{-2 \pm 2i}{2}, x = -1 \pm i$

57. $x = \dfrac{-10 \pm \sqrt{10^2 - 4(2)(13)}}{2(2)}$,

$x = \dfrac{-10 \pm \sqrt{-4}}{4}, x = \dfrac{-10 \pm 2i}{4},$
$x = \dfrac{-5}{2} \pm \dfrac{1}{2}i, x = -2.5 \pm 0.5i$

Section 4.3 (con't)

59. $(x+2)(x^2-2x+4)=0;$

$x+2=0, x=-2;\ x^2-2x+4=0,$

$x=\dfrac{-(-2)\pm\sqrt{(-2)^2-4(1)(4)}}{2(1)},$

$x=\dfrac{2\pm\sqrt{-12}}{2},\ x=\dfrac{2\pm 2i\sqrt{3}}{2},$

$x=1\pm i\sqrt{3}$

$\{-2, 1\pm i\sqrt{3}\}$

61. $x^4 - 81 = 0,\ (x^2 + 9)(x^2 - 9) = 0,$

$x^2 + 9 = 0,\ x^2 = -9,\ x = \pm\sqrt{-9},$

$x = \pm 3i;$

$x^2 - 9 = 0,\ x^2 = 9,\ x = \pm 3$

$\{\pm 3, \pm 3i\}$

63. $x^3 + 3x^2 + 2x = 0,\ x(x^2 + 3x + 2) = 0,$

$x(x + 2)(x + 1) = 0;$

$x = 0;$

$x + 2 = 0, x = -2;$

$x + 1 = 0, x = -1$

$\{-2, -1, 0\}$

65. True, use the value of the discriminant

67. True, multiply a complex number and its conjugate.

69. If $b^2 - 4ac$ is negative, *then the quadratic equation has two complex-number solutions. (c)*

71. If $b^2 - 4ac$ is positive, *then the quadratic equation has two real-number solutions. (b)*

73. $b^2 - 4ac > 0$, graphs a and e

Mid Chapter 4 Test

1. a. $(2x + 3)(2x - 3) = 4x^2 - 9$

difference of squares

b. $(2x - 3)(2x - 3) = 4x^2 + 2(2x)(-3) + 9$

$= 4x^2 - 12x + 9$, perfect square trinomial

c. $(2x + 3)(2x + 3) = 4x^2 + 2(2x)(3) + 9$

$= 4x^2 + 12x + 9$, perfect square trinomial

d. $(3x - 2)(2x + 3) = 6x^2 + 9x - 4x - 6$,

$= 6x^2 + 5x - 6$

2. a. $(x - 2)(x^2 + 2x + 4) = x^3 - 8$; difference of cubes

b. $x^3 + 27 = (x + 3)(x^2 - 3x + 9)$; sum of cubes

3. The two graphs have the same shape, $(x - 2)(x^2 + 2x + 4)$ is shifted down 8 units.

4. The two graphs have the same shape, $y = (x + 2)^2$ is shifted 2 units to the left, the vertex (and x-intercept) is shifted 2 units to the left from (0, 0) to (-2, 0), the y-intercept is at (4, 0).

5. $y = a(x + 3)(x - 4)$, $3 = a(-2 + 3)(-2 - 4)$,

$3 = -6a$, $a = -0.5$; $y = -0.5(x + 3)(x - 4)$, or $y = -0.5x^2 + 0.5x + 6$

6. a. First differences; -7, -5, -3, -1

Second differences; 2, 2, 2

Quadratic function

$2a = 2$, $a = 1$

$f(0) = 15 - (-7 - 2)$, $f(0) = 24$, $c = 24$

$f(1) = 15$, $15 = 1(1^2) + b(1) + 24$,

$15 = b + 25$, $b = -10$

$y = x^2 - 10x + 24$

b. First differences; 5, 5, 5, 5

Linear function

$a_n = 4 + (n - 1)5$, $a_n = 4 + 5n - 5$,

$a_n = 5n - 1$, $y = 5x - 1$

c. First differences; 7, 7, 7, 7

Linear function

$a_n = -3 + (n - 1)7$, $a_n = -3 + 7n - 7$,

$a_n = 7n - 10$, $y = 7x - 10$

d. First differences; -2, 2, -2, 2

Second differences; 4, -4, 4

Neither

$y = 1$ for x = odd numbers

$y = -1$ for x = even numbers

Mid Chapter 4 Test

7. Check first and second differences to determine which type of regression to use. Second differences are a constant 9.81 so quadratic regression is used.

$f(x) \approx 4.9\,x^2$

8. The negative causes the graph to open downward, the 2 makes the graph steeper.

9. The -4 shifts the graph down 4 units. The vertex remains on the y-axis but is also shifted down 4 units to (0, -4). The y-intercept remains at the vertex. There are now two x-intercepts at (-2, 0) and (2, 0).

10. a. $\sqrt{-16} = 0 + 4i$

b. $4 = 4 + 0i$

c. $3 + \sqrt{-4} = 3 + 2i$

d. $\sqrt{-56} = 0 + 2i\sqrt{14}$

11. a. $(3-2i)(3+2i) = 9 - 4i^2 = 9 + 4 = 13$

b. $(2-i)(2+i) = 4 - i^2 = 4 + 1 = 5$

c. $(x-i)(x+i) = x^2 - i^2 = x^2 + 1$

d. $(2-3i)(2+3i) = 4 - 9i^2 = 4 + 9 = 13$

12. a. Reversing a and b in (a + bi)(a - bi) gives the same real number solution.

b. $(2+i)(2-i) = 4 - i^2 = 4 + 1 = 5$

$(1+2i)(1-2i) = 1 - 4i^2 = 1 + 4 = 5$

13. $-x^2 + 2x + 3 = 0$, a = -1, b = 2, c = 3

$2^2 - 4(-1)(3) = 4 -+ 12 = 16$, there are 2 real number solutions;

$$x = \frac{-2 \pm \sqrt{16}}{2(-1)},\ x = \frac{-2 \pm 4}{-2},\ x = 1 \pm 2,$$

x = -1, or x = 3

{-1, 3}

14. $(x-2)(x^2 + 2x + 4) = 0;$

$x - 2 = 0,\ x = 2;$

$$x^2 + 2x + 4 = 0, x = \frac{-2 \pm \sqrt{4 - 4(1)(4)}}{2},$$

$$x = \frac{-2 \pm \sqrt{-12}}{2},\ x = \frac{-2 \pm 2i\sqrt{3}}{2},$$

$x = -1 \pm i\sqrt{3}$

$\{2, -2 \pm i\sqrt{3}\}$

Section 4.4

1. **a.** $y = x^2$ shifted left 2 units, equation is $y = (x + 2)^2$

b. $y = x^2$ shifted right 3 units, equation is $y = (x - 3)^2$

3. **a.** $y = x^2$ shifted up 2 units, equation is $y = x^2 + 2$

b. $y = x^2$ shifted down 1 unit, equation is $y = x^2 - 1$

5. **a.** $y = x^2$ shifted down 1 unit, equation is $y = x^2 - 1$

b. $y = -x^2$ shifted down 3 units, equation is $y = -x^2 - 3$

7. **a.** $y = x^2$ shifted left 1 unit and down 2 units, equation is $y = (x + 1)^2 - 2$

b. $y = x^2$ shifted right 2 units and up 1 unit, equation is $y = (x - 2)^2 + 1$

9. Shift the graph of $y = x^2$ left 2 units.

11. Shift the graph of $y = x^2$ right 4 units.

13. Shift the graph of $y = x^2$ right 3 units and up 4 units.

15. $y - 3 = (x + 4)^2$, $y = (x + 4)^2 + 3$; Shift the graph of $y = x^2$ left 4 units and up 3 units.

In exercises 17 to 27 refer to the vertex form of a quadratic equation.

17. The vertex is at (1, 0).

19. The vertex is at (-3, 4).

21. The vertex is at (2, -3).

23. $3 = a(-1 - 3)^2 + 1$,

$2 = a(-4)^2$, $2 = 16a$, $a = \frac{1}{8}$;

$y = \frac{1}{8}(x - 3)^2 + 1$

25. $7 = a(1 + 1)^2 - 1$,

$8 = a(2)^2$, $8 = 4a$, $a = 2$;

$y = 2(x + 1)^2 - 1$

27. $-10 = a(5 - 3)^2 - 2$,

$-8 = a(2)^2$, $-8 = 4a$, $a = -2$;

$y = -2(x - 3)^2 - 2$

29. **a.** From symmetry another point should be 5 units to the right of the vertex and 6 units down at (10, 0). The equation is $y = -0.24x^2 + 2.4x$

b. $y = -0.24(x - 5)^2 + 6$,

$y = -0.24(x^2 - 10x + 25) + 6$,

$y = -0.24x^2 + 2.4x - 6 + 6$,

$y = -0.24x^2 + 2.4x$

Section 4.4 (con't)

31. a. The result is the same as shifting the graph down 25 units.

(0, 0) becomes (0, -25)

(20, 10) becomes (20, -15)

b. Using the vertex form,

$0 = a(0 - 20)^2 - 15$, $15 = a(-20)^2$

$15 = 400a$, $a = 0.0375$;

$y = 0.0375(x - 20)^2 - 15$

c. The equations differ in their vertex coordinates.

33. Vertex is (30, 100), passes through (0, 0)

$0 = a(0 - 30)^2 + 100$, $-100 = 900a$, $a = -\frac{1}{9}$

$y = -\frac{1}{9}(x - 30)^2 + 100$

35. $y = x^2 + 10x + 25$, $y = (x + 5)^2$

37. $y = x^2 - 6x + 9$, $y = (x - 3)^2$

39. $y = x^2 + 10x + 30$, $y - 30 = x^2 + 10x$,

$y - 30 + 25 = x^2 + 10x + 25$,

$y - 5 = (x + 5)^2$, $y = (x + 5)^2 + 5$

41. $y = x^2 - 6x + 8$, $y - 8 = x^2 - 6x$,

$y - 8 + 9 = x^2 - 6x + 9$, $y + 1 = (x - 3)^2$,

$y = (x - 3)^2 - 1$

Section 4.5

1. $y = x^2 - 2x$, $y = x(x - 2)$, for x-intercepts

$y = 0$, $x(x - 2) = 0$, $x = 0$ or $x = 2$

x-intercepts are (0, 0) and (2, 0);

x value of vertex is $(0 + 2) \div 2 = 1$,

y value is $f(1) = 1(1 - 2)$, $f(1) = -1$,

vertex is at (1, -1)

3. $y = -x^2 - 4x + 21$, $y = (-x - 7)(x - 3)$

$(-x - 7)(x - 3) = 0$, $x = -7$ or $x = 3$;

x-intercepts are (-7, 0) and (3, 0);

x value of vertex is $(-7 + 3) \div 2 = -2$,

y value is $f(-2) = -(-2)^2 - 4(-2) + 21$,

$f(-2) = 25$, vertex is at (-2, 25)

5. $a = 1, b = 8, x = -\dfrac{8}{2(1)}, x = -4$

$y = f(-4)$, $y = (-4)^2 + 8(-4) + 15$, $y = -1$,

vertex is at (-4, -1)

7. $a = 1, b = -4, x = -\dfrac{-4}{2(1)}, \; x = 2$

$y = f(2)$, $y = 2^2 - 4(2) + 5$, $y = 1$,

vertex is at (2, 1)

9. $a = 2, b = -1, x = -\dfrac{-1}{2(2)}, x = \dfrac{1}{4}$

$y = f(\frac{1}{4})$, $y = 2(\frac{1}{4})^2 - \frac{1}{4} - 3$, $y = -3\frac{1}{8}$

vertex is at $(\frac{1}{4}, -3\frac{1}{8})$

11. Using $y = -\dfrac{g}{v^2}x^2 + x$,

with $g = 32.2$ ft/sec^2 and $v = 90$ ft/sec.

$y \approx -0.004x^2 + x$, $y \approx x(-0.004x + 1)$;

Find the x-intercepts; $0 = x(-0.004x + 1)$,

$x = 0$ or $x \approx 250$;

x coordinate of the vertex is

$x = -\dfrac{1}{2(-0.004)}, x \approx 125;$

y coordinate of the vertex is f(125),

$y = -0.004(125)^2 + 125$, $y \approx 62.5$

The maximum height is ≈ 62.5 ft. and horizontal distance traveled is ≈ 250 ft.

13. Using the equation from Example 5 with $v = 200$ ft/sec.; $y \approx -0.0008x^2 + x$

the x coordinate of the vertex is

$x \approx -\dfrac{1}{2(-0.0008)}, x \approx 625;$

$y = f(625)$, $y \approx -0.0008(625)^2 + 625$,

$y \approx 312.5$ ft; the flare does not get high enough to be seen over the obstacle.

Section 4.5 (con't)

15.

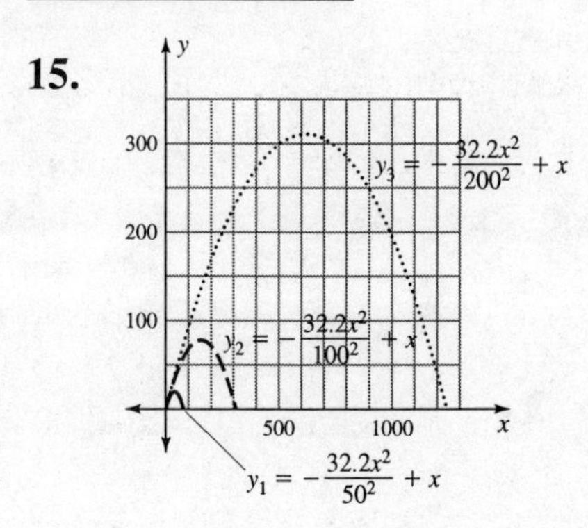

a. When the initial velocity is doubled, the height of the vertex above the ground is *quadrupled.*

b. When the initial velocity is doubled, the distance the ball travels is *quadrupled.*

17. One possible answer is a circle. If the perimeter is 80 ft to find the radius use $C = 2\pi r,\ 2\pi r = 80,\ r \approx 12.7$ ft;

Area is $A = \pi r^2,\ A \approx \pi(12.7 \text{ ft})^2,$

$A \approx 507 \text{ ft}^2$

19. The fenced perimeter is $60 = l + 2w,$

$l = 60 - 2w,\ A = lw;\ A = (60 - 2w)w,$

$A = 60w - 2w^2,$ use vertex for

maximums. $w = -\dfrac{60}{2(-2)},\ w = 15$ ft

$l = 60 - 2(15),\ l = 30$ ft

$A = 60(15) - 2(15)^2,\ A = 450 \text{ ft}^2$

21. $h = -16.1t^2 + 40t + 5.5,$

$t = -\dfrac{40}{2(-16.1)},\ t \approx 1.24$ sec

$h = -16.1(1.24)^2 + 40(1.24) + 5.5,$

$h \approx 30.3$ ft

23. L is at $x = 2000,$

$y = \frac{7}{400{,}000}(2000)^2 - \frac{1}{25}(2000) + 700,$

$y = 690$ ft, Elevation of L = 690 ft

$x = -\dfrac{-\frac{1}{25}}{2(\frac{7}{400{,}000})},\ x \approx 1143$ ft

$y = \frac{7}{400{,}000}(1143)^2 - \frac{1}{25}(1143) + 700$

$y \approx 677$ ft

Location of drain (1143, 677)

25. M is at $x = 0$, Elevation of M = 1188 ft

$x = -\dfrac{-\frac{1}{20}}{2(\frac{11}{480{,}000})},\ x \approx 1091$ ft

$y = \frac{11}{480{,}000}(1091)^2 - \frac{1}{20}(1091) + 1188,$

$y \approx 1161$ ft

Location of drain (1091, 1161)

Section 4.5 (con't)

27. $\frac{1}{60000}x^2 - \frac{3}{100}x + 1000 < 986.6$

$\frac{1}{60000}x^2 - \frac{3}{100}x + 13.4 < 0,$

Find end points by setting equation equal to zero and using quadratic formula.

$$x = \frac{-(-\frac{3}{100}) \pm \sqrt{(-\frac{3}{100})^2 - 4(\frac{1}{60000})(13.4)}}{2(\frac{1}{60000})}$$

$x \approx \{823, 978\}$

$\frac{1}{60000}x^2 - \frac{3}{100}x + 13.4 < 0,$ for

$823 < x < 978$

Length of puddle is 978 - 823 ≈ 155 ft

29. The other x-intercept will be a distance of $|h - x|$ on the opposite side of the vertex.

Chapter 4 Review

1. $(x - 6)(x - 2) = x^2 - 2x - 6x + 12$

$= x^2 - 8x + 12$; other

3. $(x + 12)(x + 1) = x^2 + x + 12x + 12$

$= x^2 + 13x + 12$; other

5. $(x - 2)^2 = x^2 + 2(-2x) + (-2)^2$

$= x^2 - 4x + 4$; perfect square trinomial

7. $225 = 15^2$, $x^2 - \mathbf{30}x + 225 = (x - \mathbf{15})^2$

9. $20 \div 2 = 10$, $x^2 + 20x + \mathbf{100} = (x + \mathbf{10})^2$

11. $x^2 - 49 = (x + 7)(x - 7)$

13. $4x^2 - 4x + 1 = (2x - 1)^2$

15. $x^2 + 16$ can not be factored

17. $4x^2 + 12x + 9 = (2x + 3)^2$

19.

Multiply	a^2	$-2ab$	b^2
a	a^3	$-2a^2b$	ab^2
$-b$	$-a^2b$	$2ab^2$	$-b^3$

$(a - b)(a^2 - 2ab + b^2) = a^3 - 3a^2b + 3ab^2 - b^3$

21. $(a + b)(a^2 - ab + b^2) = a^3 + b^3$

23. Exercise 19 is equal to $(a - b)^3$, notice that $a^2 - 2ab + b^2 = (a - b)^2$

25. $a^3 - 8 = (a - 2)(a^2 + 2a + 4)$

27. $x^3 + 27 = (x + 3)(x^2 - 3x + 9)$

29. $x^3 - 1000 = (x - 10)(x^2 + 10x + 100)$

31. a. One possible output is $y = -5$

b. The only output is $y = -4$

c. One possible output is $y = 2$

d. It is not possible to have 3 solutions

e. It is not possible to have 4 solutions

33. a. It is not possible to have no solutions

b. One possible output is $y = 5$

c. One of two possible outputs is $y = 4$

d. One possible output is $y = 2$

e. It is not possible to have 4 solutions

35. a. $2x^2 - 8x + 4 = 0$; $\approx \{0.5, 3.5\}$

b. $2x^2 - 8x + 4 = 4$; $\approx \{0, 4\}$

37. a. $x^3 + 2x^2 - 5x - 6 = 0$; $\approx \{-3, -1, 2\}$

b. $x^3 + 2x^2 - 5x - 6 = -6$; $\approx \{-3.5, 0, 1.5\}$

39. $y = (x - 4)(x + 2)$, $y = x^2 - 2x - 8$

41. $3 = a(1 - 4)(1 + 2)$, $3 = -9a$, $a = -\frac{1}{3}$

$y = -\frac{1}{3}(x^2 - 2x - 8)$, $y = -\frac{1}{3}x^2 + \frac{2}{3}x + \frac{8}{3}$

43. First differences; 9, 13, 17, 21

Second differences; 4, 4, 4

Quadratic function, $2a = 4$, $a = 2$;

$c = 6 - (9 - 4)$, $c = 1$

$6 = 2(1)^2 + b(1) + 1$, $b = 3$

$y = 2x^2 + 3x + 1$

Chapter 4 Review (con't)

45. First differences; 3, 3, 3, 3

Linear function, $y = -1 + (x - 1)3$,

$y = -1 + 3x - 3$, $y = 3x - 4$

47. First differences; -3, -5, -7, -9

Second differences; -2, -2, -2

Quadratic function, $2a = -2$, $a = -1$

$c = 15 - [-3 - (-2)]$, $c = 16$

$15 = -1(1)^2 + b(1) + 16$, $b = 0$

$y = -x^2 + 16$

49. First differences; 2, -4, -16, -34

Second differences; -6, -12, -18

Neither

51. a. Using linear regression on a calculator, $y \approx -0.000989x + 99.93$

b. Using quadratic regression on a calculator,

$y \approx 0.00000000276x^2 - 0.00107x + 100.2$

c.

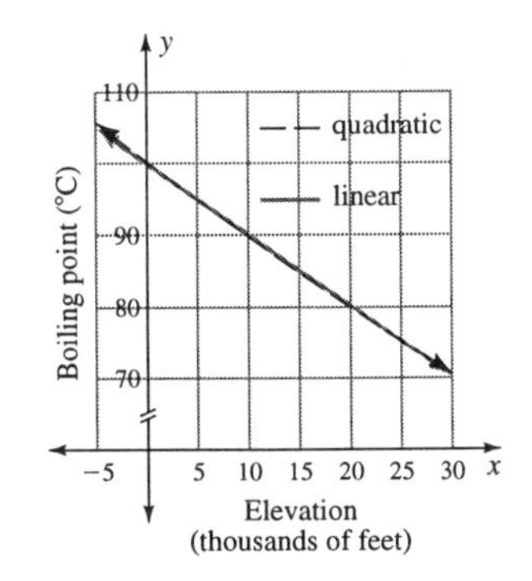

d. Linear equation, $y \approx 101.2°$ C

Quadratic equation, $y \approx 101.6°$ C

53. $y = 3x^2$ is steeper than $y = x^2$

55. $y = x^2 - 9$ is shifted down 9 units

57. $y = x^2 - 9$ crosses the x-axis twice;

$y = x^2 + 9$ does not cross the x-axis

59. The two factors are the same, this is called a double root.

61. $2^2 - 4(-1)(-1) = 4 - 4 = 0$,

1 real number solution, a double root

1 x-intercept

63. $2^2 - 4(3)(1) = 4 - 12 = -8$,

2 complex solutions,

no x-intercepts

65. If $b^2 - 4ac$ is negative, *then the quadratic equation has no real roots. (c)*

67. If $b^2 - 4ac$ is positive, *then the quadratic equation has two real roots. (b)*

69. $b^2 - 4ac > 0$, *The graph of the quadratic equation intersects the x-axis twice. (b)*

71. a. $\sqrt{-16} = i\sqrt{16} = 4i$

b. $\sqrt{-50} = i\sqrt{50} = 5i\sqrt{2}$

73. $\dfrac{8+6i}{2} = \dfrac{8}{2} + \dfrac{6i}{2} = 4 + 3i$

Chapter 4 Review (con't)

75. a. $(4+3i)(4+3i) = 4^2 + 2(4)(3i) + (3i)^2$

$= 16 + 24i + 9(-1) = 7 + 24i$

b. $(4-3i)(4+3i) = 4^2 - (3i)^2$

$= 16 - 9(-1) = 25$

c. $(3-4i)(3+4i) = 3^2 - (4i)^2$

$= 9 - 16(-1) = 25$

77. $(x - 3)(x^2 + 3x + 9) = 0,$

$x - 3 = 0, \ x = 3;$

$x^2 + 3x + 9 = 0,$

$x = \dfrac{-3 \pm \sqrt{3^2 - 4(1)(9)}}{2(1)}, \ x = \dfrac{-3 \pm \sqrt{-27}}{2},$

$x = \dfrac{-3}{2} \pm \dfrac{3i\sqrt{3}}{2}$

$\{3, \dfrac{-3}{2} \pm \dfrac{3i\sqrt{3}}{2}\}$

79. $x^2 - x - 2 = (x - 2)(x + 1),$

x-intercepts $\{-1, 2\}$

$(-1 + 2) \div 2 = 0.5$

$y = 0.5^2 - 0.5 - 2, \ y = -2.25$

vertex $(0.5, -2.25)$

81. Check determinant first;

$(-4)^2 - 4(2)(5) = 16 - 40 = -24$

no x-intercepts

x coordinate of vertex $= \dfrac{-(-4)}{2(2)} = 1$

$y = 2(1)^2 - 4(1) + 5, \ y = 3$

vertex at $(1, 3)$

83.

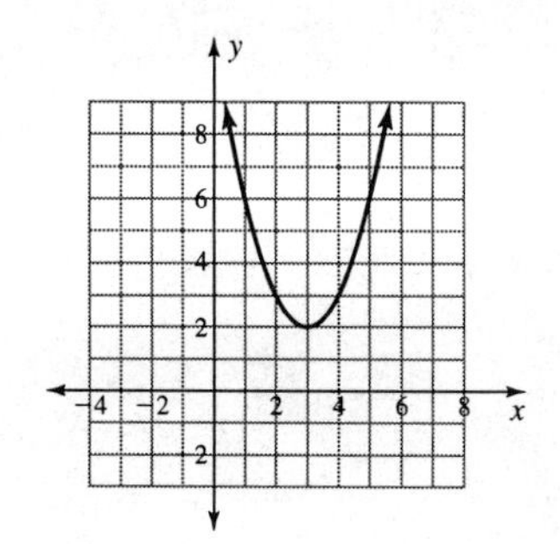

85.

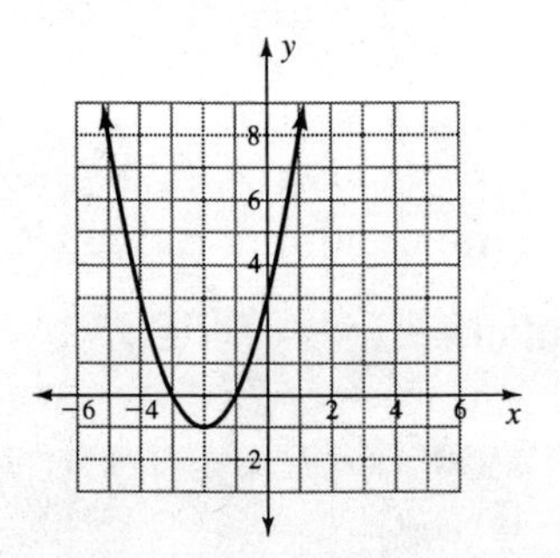

87. $y = x^2 + 4x + 9, \ y - 9 = x^2 + 4x,$

$y - 9 + 4 = x^2 + 4x + 4, \ y - 5 = (x + 2)^2,$

$y = (x + 2)^2 + 5$

89. a. $y = 0, \{-2, 4\}$

b. axis of symmetry, $x = 1$

c. $y = 3(1)^2 - 6(1) - 24, \ y = -27$

vertex is at $(1, -27)$

Chapter 4 Review (con't)

91. Using the symmetry of the path, the other x-intercept will be at (60, 0), using this point with (0, 0) and (30, 100) quadratic regression gives:

$y \approx -0.111x^2 + 6.667x$

93. a. $y = -\dfrac{32.2}{132^2}x^2 + x,\ y \approx -0.0018x^2 + x$

$x = \dfrac{-1}{2(-0.0018)},\ x \approx 278$

$y \approx -0.0018(278)^2 + 278,\ y \approx 139$

vertex at ≈ (278, 139)

b. $y = -\dfrac{32.2}{220^2}x^2 + x,\ y \approx -0.00067x^2 + x$

$x = \dfrac{-1}{2(-0.00067)},\ x \approx 746$

$y \approx -0.00067(746)^2 + 746,\ y \approx 373$

vertex at ≈ (746, 373)

c. $y = -\dfrac{32.2}{400^2}x^2 + x,\ y \approx -0.0002x^2 + x$

$x = \dfrac{-1}{2(-0.0002)},\ x \approx 2500$

$y \approx -0.0002(2500)^2 + 2500,\ y \approx 1250$

vertex at ≈ (2500, 1250)

93. d. $y = -\dfrac{32.2}{500^2}x^2 + x,\ y \approx -0.00013x^2 + x$

$x = \dfrac{-1}{2(-0.00013)},\ x \approx 3846$

$y \approx -0.00013(3846)^2 + 3846,\ y \approx 1923$

vertex at ≈ (3846, 1923)

95. $80 = 2l + 2w,\ 2l = 80 - 2w,\ l = 40 - w$

$A = lw,\ A = (40 - w)w,\ A = 40w - w^2$

$w = \dfrac{-40}{2(-1)},\ w = 20$

$A = 40(20) - 20^2,\ A = 400$

vertex is (20, 400), maximum area in a rectangular shape is a square; $l = w = 20$, $20^2 = 400$

97. $-0.111x^2 + 6.667x < 50$,

$-0.111x^2 + 6.667x - 50 < 0$,

$$x = \frac{-6.667 \pm \sqrt{6.667^2 - 4(-0.111)(-50)}}{2(-0.111)}$$

$$x \approx \frac{-6.667 \pm 4.717}{-0.222},\ x \approx 8.8 \text{ or } x \approx 51.3$$

Jet will be below 50 feet when x < 8.8 ft or x > 51.3 ft

Chapter 4 Test

1.

Multiply	x^2	$-6x$	$+9$
x	x^3	$-6x^2$	$+9x$
-3	$-3x^2$	$+18x$	-27

$(x - 3)(x^2 - 6x + 9) = x^3 - 9x^2 + 27x - 27$

2. a. $x^2 - 49 = (x + 7)(x - 7)$

b. $x^3 - 27 = (x - 3)(x^2 + 3x + 9)$

3. $y = x^2 + 2$ is shifted up 2 units, there are no x-intercepts, the vertex is at (0, 2)

4. The negative sign causes the graph to open downward.

5. $y = x - \frac{32.2}{2500}x^2$ could describe the path of a ball because its graph opens down.

6. a. First differences; 8, 10, 12, 14

Second differences; 2, 2, 2

Next number 38 + (14 + 2) = 54;

Quadratic function: 2a = 2, a = 1;

c = -6 - (8 - 2) = -12

$-6 = 1(1)^2 + b(1) - 12, \ b = 5$

$y = x^2 + 5x - 12$

b. First differences; -8, -8, -8, -8

Next number -9 + -8 = -17

Linear function: y = 23 + (x - 1)(-8)

y = - 8x + 31

c. First differences; 6, 6, 6, 6

Next number 26 + 6 = 32

Linear function: y = 2 + (x - 1)6

y = 6x - 4

d. First differences; -8, -7, -6, -5

Second differences; 1, 1, 1

Next number -1 + (-5 + 1) = -5

Quadratic function: 2a = 1, a = 0.5

c = 25 - (-8 - 1), c = 34

$25 = 0.5(1)^2 + b(1) + 34, \ b = -9.5$

$y = 0.5x^2 - 9.5x + 34$

7. $2^2 - 4(-1)(3) = 16$; 2 real number solutions, 2 x-intercepts

8. a. $\sqrt{-36} = i\sqrt{36} = 6i$

b. $\sqrt{-75} = i\sqrt{75} = 5i\sqrt{3}$

9. a. $\frac{5+10i}{5} = \frac{5}{5} + \frac{10i}{5} = 1 + 2i$

b. $\frac{8+4i}{2} = \frac{8}{2} + \frac{4i}{2} = 4 + 2i$

Chapter 4 Test (con't)

10. a. $(1+5i)(1-5i) = 1^2 - (5i)^2$

$= 1 - (25)(-1) = 26$

b. $(5-i)(5+i) = 5^2 - i^2 = 25+1 = 26$

c. $(1-5i)(1-5i) = 1^2 - 10i + 25i^2$

$= 1 - 10i - 25 = -24 - 10i$

11. $(x - 1)(x^2 + x + 1) = 0$,

$(x - 1) = 0$, $x = 1$

$(x^2 + x + 1) = 0$

$$x = \frac{-1 \pm \sqrt{1^2 - 4(1)(1)}}{2(1)},$$

$$x = \frac{-1 \pm \sqrt{-3}}{2}, \quad x = \frac{-1}{2} \pm \frac{i\sqrt{3}}{2}$$

$$\{1, \frac{-1}{2} \pm \frac{i\sqrt{3}}{2}\}$$

12. a. $2x^2 + x - 6 = (x + 2)(2x - 3)$

$(x + 2)(2x - 3) = 0$; $x = \{-2, 1.5\}$

b. $(-2 + 1.5) \div 2 = -0.25$,

axis of symmetry is $x = -0.25$

c. $y = 2(-0.25)^2 - 0.25 - 6$, $y = -6.125$

vertex is at $(-0.25, -6.125)$

d. Function is > 0 when $x < -2$ or $x > 1.5$

e. One possible answer is $y = -10$

12. f. The function will have 2 negative real number solutions when $-6.125 < y < -6$

13. $x^2 - x - 2 = (x - 2)(x + 1)$,

x-intercepts at $\{-1, 2\}$

$$x = \frac{-(-1)}{2(1)}, \quad x = \frac{1}{2}$$

$$y = \left(\frac{1}{2}\right)^2 - \frac{1}{2} - 2, \quad y = -2\frac{1}{4}$$

vertex is at $\left(\frac{1}{2}, -2\frac{1}{4}\right)$

15. $y = x^2 + 4x - 1$, $y + 1 = x^2 + 4x$,

$y + 1 + 4 = x^2 + 4x + 4$,

$y + 5 = (x + 2)^2$, $y = (x + 2)^2 - 5$

16. $y = (x - 3)^2 - 1$ is shifted right 3 units and down 1 unit

17. $y = a[x - (-1)]^2 + (-8)$, $y = a(x + 1)^2 - 8$;

$19 = a(2 + 1)^2 - 8$, $27 = 9a$, $a = 3$;

$y = 3(x + 1)^2 - 8$ or $y = 3(x^2 + 2x + 1) - 8$

$y = 3x^2 + 6x + 3 - 8$, $y = 3x^2 + 6x - 5$

Chapter 4 Test (con't)

18.

Input, x	Output, y
1	4
2	12
3	24
4	40
5	60

First differences; 8, 12, 16, 20

Second differences; 4, 4, 4

Quadratic function: 2a = 4, a = 2

c = 4 - (8 - 4), c = 0;

$4 = 2(1^2) + b(1) + 0$, b = 2;

$y = 2x^2 + 2x$

19. a. $y = lw$

b. $15 = l + 2w, l = 15 - 2w,$

$y = (15 - 2w)w, \ y = 15w - 2w^2$

c. $w = \dfrac{-15}{2(-2)}, w = \dfrac{15}{4}, w = 3\frac{3}{4}$

$y = 15\left(\dfrac{15}{4}\right) - 2\left(\dfrac{15}{4}\right)^2, \ y = \dfrac{225}{4} - \dfrac{225}{8},$

$y = \dfrac{225}{8}, \ y = 28\frac{1}{8}$

Vertex is at $\left(3\frac{3}{4}, 28\frac{1}{8}\right)$

Cumulative Review Chapters 1 to 4

1. $8 - 3(x - 2) - (x - 3)$

$= 8 - 3x + 6 - x + 3 = -4x + 17$

3. $2y = 3x + 4$, $y = 1.5x + 2$;

Parallel line will have 1.5 as a slope,

passing through origin means $b = 0$;

$y = 1.5x$

5. a. $c = nt^2, n = \dfrac{c}{t^2}$

b. $s = 2a + d(n-1),\ s - 2a = d(n-1),$

$\dfrac{s-2a}{d} = n-1,\ n = \dfrac{s-2a}{d} + 1$

7. $4.5^2 + 20^2 = 20.5^2$?

$20.25 + 400 = 420.25$ - right triangle

9. a. First differences; 7, 9, 11, 13

Second differences; 2, 2, 2

Next number $45 + (13 + 2) = 60$

Quadratic function; $2a = 2$, $a = 1$;

$c = 5 - (7 - 2)$, $c = 0$

$5 = 1(1)^2 + b(1) + 0$, $b = 4$

$y = x^2 + 4x$

b. First differences; 7, 7, 7, 7

Next number $36 + 7 = 43$

Linear function; $y = 8 + (x - 1)7$,

$y = 7x + 1$

9. c. First differences; 6, 6, 6, 6

Next number $30 + 6 = 36$

Linear function, $y = 6 + (x - 1)6$

$y = 6x$

d. First differences; 11, 15, 19, 23

Second differences; 4, 4, 4

Next number $72 + (23 + 4) = 99$

Quadratic function; $2a = 4$, $a = 2$;

$c = 4 - (11 - 4)$, $c = -3$

$4 = 2(1)^2 + b(1) - 3$, $b = 5$

$y = 2x^2 + 5x - 3$

11. a. $x^2 - 2x - 3 = 0$; $\{-1, 3\}$

b. $x^2 - 2x - 3 \le 0$ when $-1 \le x \le 3$

c. $x^2 - 2x - 3 \ge -3$ when $x \le 0$ or $x \ge 2$

d. $x^2 - 2x - 3 < 5$ when $-2 < x < 4$

13. a. $x = \dfrac{-6 \pm \sqrt{6^2 - 4(1)(10)}}{2(1)}$

$x = \dfrac{-6 \pm 2i}{2}$, $x = -3 \pm i$

b. $5x^2 + x - 4 = 0$, $(5x - 4)(x + 1) = 0$

$5x - 4 = 0$, $5x = 4$, $x = \frac{4}{5}$

$x + 1 = 0$, $x = -1$

$\{-1, \frac{4}{5}\}$

Cumulative Review (con't)

13. c. $7x^2 + 4x = 3$, $7x^2 + 4x - 3 = 0$,

$(7x - 3)(x + 1) = 0$,

$7x - 3 = 0$, $7x = 3$, $x = \frac{3}{7}$

$x + 1 = 0$, $x = -1$

$\{-1, \frac{3}{7}\}$

d. $6x^2 + 3 = 11x$, $6x^2 - 11x + 3 = 0$,

$(3x - 1)(2x - 3) = 0$

$3x - 1 = 0$, $3x = 1$, $x = \frac{1}{3}$

$2x - 3 = 0$, $2x = 3$, $x = \frac{3}{2}$

$\{\frac{1}{3}, \frac{3}{2}\}$

15. a. $16 - x^2 = (4 + x)(4 - x)$

b. $x^3 - 64 = (x - 4)(x^2 + 4x + 16)$

Section 5.0

1. **a.** $\frac{3}{5}\cdot\frac{7}{7}=\frac{21}{35}$

b. $\frac{5}{8}\cdot\frac{6}{6}=\frac{30}{48}$

c. $\frac{8}{3}\cdot\frac{16}{16}=\frac{128}{48}$

3. **a.** $\frac{3}{8}\cdot\frac{5}{5}=\frac{15}{40}$

b. $\frac{5}{6}\cdot\frac{9}{9}=\frac{45}{54}$

c. $\frac{5}{4}\cdot\frac{8}{8}=\frac{40}{32}$

5. **a.** $6=2\cdot 3;\ 10=2\cdot 5;$

$\text{LCD}=2\cdot 3\cdot 5=30$

b. $15=3\cdot 5;\ 10=2\cdot 5;$

$\text{LCD}=2\cdot 3\cdot 5=30$

c. $48=2\cdot 2\cdot 2\cdot 2\cdot 3;$

$32=2\cdot 2\cdot 2\cdot 2\cdot 2;$

$\text{LCD}=2\cdot 2\cdot 2\cdot 2\cdot 2\cdot 3=96$

7. $4=2\cdot 2;\ 10=2\cdot 5;$

$\text{LCD}=2\cdot 2\cdot 5=20;$

$\frac{1}{4}\cdot\frac{5}{5}=\frac{5}{20};\ \frac{1}{10}\cdot\frac{2}{2}=\frac{2}{20};$

$\frac{1}{4}+\frac{1}{10}=\frac{5}{20}+\frac{2}{20}=\frac{7}{20};$

$\frac{1}{4}-\frac{1}{10}=\frac{5}{20}-\frac{2}{20}=\frac{3}{20};$

$\frac{1}{4}\cdot\frac{1}{10}=\frac{1}{40};$

$\frac{1}{4}\div\frac{1}{10}=\frac{1}{4}\cdot\frac{10}{1}=\frac{10}{4}=\frac{5}{2}$

9. $4=2\cdot 2;\ 6=2\cdot 3;$

$\text{LCD}=2\cdot 2\cdot 3=12;$

$\frac{3}{4}\cdot\frac{3}{3}=\frac{9}{12};\ \frac{5}{6}\cdot\frac{2}{2}=\frac{10}{12};$

$\frac{3}{4}+\frac{5}{6}=\frac{9}{12}+\frac{10}{12}=\frac{19}{12};$

$\frac{3}{4}-\frac{5}{6}=\frac{9}{12}-\frac{10}{12}=-\frac{1}{12};$

$\frac{3}{4}\cdot\frac{5}{6}=\frac{15}{24}=\frac{5}{8};$

$\frac{3}{4}\div\frac{5}{6}=\frac{3}{4}\cdot\frac{6}{5}=\frac{18}{20}=\frac{9}{10}$

Section 5.0 (con't)

11. LCD = 3 · 2 = 6;

$$1\frac{1}{3}=\frac{3}{3}+\frac{1}{3}=\frac{4}{3},\ \frac{4}{3}\cdot\frac{2}{2}=\frac{8}{6};$$
$$2\frac{1}{2}=\frac{4}{2}+\frac{1}{2}=\frac{5}{2},\ \frac{5}{2}\cdot\frac{3}{3}=\frac{15}{6};$$
$$\frac{8}{6}+\frac{15}{6}=\frac{23}{6}=3\frac{5}{6};$$
$$\frac{8}{6}-\frac{15}{6}=-\frac{7}{6}=-1\frac{1}{6};$$
$$\frac{4}{3}\cdot\frac{5}{2}=\frac{20}{6}=\frac{10}{3}=3\frac{1}{3};$$
$$\frac{4}{3}\div\frac{5}{2}=\frac{4}{3}\cdot\frac{2}{5}=\frac{8}{15}$$

13. LCD = 4 · 5 = 20;

$$2\frac{1}{5}=\frac{10}{5}+\frac{1}{5}=\frac{11}{5},\ \frac{11}{5}\cdot\frac{4}{4}=\frac{44}{20};$$
$$1\frac{1}{4}=\frac{4}{4}+\frac{1}{4}=\frac{5}{4},\ \frac{5}{4}\cdot\frac{5}{5}=\frac{25}{20};$$
$$\frac{44}{20}+\frac{25}{20}=\frac{69}{20}=3\frac{9}{20};$$
$$\frac{44}{20}-\frac{25}{20}=\frac{19}{20};$$
$$\frac{11}{5}\cdot\frac{5}{4}=\frac{55}{20}=\frac{11}{4}=2\frac{3}{4};$$
$$\frac{11}{5}\div\frac{5}{4}=\frac{11}{5}\cdot\frac{4}{5}=\frac{44}{25}=1\frac{19}{25}$$

15. $$\frac{186\text{ cm}^2}{6\text{ cm}}=\frac{6\cdot 31\cdot\text{cm}\cdot\text{cm}}{6\text{ cm}}=31\text{ cm}$$

17. $$\frac{36\text{ in}}{1728\text{ in}^3}=\frac{36\text{ in}}{36\cdot 48\text{ in}\cdot\text{in}\cdot\text{in}}=\frac{1}{48\text{ in}^2}$$

19. $$\frac{1500\text{ foot pounds}}{25\text{ feet}}=\frac{25\cdot 60\text{ foot pounds}}{25\text{ feet}}$$

$= 60$ pounds

21. $$\frac{1200\text{ kilowatt hours}}{24\text{ hours}}=50\text{ kilowatts}$$

23. a. $d=\frac{1}{2}\cdot\frac{32.2\text{ ft}}{\text{sec}^2}\cdot(1\text{ sec})^2,\ d=16.1\text{ ft}$

b. $d=\frac{1}{2}\cdot\frac{32.2\text{ ft}}{\text{sec}^2}\cdot(3\text{ sec})^2,$

$d=\frac{16.1\text{ ft}\cdot 9\text{ sec}^2}{\text{sec}^2},\ d=144.9\text{ ft}$

c. $d=\frac{1}{2}\cdot\frac{32.2\text{ ft}}{\text{sec}^2}\cdot(9\text{ sec})^2,$

$d=\frac{16.1\text{ ft}\cdot 81\text{ sec}^2}{\text{sec}^2},\ d=1304.1\text{ ft}$

25. a. $s=\frac{9.81\text{ m}}{\text{sec}^2}\cdot 1\text{ sec},\ s=\frac{9.81\text{ m}}{\text{sec}}$

b. $s=\frac{9.81\text{ m}}{\text{sec}^2}\cdot 2\text{ sec},\ s=\frac{19.62\text{ m}}{\text{sec}}$

c. $s=\frac{9.81\text{ m}}{\text{sec}^2}\cdot 4\text{ sec},\ s=\frac{39.24\text{ m}}{\text{sec}}$

27. $$300\text{ ml}\cdot\frac{1\text{ L}}{1000\text{ ml}}=\frac{300\text{ L}}{1000}=0.3\text{ L}$$

29. $$25\text{ ft}\cdot\frac{12\text{ in}}{1\text{ ft}}\cdot\frac{1\text{ m}}{39.37\text{ in}}=\frac{25\cdot 12\text{ m}}{39.37}$$

≈ 7.62 m

31. $$150\text{ ft}^3\cdot\left(\frac{1\text{ yd}}{3\text{ ft}}\right)^3=\frac{150\text{ yd}^3}{27}\approx 5.6\text{ yd}^3$$

Section 5.0 (con't)

33. $100\text{ in}^3 \cdot \left(\frac{1\text{ ft}}{12\text{ in}}\right)^3 = \frac{100\text{ ft}^3}{1728} \approx 0.058\text{ ft}^3$

35. $200\text{ ml} \cdot \frac{1\text{ kg}}{1000\text{ ml}} \cdot \frac{1000\text{ g}}{1\text{ kg}} = \frac{200 \cdot 1000\text{ g}}{1000}$

$= 200\text{ g}$

37. $\frac{40\text{ mi}}{1\text{ hr}} \cdot \frac{5280\text{ ft}}{1\text{ mi}} \cdot \frac{1\text{ hr}}{60\text{ min}} \cdot \frac{1\text{ min}}{60\text{ sec}}$

$= \frac{40 \cdot 5280\text{ ft}}{60 \cdot 60\text{ sec}} \approx 59\text{ ft/sec}$

39. $\frac{240\text{ ml}}{12\text{ hr}} \cdot \frac{60\text{ microdrops}}{1\text{ ml}} \cdot \frac{1\text{ hr}}{60\text{ min}}$

$= \frac{240 \cdot 60\text{ microdrops}}{12 \cdot 60\text{ min}}$

$= 20\text{ microdrops/min}$

41. a. $1\text{ g fat} \cdot \frac{9\text{ cal}}{1\text{ g fat}} = 9\text{ cal}$

b. $41\text{ g carbo} \cdot \frac{4\text{ cal}}{1\text{ g carbo}} = 164\text{ cal}$

c. $5\text{ g protein} \cdot \frac{4\text{ cal}}{1\text{ g protein}} = 20\text{ cal}$

Total calories 9 + 164 + 20 = 193 cal

43. a. $8\text{ g fat} \cdot \frac{9\text{ cal}}{1\text{ g fat}} = 72\text{ cal}$

b. $27\text{ g carbo} \cdot \frac{4\text{ cal}}{1\text{ g carbo}} = 108\text{ cal}$

c. $4\text{ g protein} \cdot \frac{4\text{ cal}}{1\text{ g protein}} = 16\text{ cal}$

Total calories 71 + 108 + 16 = 196 cal

45. a. $\frac{6}{35} = \frac{3 \cdot 2}{5 \cdot 7} = \frac{3}{5} \bullet \frac{2}{7}$

b. $\frac{21}{10} = \frac{3 \cdot 7}{5 \cdot 2} = \frac{3}{5} \cdot \frac{7}{2} = \frac{3}{5} \div \frac{2}{7}$

47. a. $\frac{31}{35} = \frac{21+10}{35} = \frac{21}{35} + \frac{10}{35} = \frac{3}{5} + \frac{2}{7}$

b. $\frac{1}{21} = \frac{15-14}{21} = \frac{15}{21} - \frac{14}{21} = \frac{5}{7} - \frac{2}{3}$

49. The answer is wrong. To find the LCD, first multiply the numerator and the denominator by the same number, then add fractions.

$\frac{3}{4} + \frac{2}{3} = \frac{3 \cdot 3}{4 \cdot 3} + \frac{2 \cdot 4}{3 \cdot 4} = \frac{9}{12} + \frac{8}{12} = \frac{17}{12}$

51. This answer is right, the method always works but may give a fraction in the numerator or denominator.

53. The answer is wrong. Must find LCD before adding fractions.

$\frac{5}{6} + \frac{1}{3} = \frac{5}{6} + \frac{1 \cdot 2}{3 \cdot 2} = \frac{5}{6} + \frac{2}{6} = \frac{7}{6}$

Section 5.1

1. a. $\frac{\frac{1}{2}\text{ ft}}{2\text{ in}} = \frac{\frac{1}{2}\text{ ft}}{2\text{ in}} \cdot \frac{12\text{ in}}{1\text{ ft}} = \frac{6}{2} = \frac{3}{1}$

b. $\frac{3000\text{ g}}{6\text{ kg}} = \frac{3000\text{ g}}{6\text{ kg}} \cdot \frac{1\text{ kg}}{1000\text{ g}}$

$= \frac{3000}{6000} = \frac{1}{2}$

c. $\frac{32\text{ oz}}{6\text{ lb}} = \frac{32\text{ oz}}{6\text{ lb}} \cdot \frac{1\text{ lb}}{16\text{ oz}} = \frac{32}{96} = \frac{1}{3}$

3. a. $\frac{300\text{ ml}}{30\text{ L}} = \frac{300\text{ ml}}{30\text{ L}} \cdot \frac{1\text{ L}}{1000\text{ ml}}$

$= \frac{300}{30000} = \frac{1}{100}$

b. $\frac{2\text{ yr}}{180\text{ mo}} = \frac{2\text{ yr}}{180\text{ mo}} \cdot \frac{12\text{ mo}}{1\text{ yr}}$

$= \frac{24}{180} = \frac{2}{15}$

c. $\frac{40\text{ min}}{\frac{1}{4}\text{ hr}} = \frac{40\text{ min}}{\frac{1}{4}\text{ hr}} \cdot \frac{1\text{ hr}}{60\text{ min}}$

$= \frac{40}{15} = \frac{8}{3}$

5. a. $\frac{4.5\text{ in}}{1.68\text{ in}} = \frac{450}{168} = \frac{75}{28} \approx 2.68$ to 1

b. Area of hole: $\pi\frac{4.5^2}{4} = 5.0625\pi$

Area of ball: $\pi\frac{1.68^2}{4} = 0.7056\pi$

$\frac{5.0625\pi}{0.7056\pi} = \frac{50625}{7056} = \frac{5625}{784} \approx 7.17$ to 1

7. a. Area $= \pi\frac{1.6^2\text{ mm}^2}{4} \approx 2.01\text{ mm}^2$

b. $\frac{1\text{ cm}^2}{2.01\text{ mm}^2} \cdot \frac{(10\text{ mm})^2}{1\text{ cm}^2} = \frac{100}{2.01}$

≈ 49.7 to 1

9. Cubic feet in truck tank;

$8(12)(4) = 384\text{ ft}^3$;

Each bale expands to $4 \bullet 3 = 12\text{ ft}^3$

$384 \div 12 = 32$ bales

$\frac{384\text{ ft}^3}{1\text{ truck}} \cdot \frac{1\text{ bale}}{12\text{ ft}^3} = 32$ bales per truck

11. a. $\frac{3}{x} = \frac{5}{14}$, $5x = 3(14)$, $x = \frac{42}{5}$, $x = 8.4$

b. $\frac{4}{25} = \frac{x}{15}$, $25x = 4(15)$, $x = \frac{60}{25}$,

$x = 2.4$

c. $\frac{7}{24} = \frac{16}{x}$, $7x = 24(16)$, $x = \frac{384}{7}$,

$x \approx 54.9$

13. $\frac{1}{12} = \frac{15\text{ in}}{d}$, $1d = 12(15)$ in,

$d = 180\text{ in} \cdot \frac{1\text{ ft}}{12\text{ in}}$, $d = 15$ ft

15. $\frac{12}{9} = \frac{a}{5}$, $9a = 12(5)$, $a = \frac{60}{9}$, $a = 6\frac{2}{3}$

$\frac{12}{9} = \frac{8}{b}$, $12b = 9(8)$, $b = \frac{72}{12}$, $b = 6$

Section 5.1 (con't)

17. $\frac{6}{4}=\frac{f}{8}, 4f=48, f=12$

$\frac{6}{4}=\frac{g}{4\sqrt{3}}, 4g=24\sqrt{3}, g=6\sqrt{3}$

19. $\frac{10}{4\sqrt{3}}=\frac{e}{4}, 4\sqrt{3}e=40, e=\frac{10}{\sqrt{3}},$

$e=\frac{10\sqrt{3}}{\sqrt{3}\sqrt{3}}, e=\frac{10\sqrt{3}}{3}$

$\frac{10}{4\sqrt{3}}=\frac{f}{8}, 4\sqrt{3}f=80, f=\frac{20}{\sqrt{3}}$

$f=\frac{20\sqrt{3}}{\sqrt{3}\sqrt{3}}, f=\frac{20\sqrt{3}}{3}$

21. $\frac{\sqrt{5}}{4}=\frac{f}{8}, 4f=8\sqrt{5}, f=2\sqrt{5}$

$\frac{\sqrt{5}}{4}=\frac{g}{4\sqrt{3}}, 4g=4\sqrt{3}\sqrt{5}, g=\sqrt{15}$

23. $n=\sqrt{2(2^2)}, n=2\sqrt{2}$

$\frac{8}{2\sqrt{2}}=\frac{x}{2}, 2\sqrt{2}x=16, x=\frac{8}{\sqrt{2}},$

$x=\frac{8\sqrt{2}}{\sqrt{2}\sqrt{2}}, x=\frac{8\sqrt{2}}{2}, x=4\sqrt{2};$

$y=x=4\sqrt{2}$

25. $h_1=25$ ft, $h_2=5$ ft, let shadow = s

base of large triangle = 18 ft + s

base of small triangle = s

$\frac{5}{25}=\frac{s}{18+s}, 5(18+s)=25s,$
$90+5s=25s, 90=20s, s=4.5$

Shadow = 4.5 ft

27. h_1 = 20 ft, base of small triangle = 5 ft

base of large triangle = 12 + 5 = 17 ft

$\frac{5}{17}=\frac{h_2}{20}, 17h_2=100, h_2\approx 5.9$

Gorilla is ≈ 5.9 ft

29. a. Input, x, weight in pounds

Output, y, distance of stretch in inches

Data points; (10, 4) and (15, 6)

$m=\frac{6-4}{15-10}, m=\frac{2}{5}$

$b=4-\frac{2}{5}\cdot 10, b=0$
$y=\frac{2}{5}x$

b. Direct variation

Section 5.1 (con't)

31. a. Input, x, number of credit hrs.

Output, y, cost in \$

Data points (2, 185) and (4, 345)

$m = \dfrac{345-185}{4-2}$, $m = 80$

$b = 185 - 80(2)$, $b = 25$

$y = 80x + 25$

b. Not proportional

c. y-intercept could be student fees.

33. a. Input, x, purchase price

Output, y, tax

Data points (16, 1.20) and (30, 2.25)

$m = \dfrac{2.25-1.20}{30-16}$, $m = 0.075$

$b = 1.20 - 0.075(16)$, $b = 0$

$y = 0.075x$

b. Direct variation

35. The distance, y, varies directly with time, x;

55 mi = k(1 hr), k = 55 miles per hour

110 mi = k(2 hr), k = 55 miles per hour

165 mi = k(3 hr), k = 55 miles per hour

y = 55 mph · 8 hr, y = 440 miles

37. Cost y, varies directly with number purchased, x.

\$57.96 = k(4 CDs), k = \$14.49 per CD

\$86.94 = k(6 CDs), k = \$14.49 per CD

\$115.92 = k(8 CDs), k = \$14.49 per CD

y = 14.49(10), y = \$144.90

39. 7 inch diameter ÷ 2 = 3.5 inch radius

4.5 oz = k$(3.5\text{ in})^2$, k ≈ 0.367 oz per in^2

y = $0.367(1.5)^2$, y ≈ 0.826 oz

41. From Example 13a, k = \$0.49 per can,

y = 0.49x

From Example 13b,

$m = \dfrac{4.44-2.10}{14-5}$, $m = 0.26$

$b = 2.10 - 0.26(5)$, $b = 0.80$

$y = 0.26x + 0.80$

The y-intercept could be a connecting charge.

43. Linear equations that can be written as y = mx ***are*** linear variations.

45. If a set of data is proportional, then its equation is of the form ***y = mx.***

47. a. Constant of variation, k, is $\frac{1}{2}$

b. Constant of variation, k, is 1

c. Constant of variation, k, is $\frac{1}{3}\pi$

Section 5.1 (con't)

49. $V = \frac{1}{3}bh$, $b \approx 53{,}095\ m^2$, $h \approx 146.6\ m$;

$V \approx \frac{1}{3}(53{,}095\ m^2)(146.6\ m)$,

$V \approx 2{,}594{,}576\ m^3$

51. $C = kr$, $C = 2\pi r$, $2\pi r = kr$, $k = 2\pi$

53. $A = ks^2$, $A = s^2$, $s^2 = ks^2$, $k = 1$

55. $D = krt$, $D = rt$, $rt = krt$, $k = 1$

57. $A = ks^2$, $A = 6s^2$, $6s^2 = ks^2$, $k = 6$

59. $A = \pi\left(\frac{d}{2}\right)^2$, $A = \pi\left(\frac{d^2}{2^2}\right)$, $A = \pi\frac{d^2}{4}$

$A = \frac{\pi}{1}\cdot\frac{d^2}{4}$, $A = \frac{\pi d^2}{4}$, $A = \frac{\pi}{4}\cdot\frac{d^2}{1}$,

$A \approx 0.7854d^2$

61. The unit conversion from qt to gal is upside down.

$$\frac{3\,\text{qt}}{2\,\text{gal}}\cdot\frac{1\,\text{gal}}{4\,\text{qt}} = \frac{3}{8}$$

63. Cross multiplication result was written as a fraction instead of as an equation.

$\frac{4}{15} = \frac{6}{x}$, $4x = 90$, $x = 22.5$

Section 5.2

1. $y = \frac{1{,}000{,}000}{x}$; y = years of use,

x = yd^3 per year, k = 1,000,000 yd^3

3. $y = \frac{574{,}000{,}000}{x}$; y = years,

x = metric tons per year,

k = 5.74 X 10^8 metric tons

5. $y = \frac{100}{x}$; y = points per problem,

x = number of problems, k = 100 points

7. $160 \cdot 35 = 5600,\ 200 \cdot 28 = 5600$

Inverse variation, k = 5600 $\frac{\text{tons}}{\text{day}} \cdot$ years

9. $\frac{35.00}{160} = 0.21875,\ \frac{43.75}{200} = 0.21875$

Direct variation, k = \$0.21875 per item

11. $15 \cdot 6 = 90,\ 25 \cdot 3.6 = 90$

Inverse variation, k = 90 mi

13. $2 \cdot 6 = 12,\ 6 \cdot 2 = 12$

Inverse variation, k = 12 worker · days

15. $\frac{2200}{2} = 1100,\ \frac{3300}{3} = 1100$

Direct variation, k = \$1100 per semester

17. 35 lb · 6 ft = x · 4 ft, x = 52.5 lb

k = 35 lb · 6 ft, k = 210 lb ft

19. $80 \text{ servings} \cdot \frac{\frac{3}{4}\text{ cup}}{\text{serving}} = x \cdot \frac{1\text{ cup}}{\text{serving}}$

$x = 60$ servings,

$k = 60 \text{ servings} \cdot \frac{1\text{ cup}}{\text{serving}}$ k = 60 cups

21. 133 yrs · 900,000 metric tons per year

= x · 1,300,000 metric tons per year,

x ≈ 92 yr,

k = 133 yrs · 900,000 metric tons per yr,

k = 119,700,000 metric tons

23. First convert inches to feet,

$4\text{ in} \cdot \frac{1\text{ ft}}{12\text{ in}} = \frac{1}{3}\text{ ft},\ 3\text{ in} \cdot \frac{1\text{ ft}}{12\text{ in}} = \frac{1}{4}\text{ ft}$

$\frac{1}{3}\text{ ft} \cdot \frac{900\text{ ft}}{1\text{ gal}} = \frac{1}{4}\text{ ft} \cdot \text{x},\ \text{x} = 1200$ ft per gal

$\text{k} = \frac{1}{3}\text{ ft} \cdot \frac{900\text{ ft}}{1\text{ gal}}$, k = 300 ft^2 per gallon

25. 3 feet from the light source $y = \frac{k}{9}$,

6 feet from the light source $y = \frac{k}{36}$

$\frac{k}{9} = x\frac{k}{36},\ \frac{k}{9} \cdot \frac{36}{k} = x,\ x = 4$

Intensity is 4 times greater

27. $\frac{x_1 y_1}{x_2 y_1} = \frac{x_2 y_2}{x_2 y_1},\ \frac{x_1}{x_2} = \frac{y_2}{y_1};\ \frac{y_1}{y_2}$ has been inverted.

Section 5.2 (con't)

29.

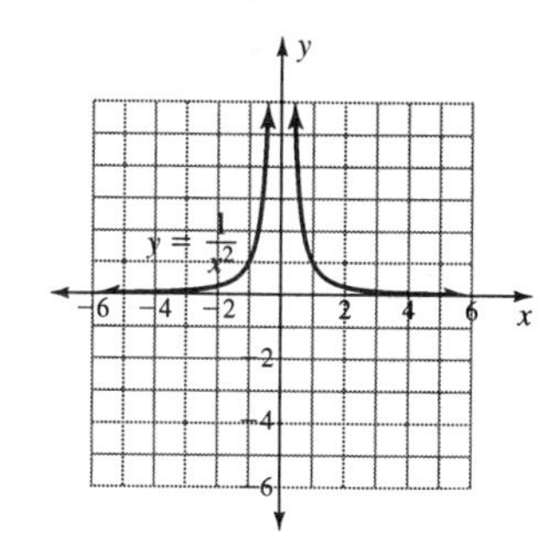

a. f(0) is undefined

b. Axis of symmetry is x = 0

c. The graph approaches y = 0 (the x-axis)

31. a.

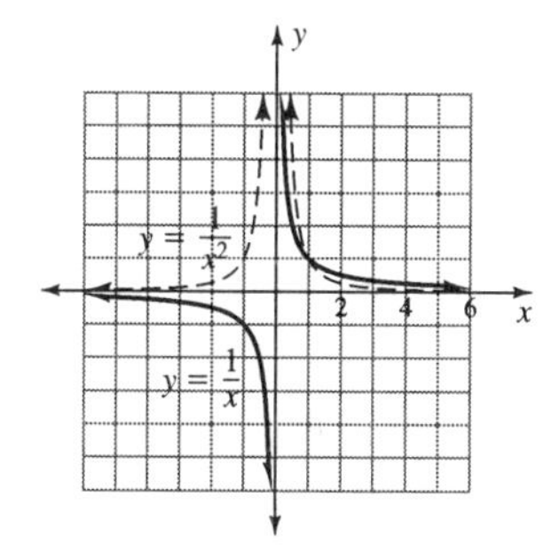

b. $y=\frac{1}{x}$ is above $y=\frac{1}{x^2}$ for x > 1

c. Graphs are equal when x = 1

d. When x < 0 $\frac{1}{x^2} > 0$, $\frac{1}{x} < 0$

33. a.

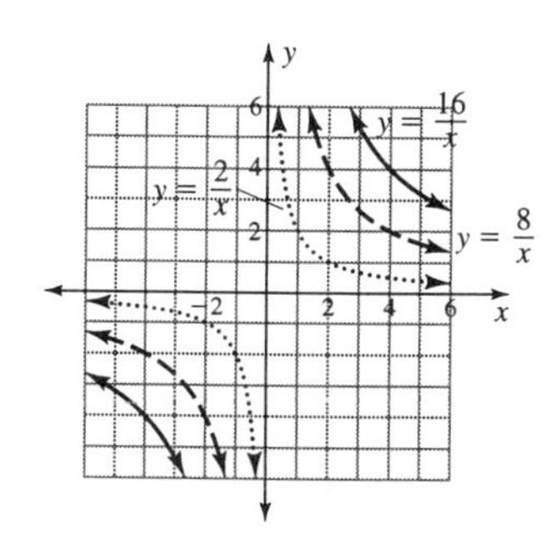

b. All graphs are in the 1st and 3rd quadrants.

33. c. $\frac{16}{x}$ is flatter than $\frac{8}{x}$ which is flatter than $\frac{2}{x}$; in the 1st quadrant $\frac{16}{x}$ is above $\frac{8}{x}$ which is above $\frac{2}{x}$, in the 3rd quadrant the order is reversed.

d. The numerator is the constant of variation for each equation.

e. Increasing the constant of variation increases the steepness and increases the distance from the x-axis.

f. $y=\frac{1}{x}$, will be closer to the x-axis.

g. $y=\frac{32}{x}$, will be farther from the x-axis.

h. As x gets larger the graphs get closer to the x-axis.

Section 5.2 (con't)

35.

Input, x mph	Output, t, hrs
$6 \div 1$	6
$6 \div 2$	3
$6 \div 3$	2
$6 \div 4$	1.5
$6 \div 5$	1.2
$6 \div 6$	1
$6 \div 7$	≈ 0.86
$6 \div 8$	0.75
$6 \div 9$	≈ 0.67
$6 \div 10$	0.6

A reasonable speed would be 6 mph and above.

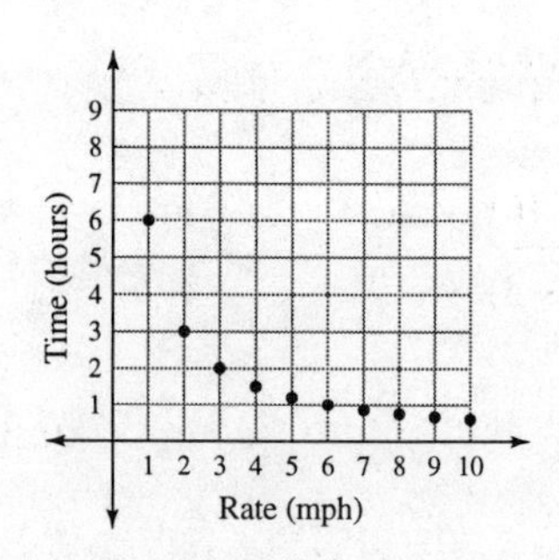

37. a. $\frac{20 \text{ mi}}{65 \text{ mph}} \cdot \frac{60 \text{ min}}{1 \text{ hr}} \approx 18.5 \text{ min}$

b.

Input, x mph	Output, y min
60	$\frac{20}{60} \cdot \frac{60}{1} = 20$
65	$\frac{20}{65} \cdot \frac{60}{1} \approx 18.5$
70	$\frac{20}{70} \cdot \frac{60}{1} \approx 17.1$
75	$\frac{20}{75} \cdot \frac{60}{1} = 16$
80	$\frac{20}{80} \cdot \frac{60}{1} = 15$

c. $18.5 - 5 = 13.5$;

$$\frac{20 \text{ mi} \cdot 60 \text{ min}}{x \text{ mph} \cdot 1 \text{hr}} = 13.5 \text{ min}, \ x = \frac{1200 \text{ mi}}{13.5 \text{ hr}}$$

$x \approx 88.9$ mph

39. a. $\frac{120}{12} = \frac{100 \text{ mA}}{I_1}, \ 120 I_1 = 1200 \text{ mA}$

$I_1 = 10 \text{ mA}$

b. Voltage and current are inversely proportional; $V_1 I_1 = V_2 I_2$.

c. 120 volt • 10 mA = 1200 volt mA;

Note: 1 Amp = 1000 mA

$$1200 \text{ volt mA} \cdot \frac{1 \text{ Amp}}{1000 \text{ mA}} = 1.2 \text{ volt Amp}$$

Section 5.3

1. Undefined when x + 3 = 0,

x + 3 = 0, x = -3

3. Undefined when 4 - a = 0,

4 - a = 0, 4 = a

5. Undefined when $x^2 - 3x + 28 = 0$

$$x = \frac{-(-3) \pm \sqrt{(-3)^2 - 4(1)(28)}}{2(1)},$$

$$x = \frac{3 \pm \sqrt{-103}}{2}, \; x = \frac{3 \pm i\sqrt{103}}{2}$$

solutions are complex numbers, therefore the expression is defined for all real numbers

7. a. Additive inverse of x + y is -x - y

b. Additive inverse of -x + y is x - y

c. Additive inverse of y - x is x - y

9. a. $\frac{x-3}{x-3} = 1, \; x \neq 3$

b. $\frac{x+2}{-x-2} = -1, \; x \neq -2$

c. $\frac{b-a}{b-a} = 1, \; b \neq a$

d. $\frac{3-x}{x-3} = -1, \; x \neq 3$

11. a. Common factor is $2a$

$$\frac{16a^2}{10a} = \frac{2a \cdot 8a}{2a \cdot 5} = \frac{8a}{5}; \; a \neq 0$$

b. Common factor is $2xy$

$$\frac{2x^2y}{10xy^2} = \frac{2xy \cdot x}{2xy \cdot 5y} = \frac{x}{5y}; \; x \neq 0, \; y \neq 0$$

13. a. No common factors, expression is already simplified; $y \neq -2x$

b. Common factor is x

$$\frac{xy}{2x + xy} = \frac{x \cdot y}{x(2+y)} = \frac{y}{2+y}, x \neq 0, y \neq -2$$

15. a. Notice the numerator is the additive inverse of a factor in the denominator, rewrite the expression:

$$\frac{3-x}{(x+3)(x-3)} = \frac{-1(x-3)}{(x+3)(x-3)}$$

Common factor is (x - 3),

$$\frac{-1(x-3)}{(x+3)(x-3)} = -\frac{1}{x+3}, \; x \neq \pm 3$$

b. Factor numerator and denominator;

$$\frac{x^2 - 9}{x^2 + 5x + 6} = \frac{(x+3)(x-3)}{(x+3)(x+2)}$$

Common factor is (x + 3),

$$\frac{(x+3)(x-3)}{(x+3)(x+2)} = \frac{x-3}{x+2}, \; x \neq -3, -2$$

Section 5.3 (con't)

17. a. Factor numerator and denominator;

$$\frac{2ac+4bc}{4ad+8bd}=\frac{2c(a+2b)}{4d(a+2b)}=\frac{2c(a+2b)}{2\cdot 2d(a+2b)}$$

Common factor is $2(a + 2b)$,

$$\frac{2c(a+2b)}{4d(a+2b)}=\frac{c}{2d},\ a\neq -2b,\ d\neq 0$$

b. Factor numerator and denominator;

$$\frac{6x^2+3x}{12x^2-6x}=\frac{3x(2x+1)}{6x(2x-1)}$$

Common factor is 3x,

$$\frac{3x(2x+1)}{6x(2x-1)}=\frac{2x+1}{2(2x-1)},x\neq 0,\ \tfrac{1}{2}$$

19. a. Factor numerator;

$$\frac{x^2+x-6}{2-x}=\frac{(x+3)(x-2)}{2-x}=\frac{(x+3)(x-2)}{-1(x-2)}$$

Common factor is (x - 2),

$$\frac{(x+3)(x-2)}{-1(x-2)}=-(x+3),\ x\neq 2$$

b. Factor denominator;

$$\frac{x-3}{6-2x}=\frac{x-3}{-2(-3+x)}=\frac{x-3}{-2(x-3)}$$

Common factor is (x - 3),

$$\frac{x-3}{-2(x-3)}=-\frac{1}{2},\ x\neq 3$$

21. a. $\frac{1}{x}\cdot\frac{x^2}{1}=\frac{x^2}{x}=\frac{x\cdot x}{x}=x$

b. $\frac{1}{a}\div\frac{a^2b^2}{1}=\frac{1}{a}\cdot\frac{1}{a^2b^2}=\frac{1}{a^3b^2}$

c. $\frac{b}{a}\div\frac{a^2}{b^2}=\frac{b}{a}\cdot\frac{b^2}{a^2}=\frac{b^3}{a^3}$

d. $\frac{x}{y}\div\frac{x^3}{y^2}=\frac{x}{y}\cdot\frac{y^2}{x^3}=\frac{y}{x^2}$

23. a. $\frac{a^2+7a+12}{a^2-4}\div\frac{a^2+4a}{a-2}$

$$=\frac{(a+3)(a+4)}{(a-2)(a+2)}\div\frac{a(a+4)}{(a-2)}$$

$$=\frac{(a+3)(a+4)}{(a-2)(a+2)}\cdot\frac{(a-2)}{a(a+4)}=\frac{(a+3)}{a(a+2)}$$

b. $\frac{x^2-2x}{x}\cdot\frac{x^2-1}{x^2-3x+2}$

$$=\frac{x(x-2)}{x}\cdot\frac{(x+1)(x-1)}{(x-1)(x-2)}=x+1$$

25. a. $\frac{x^2+3x}{x}\cdot\frac{x^2-x-6}{x^2-9}$

$$=\frac{x(x+3)}{x}\cdot\frac{(x-3)(x+2)}{(x-3)(x+3)}=x+2$$

b. $\frac{4a^2+4a+1}{4-9a^2}\div\frac{4a^2+2a}{3a-3}$

$$=\frac{(2a+1)(2a+1)}{(2-3a)(2+3a)}\cdot\frac{(3a-2)}{2a(2a+1)}$$

$$=\frac{-1(2a+1)}{2a(2+3a)}=\frac{-2a-1}{2a(2+3a)}$$

Section 5.3 (con't)

27. a. $\dfrac{x+2}{x^2-4x+4}\cdot\dfrac{x^2-2x}{x+2}$

$=\dfrac{(x+2)}{(x-2)(x-2)}\cdot\dfrac{x(x-2)}{(x+2)}=\dfrac{x}{x-2}$

b. $\dfrac{a^2-5a}{a^2+5a}\div\dfrac{a^2-10a+25}{a}$

$=\dfrac{a(a-5)}{a(a+5)}\cdot\dfrac{a}{(a-5)^2}=\dfrac{a}{(a+5)(a-5)}$

29. a. $\dfrac{a^2-b^2}{a^2+2ab+b^2}\cdot\dfrac{a+b}{a-b}$

$=\dfrac{(a-b)(a+b)}{(a+b)^2}\cdot\dfrac{(a+b)}{(a-b)}=1$

b. $\dfrac{x^2-2xy+y^2}{x^2-y^2}\cdot\dfrac{x-y}{y-x}$

$=\dfrac{(x-y)^2}{(x+y)(x-y)}\cdot -1=-\dfrac{x-y}{x+y}$

31. a. $\dfrac{a^3-b^3}{a+b}\cdot\dfrac{a^2+2ab+b^2}{a^2-b^2}$

$=\dfrac{(a-b)(a^2+ab+b^2)}{(a+b)}\cdot\dfrac{(a+b)^2}{(a+b)(a-b)}$

$=a^2+ab+b^2$

31. b. $\dfrac{x^2+2xy+y^2}{x^3+y^3}\cdot\dfrac{x^2-y^2}{x+y}$

$=\dfrac{(x+y)^2}{(x+y)(x^2-xy+y^2)}\cdot\dfrac{(x+y)(x-y)}{(x+y)}$

$=\dfrac{(x+y)(x-y)}{x^2-xy-y^2}$

33. a. $\dfrac{\frac{1}{a}}{\frac{1}{b}}=\dfrac{1}{a}\cdot\dfrac{b}{1}=\dfrac{b}{a}$

b. $\dfrac{\frac{1}{b}}{a}=\dfrac{1}{b}\cdot\dfrac{1}{a}=\dfrac{1}{ab}$

c. $\dfrac{\frac{1}{b}}{b}=\dfrac{1}{b}\cdot\dfrac{1}{b}=\dfrac{1}{b^2}$

d. $\dfrac{\frac{a}{b}}{\frac{1}{b}}=\dfrac{a}{b}\cdot\dfrac{b}{1}=a$

35. $\dfrac{\dfrac{x^2-3x+2}{x}}{\dfrac{x^2-1}{x^2}}$

$=\dfrac{(x-2)(x-1)}{x}\cdot\dfrac{x^2}{(x+1)(x-1)}$

$=\dfrac{x(x-2)}{x+1}$

37. $\dfrac{\dfrac{x-2}{x^2+3x-4}}{\dfrac{2-x}{x-1}}=\dfrac{(x-2)}{(x+4)(x-1)}\cdot\dfrac{x-1}{2-x}$

$=\dfrac{-1}{x+4}$

Section 5.3 (con't)

39. $\dfrac{5280 \text{ feet}}{88 \text{ feet per sec}} = \dfrac{5280 \text{ feet}}{1} \cdot \dfrac{1 \text{ sec}}{88 \text{feet}}$

$= 60$ sec

41. $\dfrac{65 \text{ mi per hr}}{15 \text{ mi per gal}} = \dfrac{65 \text{ mi}}{1 \text{ hr}} \cdot \dfrac{1 \text{ gal}}{15 \text{ mi}}$

$= 4\frac{1}{3}$ gal per hr

43. $\dfrac{24 \text{ cans per case}}{\$3.98 \text{ per case}} = \dfrac{24 \text{ cans}}{\text{case}} \cdot \dfrac{1 \text{ case}}{\$3.98}$

≈ 6 cans per $

45. $\dfrac{8 \text{ stitches per inch}}{\dfrac{1 \text{ foot}}{12 \text{ inches}}} = \dfrac{8 \text{ stitches}}{1 \text{in}} \cdot \dfrac{12 \text{ in}}{1 \text{ ft}}$

$= 96$ stitches per foot

47. $\dfrac{95 \text{ words per min}}{300 \text{ words per page}}$

$= \dfrac{95 \text{ words}}{\text{min}} \cdot \dfrac{1 \text{ page}}{300 \text{ words}}$

≈ 0.3 page per minute

49. $x^2 + x + 2$ does not factor; x is a term, not a factor; this expression is already simplified.

51. $x^2 - 2x + 1$ factors to $(x - 1)(x - 1)$; the numerator and denominator have no common factors; this expression is already simplified.

53. When we multiply $a \cdot b \cdot c$ we do not multiply $a \cdot b$ and $a \cdot c$ because multiplication is not distributive over multiplication.

Mid-Chapter 5 Test

1. $\frac{32}{24} = \frac{8(4)}{8(3)} = \frac{4}{3}$

2. $\frac{3}{10} + \frac{5}{18} = \frac{3(9)}{10(9)} + \frac{5(5)}{18(5)} = \frac{27+25}{90}$

$= \frac{52}{90} = \frac{26(2)}{45(2)} = \frac{26}{45}$

3. $\frac{abc}{aces} = \frac{b}{es}$

4. $\frac{3}{10} \div \frac{5}{18} = \frac{3}{10} \cdot \frac{18}{5} = \frac{3(2)(9)}{5(2)(5)}$

$\frac{27}{25} = 1\frac{2}{25}$

5. $\frac{12cd^2}{8c^2d} = \frac{4(3)cdd}{4(2)ccd} = \frac{3d}{2c}$

6. $\frac{(x-2)(x+3)}{(x+3)} = x-2$

7. $\frac{a-6}{a^2-3a-18} = \frac{a-6}{(a-6)(a+3)} = \frac{1}{a+3}$

8. $\frac{b+bc}{b} = \frac{b(1+c)}{b} = 1+c$

9. $\frac{a^2b}{c} \cdot \frac{c^2}{ab^2} = \frac{aabcc}{cabb} = \frac{ac}{b}$

10. $\frac{ab^2}{c^2} \div \frac{ac}{b} = \frac{ab^2}{c^2} \cdot \frac{b}{ac} = \frac{b^3}{c^3}$

11. $\frac{x^2-16}{x^2+6x+8} \cdot \frac{x+2}{x-2}$

$= \frac{(x+4)(x-4)}{(x+4)(x+2)} \cdot \frac{(x+2)}{(x-2)} = \frac{x-4}{x-2}$

12. $\dfrac{\frac{x^2}{x+1}}{\frac{x}{x^2-1}} = \frac{x^2}{x+1} \cdot \frac{(x+1)(x-1)}{x}$

$= x(x-1)$

13. $\frac{x^2-16}{x^2+6x+8} \div \frac{4-x}{x+2}$

$\frac{(x+4)(x-4)}{(x+4)(x+2)} \cdot \frac{(x+2)}{-1(x-4)} = -1$

14. $\frac{2\frac{2}{3}\text{ yards}}{5\text{ feet}} = \frac{\frac{8}{3}\text{yards}}{5\text{ feet}} \cdot \frac{3\text{ feet}}{1\text{ yard}}$

$= \frac{\frac{8}{3}(3)}{5} = \frac{8}{5}$

15. $\frac{150\text{ months}}{15\text{ years}} = \frac{150\text{ mo.}}{15\text{ yr}} \cdot \frac{1\text{ yr}}{12\text{ mo.}}$

$= \frac{150}{180} = \frac{3(5)(10)}{3(6)(10)} = \frac{5}{6}$

16. Expression is undefined when the denominator = 0

(x + 4) = 0, x = -4

or (x + 2) = 0, x = -2

or (x - 2) = 0, x = 2

Mid-Chapter 5 Test (con't)

17.

$$1{,}000{,}000\text{ tbsp}\cdot\frac{1\text{ cup}}{16\text{ tbsp}}\cdot\frac{1\text{ pt}}{2\text{ cup}}\cdot\frac{1\text{ qt}}{2\text{ pt}}\cdot\frac{0.25\text{ gal}}{1\text{ qt}}$$

= 3906.25 gallons

18. 5 feet 2 inches = 62 inches, 62 - 15 = 47

rate is 47 inches in 14 years, convert this to miles per hour;

$$\frac{47\text{ in}}{14\text{ yr}}\cdot\frac{1\text{ ft}}{12\text{ in}}\cdot\frac{1\text{ mi}}{5280\text{ ft}}\cdot\frac{1\text{ yr}}{365\text{ day}}\cdot\frac{1\text{ day}}{24\text{ hr}}$$

$\approx 6.05 \text{ X } 10^{-9}$ miles per hour

19. $\frac{a}{15}=\frac{12}{35},\ 35a=180,\ a\approx 5.14$

20. $\frac{x+1}{8}=\frac{2x-1}{14},\ 14(x+1)=8(2x-1)$

$14x+14=16x-8,\ 2x=22,\ x=11$

21. $\frac{5}{6}=\frac{n}{9.6},\ 6n=48,\ n=8$

$\frac{6}{5}=\frac{p}{11},\ 5p=66,\ p=13.2$

22. $m=\frac{15-(-6)}{10-(-4)}=\frac{21}{14}=\frac{3}{2},$

$b=15-\frac{3}{2}(10),\ b=0,$

$y=\frac{3}{2}x$

proportional

23. $m=\frac{1.50-0.60}{3-1},\ m=0.45;$

$b=0.60-0.45(1),\ b=0.15;$

$y=0.45x+0.15$

not proportional

24. a. k = 100 lb · 4 ft = 400 lb ft

or k = 50 lb · 8 ft = 400 lb ft

This is an inverse proportion

b. 80x=400, x = 5 ft (from the pivot point)

25. $y=\frac{100}{x}$, where y = number of cookies per child, x = number of children;

k = 100 cookies

26. $10 = k(13)^2,\ k\approx 0.05917$ \$ per in^2

$p = 0.05917(16)^2,\ p\approx \$15.15$

27. $1350=\frac{k}{1^2},\ k=1350;$

$n=\frac{1350}{4^2},\ n\approx 84$

Section 5.4

1. a. $\frac{3}{4} - \frac{x}{4} = \frac{3-x}{4}$

b. $\frac{2}{5x} - \frac{7}{5x} = \frac{2-7}{5x} = \frac{-5}{5x} = -\frac{1}{x}, x \neq 0$

c. $\frac{4}{x+3} - \frac{x^2}{x+3} = \frac{4-x^2}{x+3}, x \neq -3$

d. $\frac{2}{x^2+1} - \frac{x-1}{x^2+1} = \frac{2-(x-1)}{x^2+1}$

$= \frac{2-x+1}{x^2+1} = \frac{3-x}{x^2+1}$

3. a. Common denominator is 12

b. Common denominator is 2a

c. Common denominator is a^2b^2

d. $x^2 - 2x = x(x - 2)$;

$x^2 - 4 = (x - 2)(x + 2)$;

Common denominator is $x(x - 2)(x + 2)$

e. $x^2 - 6x + 9 = (x - 3)^2$;

$x^2 - 9 = (x - 3)(x + 3)$;

Common denominator is $(x - 3)^2(x + 3)$

5. a. $\frac{1}{4} + \frac{7}{12} = \frac{1 \cdot 3}{4 \cdot 3} + \frac{7}{12} = \frac{3+7}{12}$

$= \frac{10}{12} = \frac{5}{6}$

5. b. $\frac{2}{a} - \frac{5}{2a} = \frac{2 \cdot 2}{2a} - \frac{5}{2a} = \frac{4-5}{2a}$

$= -\frac{1}{2a}, \ a \neq 0$

c. $\frac{b}{a^2} - \frac{c}{ab^2} = \frac{b \cdot b^2}{a^2b^2} - \frac{a \cdot c}{a^2b^2}$

$= \frac{b^3 - ac}{a^2b^2}, a \neq 0, b \neq 0$

d. $\frac{5}{x^2 - 2x} + \frac{3}{x^2 - 4}$

$= \frac{5(x+2)}{x(x-2)(x+2)} + \frac{3x}{x(x-2)(x+2)}$

$= \frac{5x+10+3x}{x(x-2)(x+2)} = \frac{8x+10}{x(x-2)(x+2)}$

$x \neq 0, x \neq \pm 2$

e. $\frac{5x}{x^2 - 6x + 9} - \frac{3x}{x^2 - 9}$

$= \frac{5x(x+3)}{(x-3)^2(x+3)} - \frac{3x(x-3)}{(x-3)^2(x+3)}$

$\frac{5x^2 + 15x - 3x^2 + 9x}{(x-3)^2(x+3)} = \frac{2x^2 + 24x}{(x-3)^2(x+3)}$

$x \neq \pm 3$

7. $\frac{5}{4a} + \frac{3}{6b} = \frac{5(3b)}{4a(3b)} + \frac{3(2a)}{6b(2a)}$

$= \frac{15b + 6a}{12ab} = \frac{3(5b+2a)}{3(4ab)} = \frac{2a+5b}{4ab}$

$a \neq 0, \ b \neq 0$

Section 5.4 (con't)

9. $\dfrac{x}{x-3}+\dfrac{1}{x}=\dfrac{x(x)}{x(x-3)}+\dfrac{1(x-3)}{x(x-3)}$

$=\dfrac{x^2+x-3}{x(x-3)},\ x\neq 0, 3$

11. $\dfrac{5x}{x-1}-\dfrac{8+x}{x}=\dfrac{5x(x)}{(x-1)x}-\dfrac{(8+x)(x-1)}{x(x-1)}$

$=\dfrac{5x^2-(x^2+7x-8)}{x(x-1)}=\dfrac{4x^2-7x+8}{x(x-1)}$

$x\neq 0, 1$

13. $\dfrac{x}{x-3}+\dfrac{3}{x^2-6x+9}=\dfrac{x}{x-3}+\dfrac{3}{(x-3)^2}$

$\dfrac{x(x-3)}{(x-3)^2}+\dfrac{3}{(x-3)^2}=\dfrac{x^2-3x+3}{(x-3)^2}.\ x\neq 3$

15. $\dfrac{2b}{b^2-1}-\dfrac{3}{1-b}=\dfrac{2b}{(b+1)(b-1)}-\dfrac{3}{-1(b-1)}$

$\dfrac{2b}{(b+1)(b-1)}+\dfrac{3(b+1)}{(b+1)(b-1)}=\dfrac{2b+3b+3}{(b+1)(b-1)}$

$=\dfrac{5b+3}{(b+1)(b-1)},\ b\neq -1, 1$

17. $\dfrac{2}{2a+ab}-\dfrac{3}{2b+b^2}$

$=\dfrac{2(b)}{a(2+b)(b)}-\dfrac{3(a)}{b(2+b)(a)}$

$=\dfrac{2b-3a}{ab(2+b)},\ a\neq 0,\ b\neq -2, 0$

19. $\dfrac{x}{x^2-6x+9}+\dfrac{3}{x^2-3x}=\dfrac{x}{(x-3)^2}+\dfrac{3}{x(x-3)}$

$\dfrac{x(x)}{x(x-3)^2}+\dfrac{3(x-3)}{x(x-3)^2}=\dfrac{x^2+3x-9}{x(x-3)^2}$

$x\neq 0, 3$

21. $\dfrac{1}{t}=\dfrac{1}{t_1}+\dfrac{1}{t_2},\ \dfrac{1}{t}=\dfrac{t_2}{t_1t_2}+\dfrac{t_1}{t_1t_2},\ \dfrac{1}{t}=\dfrac{t_2+t_1}{t_1t_2}$

23. $\dfrac{1}{m}=\dfrac{1}{m_1}+\dfrac{1}{m_2},\dfrac{1}{m}=\dfrac{m_2}{m_1m_2}+\dfrac{m_1}{m_1m_2},$

$\dfrac{1}{m}=\dfrac{m_2+m_1}{m_1m_2}$

25. $t_1=5,\ t_2=6,\ \dfrac{6+5}{5(6)}=\dfrac{11}{30}\approx 0.367$

$t=3,\ \dfrac{1}{3}\approx 0.333$; the two fans will be able to meet code.

27. $\dfrac{367}{45}\approx 8.156,\ \dfrac{367}{60}\approx 6.117,$

total time $8.156+6.117\approx 14.3$ hrs

average rate $2(367)\div 14.3\approx 51.4$ mph

29. $\dfrac{x}{55}+\dfrac{x}{65}=\dfrac{13x}{13(55)}+\dfrac{11x}{11(65)}=\dfrac{24x}{715}$

$2x\div\dfrac{24x}{715}=2x\cdot\dfrac{715}{24x}=\dfrac{715}{12}\approx 59.6$ mph

Section 5.4 (con't)

31. $m_1 = 5,\ m_2 = 9,\ \frac{1}{m} = \frac{9+5}{9(5)},$

$\frac{1}{m} = \frac{14}{45},\ m = \frac{45}{14},\ m \approx 3.2 \text{ min}$

33. $1 + x + \frac{x^2}{2} + \frac{x^3}{6} + \frac{x^4}{24}$

$= \frac{24}{24} + \frac{24x}{24} + \frac{12x^2}{24} + \frac{4x^3}{24} + \frac{x^4}{24}$

$= \frac{x^4 + 4x^3 + 12x^2 + 24x + 24}{24}$

35. $\frac{RT}{v-b} - \frac{a}{v^2} = \frac{RTv^2}{v^2(v-b)} - \frac{a(v-b)}{v^2(v-b)}$

$= \frac{RTv^2 - av + ab}{v^2(v-b)}$

37. $\frac{1}{t} = \frac{1}{t_1} + \frac{1}{t_2} + \frac{1}{t_3},\ \frac{1}{t} = \frac{t_2t_3}{t_1t_2t_3} + \frac{t_1t_3}{t_1t_2t_3} + \frac{t_1t_2}{t_1t_2t_3}$

$\frac{1}{t} = \frac{t_2t_3 + t_1t_3 + t_1t_2}{t_1t_2t_3},\ t = \frac{t_1t_2t_3}{t_2t_3 + t_1t_3 + t_1t_2}$

The time required is not the sum of the individual times.

39. $\frac{7+1}{\frac{1}{7}+1} = \frac{8}{\frac{8}{7}} = \frac{8}{1}\cdot\frac{7}{8} = 7$

41. $a = \frac{V}{\frac{4}{3}\pi b^2},\ a = \frac{V}{\frac{4\pi b^2}{3}},$

$a = \frac{V}{1}\cdot\frac{3}{4\pi b^2},\ a = \frac{3V}{4\pi b^2}$

43. $\frac{x - \frac{x}{2}}{2 + \frac{x}{3}} = \frac{\frac{2x-x}{2}}{\frac{6+x}{3}} = \frac{x}{2}\cdot\frac{3}{(x+6)} = \frac{3x}{2(x+6)}$

45. $I = \frac{E}{R + \frac{r}{2}},\ I = \frac{E}{\frac{2R+r}{2}},$

$I = \frac{E}{1}\cdot\frac{2}{2R+r},\ I = \frac{2E}{2R+r}$

47. a. $\frac{1}{b} = \frac{a}{b} \bullet \frac{1}{a}$

b. $\frac{a^2}{b} = \frac{a}{b}\cdot\frac{a}{1} = \frac{a}{b} \div \frac{1}{a}$

49. a. $\frac{a+b}{ab} = \frac{a}{ab} + \frac{b}{ab} = \frac{1}{b} + \frac{1}{a}$

b. $\frac{1}{ab} = \frac{1}{a}\cdot\frac{1}{b}$

51. $\frac{1}{\frac{2}{3}+\frac{4}{7}} = \frac{1}{\frac{14}{21}+\frac{12}{21}} = \frac{1}{\frac{26}{21}} = 1\cdot\frac{21}{26} = \frac{21}{26}$

$\frac{3}{2} + \frac{7}{4} = \frac{6}{4} + \frac{7}{4} = \frac{13}{4} \neq \frac{21}{26}$

53. $\frac{a}{b} + \frac{c}{d} = \frac{ad}{bd} + \frac{bc}{bd} = \frac{ad+bc}{bd}$

The reciprocal is $\frac{bd}{ad+bc}$

Section 5.5

1. Defined for all real numbers, $x \neq 4$

$$\begin{array}{r} x+1 \\ x-4\overline{)\,x^2-3x-4} \\ -(x-4x) \\ \hline x-4 \\ -(x-4) \\ \hline 0 \end{array} = x+1$$

3. Defined for all real numbers, $x \neq 3$

$$\begin{array}{r} x \\ x-3\overline{)\,x^2-3x-4} \\ -(x^2-3x) \\ \hline -4 \end{array} = x-\frac{4}{x-3}$$

5. Defined for all real numbers, $x \neq 2$

$$\begin{array}{r} x-1 \\ x-2\overline{)\,x^2-3x-4} \\ -(x^2-2x) \\ \hline -x-4 \\ -(-x+2) \\ \hline -6 \end{array} = x-1-\frac{6}{x-2}$$

7. Defined for all real numbers, $x \neq 1$

$$\begin{array}{r} x-2 \\ x-1\overline{)\,x^2-3x-4} \\ -(x^2-x) \\ \hline -2x-4 \\ -(-2x+2) \\ \hline -6 \end{array} = x-2-\frac{6}{x-1}$$

9. Defined for all real numbers, $x \neq 0$

$$\begin{array}{r} x-3 \\ x\overline{)\,x^2-3x-4} \\ -x^2 \\ \hline -3x \\ -(-3x) \\ \hline -4 \end{array} = x-3-\frac{4}{x}$$

11. Defined for all real numbers, $x \neq -1$

$$\begin{array}{r} x-4 \\ x+1\overline{)\,x^2-3x-4} \\ -(x^2+x) \\ \hline -4x-4 \\ -(-4x-4) \\ \hline 0 \end{array} = x-4$$

13. $x^2 - 3x - 4 = (x + 1)(x - 4)$, the remainder was zero when the denominator was a factor.

15.

x	$x^2 - 3x - 4$
-1	$(-1)^2 - 3(-1) - 4 = 0$
0	$(0)^2 - 3(0) - 4 = -4$
1	$(1)^2 - 3(1) - 4 = -6$
2	$(2)^2 - 3(2) - 4 = -6$
3	$(3)^2 - 3(3) - 4 = -4$
4	$(4)^2 - 3(4) - 4 = 0$

The answers and remainders are the same.

Section 5.5 (con't)

17.

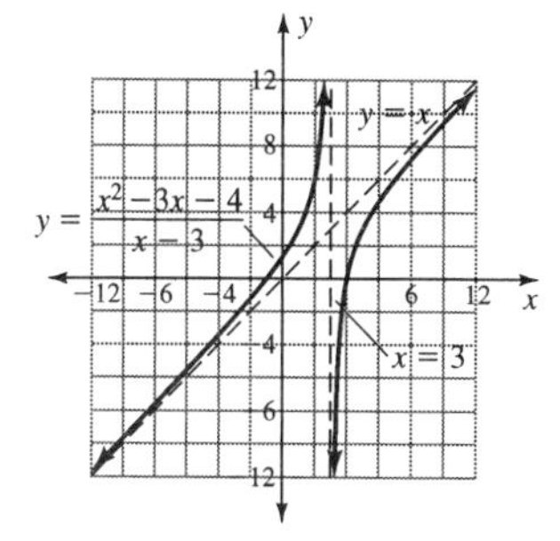

a. The function is undefined at x = 3, nearly vertical portion is near x = 3

b. The graph approaches the function y = x.

c.

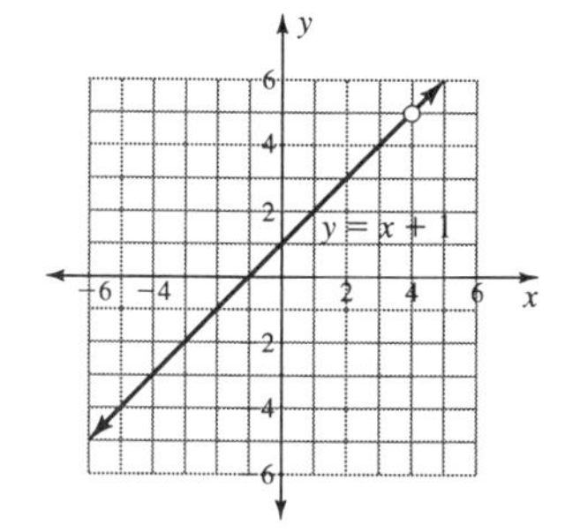

The graph is different because the remainder was zero. The undefined value is a hole in the graph.

d.

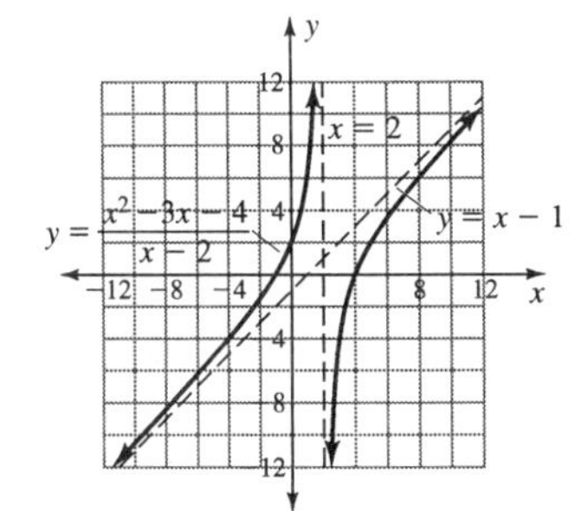

The graphs are similar because the remainder was not zero. As the graph approaches the undefined value it is nearly vertical.

19. $\dfrac{x^3-1}{x-2} = x^2+2x+4+\dfrac{7}{x-2}$

$$
\begin{array}{r}
x^2+2x+4 \\
x-2\overline{)\,x^3+0x^2+0x\;\;-1} \\
\underline{-(x^3-2x^2)\quad\quad\quad\quad} \\
2x^2+0x\quad\quad \\
\underline{-(2x^2-4x)\quad\quad} \\
4x\;\;-1 \\
\underline{-(4x\;\;-8)} \\
7
\end{array}
$$

21. $\dfrac{x^3-1}{x-1} = x^2+x+1$, denominator is a factor

$$
\begin{array}{r}
x^2\;+x+1 \\
x-1\overline{)\,x^3+0x^2+0x-1} \\
\underline{-(x^3-\;x^2)\quad\quad\quad\quad} \\
x^2+0x\quad\quad \\
\underline{-(x^2-\;x)\quad\quad} \\
x-1 \\
\underline{-(x-1)} \\
0
\end{array}
$$

23. $\dfrac{x^3-1}{x} = x^2-\dfrac{1}{x}$

$$
\begin{array}{r}
x^2\quad\quad\quad\quad \\
x\overline{)\,x^3+0x^2+0x-1} \\
\underline{-x^3\quad\quad\quad\quad\quad\quad\quad} \\
-1
\end{array}
$$

Section 5.5 (con't)

25. $\dfrac{x^3-1}{x+1}=x^2-x+1-\dfrac{2}{x+1}$

$$\begin{array}{r}
x^2-x+1 \\
x+1\overline{)x^3+0x^2+0x-1} \\
\underline{-(x^3+x^2)} \\
-x^2+0x \\
\underline{-(-x^2-x)} \\
x-1 \\
\underline{-(x+1)} \\
-2
\end{array}$$

27.

x	x^3 - 1
-1	$(-1)^3 - 1 = -2$
0	$(0)^3 - 1 = -1$
1	$(1)^3 - 1 = 0$
2	$(2)^3 - 1 = 7$

Solutions and remainders are the same.

29. $\dfrac{x^3+1}{x+1}=x^2-x+1$, denominator is a factor of numerator

$$\begin{array}{r}
x^2-x+1 \\
x+1\overline{)x^3+0x^2+0x+1} \\
\underline{-(x^3+x^2)} \\
-x^2+0x \\
\underline{-(-x^2-x)} \\
x+1 \\
\underline{-(x+1)} \\
0
\end{array}$$

31. $\dfrac{x^4+1}{x+1}=x^3-x^2+x-1+\dfrac{2}{x+1}$

$$\begin{array}{r}
x^3-x^2+x-1 \\
x+1\overline{)x^4+0x^3+0x^2+0x+1} \\
\underline{-(x^4+x^3)} \\
-x^3+0x^2 \\
\underline{-(-x^3-x^2)} \\
x^2+0x \\
\underline{-(x^2+x)} \\
-x+1 \\
\underline{-(-x-1)} \\
2
\end{array}$$

33. $\dfrac{x^4-4x^3+6x^2-4x+1}{x-1}=x^3-3x^2+3x-1$

denominator is a factor of numerator

$$\begin{array}{r}
x^3-3x^2+3x-1 \\
x-1\overline{)x^4-4x^3+6x^2-4x+1} \\
\underline{-(x^4-x^3)} \\
-3x^3+6x^2 \\
\underline{-(-3x^3+3x^2)} \\
3x^2-4x \\
\underline{-(3x^2-3x)} \\
-x+1 \\
\underline{-(-x+1)} \\
0
\end{array}$$

Section 5.5 (con't)

35. $\dfrac{3x^3 + x^2 + 0x - 1}{x+1} = 3x^2 - 2x + 2 - \dfrac{3}{x+1}$

$$\begin{array}{r} 3x^2 - 2x + 2 \\ x+1 \overline{)\,3x^3 + x^2 + 0x - 1} \\ \underline{-(3x^3 + 3x^2)} \\ -2x^2 + 0x \\ \underline{-(-2x^2 - 2x)} \\ 2x - 1 \\ \underline{-(2x+2)} \\ -3 \end{array}$$

37. Subtract the degree of the denominator from the degree of the numerator.

39. Hole is at (-1, 3)

41. Hole is at (1, 0)

43. $\dfrac{x^3 + y^3}{x + y} = x^2 - xy + y^2$

$$\begin{array}{r} x^2 - \quad xy + y^2 \\ x+y \overline{)\,x^3 + 0x^2y + 0xy^2 + y^3} \\ \underline{-(x^3 + x^2y)} \\ -x^2y + 0xy^2 \\ \underline{-(-x^2y - \quad xy^2)} \\ xy^2 + y^3 \\ \underline{-(xy^2 + y^3)} \\ 0 \end{array}$$

45. True

47. f(-2) = $(-2)^2$ - 3(-2) - 4, f(-2) = 6

f(5) = $(5)^2$ - 3(5) - 4, f(5) = 6

$$\begin{array}{r} x - 5 \\ x+2 \overline{)\,x^2 - 3x - 4} \\ \underline{-(x^2 + 2x)} \\ -5x - 4 \\ \underline{-(-5x - 10)} \\ 6 \end{array}$$

$$\begin{array}{r} x + 2 \\ x-5 \overline{)\,x^2 - 3x - 4} \\ \underline{-(x^2 - 5x)} \\ 2x - 4 \\ \underline{-(2x - 10)} \\ 6 \end{array}$$

Remainders equal 6

49. The output at $x = a$, $f(a)$, is the ***remainder*** upon division of $f(x)$ by $(x - a)$.

51. Using (0, 1), (1, 4), (2, 9) gives the same equation as the quotient: $x^2 + 2x + 1$

Section 5.6

1. $\frac{3}{4}x+5=23, \frac{3}{4}x=18, x=\frac{18\cdot 4}{3}, x=24$

3. $L=\frac{\pi r\theta}{180}, r=\frac{180L}{\pi\theta}, \theta\neq 0$

5. $x(\frac{1}{10}+\frac{1}{12})=1, x(\frac{6}{60}+\frac{5}{60})=1$

$x(\frac{11}{60})=1, x=\frac{60}{11}, x=5\frac{5}{11}$

7. $\frac{2n+4}{7}=\frac{3n-7}{4}, 4(2n+4)=7(3n-7)$

$8n+16=21n-49, 13n=65, n=5$

9. There were no variables in the denominators.

11. $\frac{7}{11+3}=\frac{11-7}{8}, \frac{7}{14}=\frac{4}{8}, \frac{1}{2}=\frac{1}{2}$

13. $\frac{3}{4}+\frac{1}{5}=\frac{1}{\frac{20}{19}}, \frac{15}{20}+\frac{4}{20}=\frac{19}{20}$

15. $\frac{2}{-3+1}+\frac{3}{-3}=-2, \frac{2}{-2}-1=-2,$

-1 - 1 = -2

17. $\frac{x+1}{2}=\frac{x-3}{1}, 2(x-3)=x+1,$

$2x-6=x+1, x=7$

19. $\frac{x-8}{5}=\frac{-6}{x+5}, (x-8)(x+5)=-30,$

$x^2-3x-40=-30, x^2-3x-10=0,$

$(x+2)(x-5)=0, \{-2, 5\}, x\neq -5$

21. $\frac{x-3}{6}=\frac{3}{2x-1}, (x-3)(2x-1)=18,$

$2x^2-7x+3=18, 2x^2-7x-15=0$

$(2x+3)(x-5)=0, \{-1.5, 5\}, x\neq\frac{1}{2}$

23. $\frac{x}{32}+\frac{x}{8}=10, \frac{x}{32}+\frac{4x}{32}=10,$

$\frac{5x}{32}=10, 5x=320, x=64$

25. $12x^2\left(\frac{1}{3x^2}+\frac{1}{4x}\right)=\frac{12x^2}{3x^2}+\frac{12x^2}{4x}=4+3x$

27. $2x^2\left(\frac{1}{2x^2}+\frac{3}{x}\right)=\frac{2x^2}{2x^2}+\frac{6x^2}{x}=1+6x$

29. $x(x+1)\left(\frac{1}{x+1}-\frac{2}{x}\right)=\frac{x(x+1)}{x+1}-\frac{2x(x+1)}{x}$

$=x-2(x+1)=x-2x-2=-x-2$

31. $(x+3)(x-1)\left(\frac{1}{x+3}+\frac{2}{x-1}\right)$

$=\frac{(x+3)(x-1)}{x+3}+\frac{2(x+3)(x-1)}{x-1}$

$=(x-1)+2(x+3)=x-1+2x+6=3x+5$

33. $x\neq 0$ $\frac{3}{5}+\frac{2}{3}=\frac{1}{x}, \frac{9}{15}+\frac{10}{15}=\frac{1}{x},$

$=\frac{1}{x}, \frac{19}{15}=\frac{1}{x}, x=\frac{15}{19}$

Section 5.6 (con't)

35. $x \neq 0,\ \frac{1}{3}+\frac{1}{x}=\frac{1}{2},$

$$(6x)\left(\frac{1}{3}+\frac{1}{x}\right)=(6x)\frac{1}{2},2x+6=3x,\ x=6$$

37. $x \neq 0,\ x\left(\frac{5}{x}+\frac{2}{x}\right)=x\left(\frac{4}{x}\right),\ 5+2=4$

No solution { }

39. $x \neq 0\ 2x\left(\frac{1}{2}-\frac{1}{x}\right)=2x\left(\frac{1}{2x}\right),\ x-2=1,$

$x=3$

41. $x \neq \frac{1}{2},\ 1;\ 4(x-1)=2x-1,$

$4x-4=2x-1,\ 2x=3,\ x=\frac{3}{2}$

43. $x \neq -2;\ (x+1)(x+2)=30,$

$x^2+3x+2=30,\ x^2+3x-28=0$

$(x+7)(x-4)=0,\ \{-7,\ 4\}$

45. $x \neq 1;\ (x-1)(x+2)=40,$

$x^2+x-2=40,\ x^2+x-42=0$

$(x+7)(x-6)=0,\ \{-7,\ 6\}$

47. $x \neq 0;\ x^2\left(\frac{18}{x^2}+\frac{9}{x}-2\right)=x^2(0),$

$18+9x-2x^2=0,\ -2x^2+9x+18=0,$
$(-2x-3)(x-6)=0,\ \{-1.5,\ 6\}$

49. $x \neq 2;\ (x-2)\left(\frac{1}{x-2}-4\right)=(x-2)\left(\frac{3-x}{x-2}\right)$

$1-4(x-2)=3-x,\ 1-4x+8=3-x,$
$9-4x=3-x,\ 6=3x,\ x=2,$ violates $x \neq 2$

no solution, { }

51. $x \neq \pm 1,$

$$2(x+1)(x-1)\left(\frac{3}{x-1}+\frac{5}{2(x+1)}\right)=2(x+1)(x-1)\frac{3}{2}$$

$6(x+1)+5(x-1)=3(x+1)(x-1)$
$6x+6+5x-5=3(x^2-1),$
$11x+1=3x^2-3,\ 3x^2-11x-4=0,$
$(3x+1)(x-4)=0,\ \{-\frac{1}{3},\ 4\}$

53. $x \neq 3,\ 6;\ x(x-3)=2(6-x),$

$x^2-3x=12-2x,\ x^2-x-12=0,$
$(x-4)(x+3)=0,\ \{-3,4\}$

55. $x \neq 0,\ 2;$

$$2x(x-2)\left(\frac{1}{x-2}+\frac{x-1}{x}\right)=2x(x-2)\left(\frac{2x+1}{2x}\right)$$

$2x+2(x-2)(x-1)=(x-2)(2x+1),$
$2x+2(x^2-3x+2)=2x^2-3x-2,$
$2x+2x^2-6x+4=2x^2-3x-2,$
$2x^2-4x+4=2x^2-3x-2,$
$-4x+4=-3x-2,\ x=6$

Section 5.6 (con't)

57. $x \neq -3, 0;$

$$8x(x+3)\left(\frac{x-3}{x}+\frac{x-2}{x+3}\right)=8x(x+3)\frac{6x+1}{8x}$$

$8(x+3)(x-3)+8x(x-2)=(x+3)(6x+1),$

$8(x^2-9)+8x^2-16x=6x^2+19x+3,$

$8x^2-72+8x^2-16x=6x^2+19x+3,$

$16x^2-16x-72=6x^2+19x+3,$

$10x^2-35x-75=0,\ 2x^2-7x-15=0,$

$(2x+3)(x-5)=0,\ \{-1.5, 5\}$

59. $x \neq \pm 1,$

$$3(x-1)(x+1)\left(\frac{2}{x-1}+\frac{x-2}{x+1}=\frac{4x}{3(x+1)}\right)$$

$6(x+1)+3(x-1)(x-2)=4x(x-1),$

$6x+6+3(x^2-3x+2)=4x^2-4x,$

$6x+6+3x^2-9x+6=4x^2-4x,$

$3x^2-3x+12=4x^2-4x,$

$x^2-x-12=0,\ (x-4)(x+3)=0,\ \{-3,4\}$

61. $8(x+3)\frac{7}{x+3}=8(x+3)\frac{x-7}{8},$

$56=(x+3)(x-7),$

$56=x^2-4x-21,\ x^2-4x-77=0,$

$(x+7)(x-11)=0,\ \{-7,11\},\ x\neq -3$

63. From the graph, $\{\frac{1}{2}, 2\}$, $x \neq 0, 1$

65. $\frac{1}{6}+\frac{1}{x}=\frac{1}{2},$

$6x\left(\frac{1}{6}+\frac{1}{x}\right)=6x\left(\frac{1}{2}\right),\ x+6=3x,$

$2x=6,\ x=3$ min

67. $\frac{1}{20}+\frac{1}{x}=\frac{1}{12},\ 60x\left(\frac{1}{20}+\frac{1}{x}\right)=60x\frac{1}{12}$

$3x+60=5x,\ 2x=60,\ x=30$ min

69. $\frac{1}{10{,}000}+\frac{1}{4000}=\frac{1}{R},$

$\frac{2}{20{,}000}+\frac{5}{20{,}000}=\frac{1}{R},$

$\frac{7}{20{,}000}=\frac{1}{R},\ R=\frac{20{,}000}{7}\approx 2857$ ohms

71. $\frac{1}{5}+\frac{1}{3}+\frac{1}{3}=\frac{1}{x},\ 15x\left(\frac{1}{5}+\frac{1}{3}+\frac{1}{3}\right)=15x\frac{1}{x}$

$3x+5x+5x=15,\ 13x=15,\ x=\frac{15}{13}$

$x\approx 1.15$ min

73. $\frac{l}{w}=\frac{5}{2},\ 2l=5w,\ l=\frac{5w}{2};$

$2l+2w=39.9,\ 2\left(\frac{5w}{2}\right)+2w=39.9,$

$5w+2w=39.9,\ 7w=39.9,\ w=5.7,$

$l=\frac{5(5.7)}{2},\ l\approx 14.3$

length $\approx$ 14.3 cm, width = 5.7 cm

Section 5.6 (con't)

75. $\frac{l}{w} = \frac{7}{3}, l = \frac{7w}{3}, 2l + 2w = 44.4$

$2\left(\frac{7w}{3}\right) + 2w = 44.4, \frac{14w}{3} + 2w = 44.4,$

$3\left(\frac{14w}{3} + 2w\right) = 3(44.4),$

$14w + 6w = 133.2,\ 20w = 133.2,$

$w \approx 6.7, l = \frac{7(6.7)}{3}, l \approx 15.6$

length ≈ 15.6 cm, width ≈ 6.7 cm

77. $\frac{1}{a} + \frac{1}{b} = \frac{1}{c}, abc\left(\frac{1}{a} + \frac{1}{b}\right) = abc\frac{1}{c},$

$bc + ac = ab, bc = ab - ac,$

$a(b - c) = bc, a = \frac{bc}{b - c}$

79. $abc\left(\frac{1}{a} + \frac{1}{b}\right) = abc\frac{1}{c},$

$bc + ac = ab, c(a + b) = ab,$

$c = \frac{ab}{a + b}$

81. $I = \frac{E}{R + r}, R + r = \frac{E}{I}$

$R = \frac{E}{I} - r$

83. $15x(x + 2)$ was not distributed over $\frac{2}{x}$,

$15x + 30(x + 2) = 11x(x + 2)$

85 Forgot to multiply by 2 in the 2nd term,

$15x + 30(x + 2) = 11x(x + 2)$

Chapter 5 Review

1. $\frac{2}{3}+\frac{4}{3}=\frac{6}{3}=2,\ \frac{2}{3}-\frac{4}{3}=-\frac{2}{3},$

$\frac{2}{3}\cdot\frac{4}{3}=\frac{8}{9},\ \frac{2}{3}\div\frac{4}{3}=\frac{2}{3}\cdot\frac{3}{4}=\frac{1}{2}$

3. $\frac{a}{b}+\frac{a}{c}=\frac{ac}{bc}+\frac{ab}{bc}=\frac{ac+ab}{bc},$

$\frac{a}{b}-\frac{a}{c}=\frac{ac}{bc}-\frac{ab}{bc}=\frac{ac-ab}{bc},$

$\frac{a}{b}\cdot\frac{a}{c}=\frac{a^2}{bc},\ \frac{a}{b}\div\frac{a}{c}=\frac{a}{b}\cdot\frac{c}{a}=\frac{c}{b}$

5. **a.** $\frac{27xy^2}{15x^2y}=\frac{3(9)xy^2}{3(5)x^2y}=\frac{9y}{5x}$

b. $\frac{2-x}{x-2}=\frac{-1(x-2)}{x-2}=-1$

c. $\frac{a-ac}{a}=\frac{a(1-c)}{a}=1-c$

7. **a.** $\frac{3x^2-12}{x+2}=\frac{3(x^2-4)}{x+2}$

$\frac{3(x+2)(x-2)}{x+2}=3(x-2)$

b. $\frac{(a-b)}{(a+b)(a-b)}=\frac{1}{a+b}$

c. $8x\left(\frac{1}{4x}+\frac{1}{2}\right)=\frac{8x}{4x}+\frac{8x}{2}=2+4x$

9. **a.** $\frac{x^2-5x-6}{x^2-4x-5}=\frac{(x+1)(x-6)}{(x+1)(x-5)}=\frac{x-6}{x-5}$

b. $\frac{4x^2-1}{2x^2+5x+2}=\frac{(2x+1)(2x-1)}{(2x+1)(x+2)}$

$=\frac{2x-1}{x+2}$

c. $\frac{x^2+3x-4}{x^2-16}=\frac{(x-1)(x+4)}{(x-4)(x+4)}=\frac{x-1}{x-4}$

11. $\frac{6-x}{x-6}=-1$

13. $4\text{ gal}\cdot\frac{4\text{ qt}}{1\text{ gal}}=16\text{ qt},$

3 quarts to 16 quarts = 3 to 16

15. Undefined with denominator = 0,

(x - 1)(x + 3) = 0,

x - 1 = 0, x = 1 or x + 3 = 0, x = -3

17. $\frac{4y^2}{9x^2}\cdot\frac{3x}{8y}=\frac{y}{6x}$

19. $15x\left(\frac{1}{3x}+\frac{2}{5x}\right)=5+6=11$

21. $\frac{x^2+5x+4}{x^2-16}\cdot\frac{2x-8}{1-x^2}$

$=\frac{(x+1)(x+4)}{(x-4)(x+4)}\cdot\frac{2(x-4)}{(1+x)(1-x)}=\frac{2}{1-x}$

Chapter 5 Review (con't)

23. $\frac{1}{x-3}+\frac{x}{x-3}=\frac{x+1}{x-3}$

25. $\frac{x}{x-1}-\frac{1}{1-x}=\frac{x}{x-1}+\frac{1}{x-1}=\frac{x+1}{x-1}$

27. $\frac{x-2}{x^2-1}-\frac{2}{x-1}=\frac{x-2}{(x+1)(x-1)}-\frac{2(x+1)}{(x-1)(x+1)}$

$=\frac{x-2-2(x+1)}{(x-1)(x+1)}=\frac{x-2-2x-2}{(x-1)(x+1)}$

$=\frac{-x-4}{(x-1)(x+2)}$

29. As x approaches 1 from the left, y approaches - ∞.

31. As x approaches -3 from the right, y approaches - ∞.

33. $1{,}000{,}000\,\text{hrs}\cdot\frac{1\,\text{day}}{24\,\text{hrs}}\cdot\frac{1\,\text{yr}}{365\,\text{days}}\approx 114.2\,\text{yr}$

35. a. $\frac{21\,\text{in}}{9\,\text{mo}}\cdot\frac{1\,\text{ft}}{12\,\text{in}}\cdot\frac{1\,\text{mi}}{5280\,\text{ft}}\cdot\frac{1\,\text{mo}}{30\,\text{day}}\cdot\frac{1\,\text{day}}{24\,\text{hr}}$

$\approx 5.115 \text{ X } 10^{-8}$ mph

b. 18 yrs (12 mo/yr) = 216 mo

$\frac{x}{216\,\text{mo}}=\frac{21\,\text{in}}{9\,\text{mo}},\ x=\frac{216(21\,\text{in})}{9},$

$x\approx 504\,\text{in};\ +21\,\text{in at birth}=525\,\text{in};$

$525\,\text{in}\cdot\frac{1\,\text{ft}}{12\,\text{in}}\approx 43.75\,\text{ft}$

37. $\frac{4}{5}=\frac{x}{6},\ 5x=24,\ x=4.8$

$\frac{4}{5}=\frac{y}{7},\ 5y=28,\ y=5.6$

39. $\frac{8}{6}=\frac{r}{7.5},\ 6r=60,\ r=10$

$s=\sqrt{6^2+8^2},\ s=10$

$t=\sqrt{7.5^2+10^2},\ t=12.5$

41. $m=\frac{12-9}{24-12},\ m=0.25$

$b=12-0.25(24),\ b=6$

$y=0.25x+6.00,$

y-intercept is cost of developing

43. $m=\frac{121.55-74.80}{65-40},\ m=1.87$

$b=74.80-1.87(40),\ b=0$

$y=1.87x,$ direct variation

45. D = distance, h = height, $D=k\sqrt{h}$

47. V = volume, l = length, w = width, h = height; V = klwh; k = 1

49. $d=ks^2$

51. $s=k\sqrt{df}$

53. $y=\frac{90}{x},\ k=90\,\text{credits}$

Chapter 5 Review (con't)

55. $y = \dfrac{10{,}000}{x}$, $k = \$10{,}000$

57. a. If t is the total time for the trip then the 3 hours spent not traveling must be subtracted to obtain a meaningful rate.

b.

Total Trip	Rate
4	300 ÷ (4 - 3) = 300
5	300 ÷ (5 - 3) = 150
6	300 ÷ (6 - 3) = 100
7	300 ÷ (7 - 3) = 75
8	60

c. Replacing t with t - 3 causes the graph to shift 3 units to the right.

59. a 300 ÷ (3.5 - 3) = 600 mph

b. 300 ÷ (3.1 - 3) = 3000 mph

c. 300 ÷ (3.01 - 3) = 30,000 mph

d. 300 ÷ (3 - 3), division by zero, not possible

61. $h = \dfrac{V}{\frac{1}{3}\pi r^2}, h = \dfrac{V}{\frac{\pi r^2}{3}}, h = V \cdot \dfrac{3}{\pi r^2}$

$$h = \frac{3V}{\pi r^2}$$

63. $40 \text{ mo} \cdot \dfrac{\frac{1}{150}\text{ grain}}{150 \text{ mo}} = 40 \cdot \dfrac{1}{150} \cdot \dfrac{1}{150}\text{ grain}$

≈ 0.00178 grain

65. $\dfrac{x - \frac{4}{x}}{1 - \frac{2}{x}} = \dfrac{\frac{x^2-4}{x}}{\frac{x-2}{x}} = \dfrac{x^2-4}{x} \cdot \dfrac{x}{x-2}$

$= \dfrac{(x+2)(x-2)}{x-2} = x + 2$

67. $\dfrac{1}{5.5} + \dfrac{1}{7} = \dfrac{1}{3}?, \dfrac{7+5.5}{7(5.5)} = \dfrac{12.5}{38.5} \approx 0.3247$

$\dfrac{1}{3} \approx 0.3333$, fans will not meet code

69. $x \neq 0$

$\dfrac{x^3 - 3x^2 + 3x - 1}{x} = \dfrac{x^3}{x} - \dfrac{3x^2}{x} + \dfrac{3x}{x} - \dfrac{1}{x}$

$= x^2 - 3x + 3 - \dfrac{1}{x}$

71. $x \neq 1$

$$\begin{array}{r} x^2 - 2x + 1 \\ x-1 \overline{) x^3 - 3x^2 + 3x - 1} \\ \underline{-(x^3 - x^2)} \\ -2x^2 + 3x \\ \underline{-(-2x^2 + 2x)} \\ x - 1 \\ \underline{-(x-1)} \\ 0 \end{array}$$

$= x^2 - 2x + 1$

Chapter 5 Review (con't)

73. $x \neq 2$

$$\begin{array}{r} x^2 - x + 1 \\ x-2\overline{)\,x^3 - 3x^2 + 3x - 1} \\ -(x^3 - 2x^2) \qquad\qquad \\ \hline -x^2 + 3x \qquad \\ -(-x^2 + 2x) \qquad \\ \hline x - 1 \\ -(x-2) \\ \hline 1 \end{array}$$

$$= x^2 - x + 1 + \frac{1}{x-2}$$

75. $x \neq 3$

$$\begin{array}{r} x^2 + 0x + 3 \\ x-3\overline{)\,x^3 - 3x^2 + 3x - 1} \\ -(x^3 - 3x^2) \qquad\qquad \\ \hline 0x^2 + 3x \qquad \\ -(0x^2 + 0x) \qquad \\ \hline +3x - 1 \\ -(3x-9) \\ \hline 8 \end{array}$$

$$= x^2 + 3 + \frac{8}{x-3}$$

77.

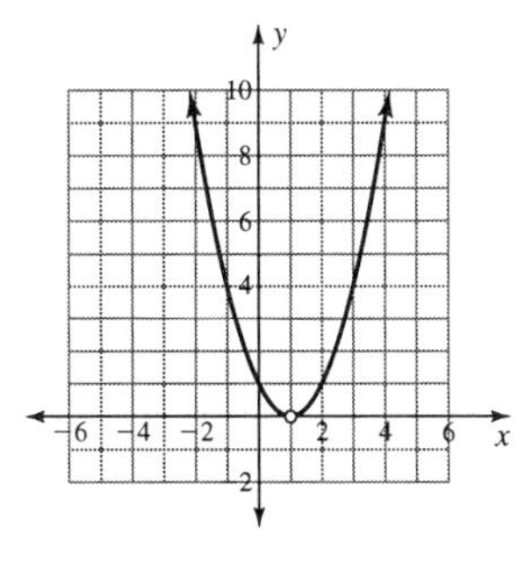

The hole is created because the denominator is a factor of the numerator.

79. $\frac{6}{x} = \frac{15}{32}$, $15x = 192$, $x = 12.8$, $x \neq 0$

81. $\frac{x}{2} = \frac{7}{x+5}$, $x(x+5) = 14$, $x^2 + 5x - 14 = 0$,

$(x+7)(x-2) = 0$, $\{-7, 2\}$, $x \neq -5$

83. $\frac{x-2}{4} = \frac{x+1}{3} - 3$, $(12)\frac{x-2}{4} = (12)\left(\frac{x+1}{3} - 3\right)$

$3(x-2) = 4(x+1) - 36$,

$3x - 6 = 4x + 4 - 36$, $3x - 6 = 4x - 32$

$x = 26$,

85. $3x(x+2)\left(\frac{1}{x} + \frac{5}{x+2}\right) = 3x(x+2)\frac{13}{3x}$,

$3(x+2) + 15x = 13(x+2)$,

$3x + 6 + 15x = 13x + 26$,

$18x + 6 = 13x + 26$, $5x = 20$

$x = 4$, $x \neq -2, 0$

87. $6x(x-1)\left(\frac{1}{x-1} + \frac{2}{x}\right) = 6x(x-1)\frac{7}{6}$

$6x + 12(x-1) = 7x(x-1)$,

$6x + 12x - 12 = 7x^2 - 7x$,

$7x^2 - 25x + 12 = 0$,

$(7x-4)(x-3) = 0$, $\{\frac{4}{7}, 3\}$, $x \neq 0, 1$

Chapter 5 Review (con't)

89. a. $m = \frac{8-4}{5-10}, m = -\frac{4}{5}$

$b = 8 - (-\frac{4}{5})(5), b = 12$

$y = -\frac{4}{5}x + 12$

b. For an inverse variation, k = xy,

$5(8) = 40;\ 10(4) = 40;$

$y = \frac{40}{x}$

89. c.

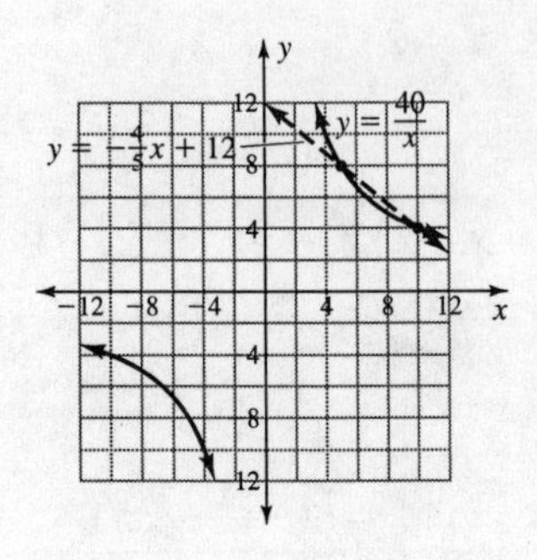

d. Both graphs contain the points (5, 8) and (10, 4). Both graphs are decreasing. One is linear, the other is curved.

Chapter 5 Test

1. $\dfrac{ab^2c}{a^2bc^2} = \dfrac{b}{ac}$, $a \neq 0, b \neq 0, c \neq 0$

2. $\dfrac{b+3}{b-3}$, $b \neq 3$, already simplified

3. $\dfrac{3ac}{15ac^2} = \dfrac{3ac}{3(5)ac^2} = \dfrac{1}{5c}, a \neq 0, c \neq 0$

4. $\dfrac{15b^2c^3}{10b^3c} = \dfrac{3c^2}{2b}, b \neq 0, c \neq 0$

5. $\dfrac{x^2-25}{20+x-x^2} = \dfrac{(x-5)(x+5)}{(5-x)(4+x)}$

$= \dfrac{(x-5)(x+5)}{-1(x-5)(x+4)} = -\dfrac{x+5}{x+4}, x \neq -4, 5$

6. $\dfrac{12-2a}{18+3a-a^2} = \dfrac{2(6-a)}{(3+a)(6-a)}$

$= \dfrac{2}{3+a}, a \neq -3, 6$

7. $\dfrac{3x^2-7x+2}{2x^2-5x+2} = \dfrac{(3x-1)(x-2)}{(2x-1)(x-2)}$

$= \dfrac{3x-1}{2x-1}, x \neq \frac{1}{2}, 2$

8. $\dfrac{2x^2+7x+6}{x^2+4x+4} = \dfrac{(2x+3)(x+2)}{(x+2)^2}$

$\dfrac{2x+3}{x+2}, x \neq -2$

9. $\dfrac{12xy^2}{7y-y^2} \div \dfrac{6x^2}{7-y} = \dfrac{12xy^2}{y(7-y)} \cdot \dfrac{7-y}{6x^2}$

$\dfrac{2y}{x}, y \neq 0, 7; x \neq 0$

10. $\dfrac{x^2-9}{x^2+4x+3} \cdot \dfrac{x-3}{x+1} = \dfrac{(x+3)(x-3)}{(x+1)(x+3)} \cdot \dfrac{x-3}{x+1}$

$\dfrac{(x-3)^2}{(x+1)^2}, x \neq -3, -1$

11. $\dfrac{x^2+8x+16}{x-2} \cdot \dfrac{x^2-4}{x+4} = \dfrac{(x+4)^2}{x-2} \cdot \dfrac{(x+2)(x-2)}{x+4}$

$(x+4)(x+2), x \neq -4, 2$

12. $x(x+1)\left(\dfrac{3}{x} + \dfrac{2}{x+1}\right) = 3(x+1) + 2x$

$= 3x+3+2x = 5x+3, x \neq -1, 0$

13. $\dfrac{2-x}{x+2} + \dfrac{x+2}{x-2}$

$= \dfrac{(x-2)(2-x)}{(x-2)(x+2)} + \dfrac{(x+2)(x+2)}{(x-2)(x+2)}$

$= \dfrac{-(x-2)^2 + (x+2)^2}{(x-2)(x+2)}$

$= \dfrac{-(x^2-4x+4) + x^2+4x+4}{(x-2)(x+2)}$

$= \dfrac{8x}{(x-2)(x+2)}, x \neq \pm 2$

Chapter 5 Test (con't)

14. $\frac{x-2}{x+2}-\frac{x}{x(x+2)}=\frac{x(x-2)}{x(x+2)}-\frac{x}{x(x+2)}$

$\frac{x^2-2x-x}{x(x+2)}=\frac{x(x-3)}{x(x+2)}=\frac{x-3}{x+2}, x\neq -2,0$

15. $\frac{10000\text{ cm}^3}{10\text{ cm}}=1000\text{ cm}^2$

16. $\frac{30\text{ miles per gallon}}{60\text{ miles per hour}}=\frac{30\text{ miles}}{\text{gallon}}\cdot\frac{1\text{ hour}}{60\text{ miles}}$

$\frac{1}{2}$ hour per gallon

17. $\frac{\frac{3}{x}-x}{x+\frac{2}{x}}=\frac{\frac{3-x^2}{x}}{\frac{x^2+2}{x}}=\frac{3-x^2}{x}\cdot\frac{x}{x^2+2}$

$\frac{3-x^2}{x^2+2}, x\neq 0$

18. $\frac{\frac{x^2-3x}{3}}{\frac{9-x^2}{x}}=\frac{x(x-3)}{3}\cdot\frac{x}{(3-x)(3+x)}$

$\frac{x^2(x-3)}{-3(x-3)(3+x)}=-\frac{x^2}{3(3+x)}, x\neq 0,\pm 3$

19. $\frac{7\text{ lb}}{9\text{ mo}}\cdot\frac{1\text{ mo}}{30\text{ day}}\cdot\frac{1\text{ day}}{24\text{ hr}}\cdot\frac{16\text{ oz}}{1\text{ lb}}$

≈ 0.0173 oz per hour

20. $\frac{133\text{ lb}}{18\text{ yr}}\cdot\frac{1\text{ yr}}{365\text{ day}}\cdot\frac{1\text{ day}}{24\text{ hr}}\cdot\frac{16\text{ oz}}{1\text{ lb}}$

≈ 0.0135 oz per hour

21. $d=\frac{45^2}{30(0.6)}, d=112.5\text{ ft}$

22. $208=\frac{s^2}{30(0.4)}, s^2=2496,$

$s\approx 50$ mph

23. The volume varies directly with the cube of the radius. $k=\frac{4}{3}\pi$

24. $\frac{4}{6}=\frac{EF}{15}$, 6(EF) = 60, EF = 10

$AB=\sqrt{6^2+15^2}$, $AB=\sqrt{261}$,
$AB=3\sqrt{29}$,

$DE=\sqrt{4^2+10^2}$, $DE=\sqrt{116}$,
$DE=2\sqrt{29}$

25. a. The phrase "cover the same distance" indicates an inverse variation.

b. 6(60) = 360, k = 360

c. $60(6)=55(t), t=\frac{360}{55}, t\approx 6.5\text{ hr}$

26. $\frac{x-2}{10}=\frac{2}{x-1}, (x-2)(x-1)=20,$

$x^2-3x+2=20, x^2-3x-18=0$

$(x-6)(x+3)=0, \{-3, 6\}, x\neq 1$

Chapter 5 Test (con't)

27. $3x\left(\frac{1}{3}-\frac{1}{x}\right)=3x\frac{2x}{3},\ x-3=2x^2$

$2x^2-x+3=0,$

$x=\frac{-(-1)\pm\sqrt{(-1)^2-4(2)(3)}}{2(2)},$

$x=\frac{1\pm\sqrt{-23}}{4},\ x=\frac{1\pm i\sqrt{23}}{4}$

$x\neq 0$

28. $3x(x-4)\left(\frac{1}{x-4}+\frac{1}{x}\right)=3x(x-4)\frac{10}{3x}$

$3x+3(x-4)=10(x-4),$
$3x+3x-12=10x-40,$
$6x-12=10x-40,\ 4x=28,\ x=7$
$x\neq 0,\ 4$

29. $10(x-1)=9(x+3),\ 10x-10=9x+27,$

$x=37$

30. $x(x+1)\left(\frac{3}{x}+\frac{2}{x+1}\right)=x(x+1)4$

$3(x+1)+2x=4x^2+4x,$
$3x+3+2x=4x^2+4x,$
$4x^2-x-3=0,\ (4x+3)(x-1)=0,$
$\{-\frac{3}{4},1\},\ x\neq -1,0$

31. Defined for all real numbers, x ≠ -2

$$\begin{array}{r}
x^2+4x+4 \\
x+2\overline{)x^3+6x^2+12x+8} \\
-(x^3+2x^2) \\
\hline
4x^2+12x \\
-(4x^2+8x) \\
\hline
4x+8 \\
-(4x+8) \\
\hline
0
\end{array}$$

$=x^2+4x+4$

32. Defined for all real numbers, x ≠ -4

$$\begin{array}{r}
x^2+2x+4 \\
x+4\overline{)x^3+6x^2+12x+8} \\
-(x^3+4x^2) \\
\hline
2x^2+12x \\
-(2x^2+8x) \\
\hline
4x+8 \\
-(4x+16) \\
\hline
-8
\end{array}$$

$=x^2+2x+4-\frac{8}{x+4}$

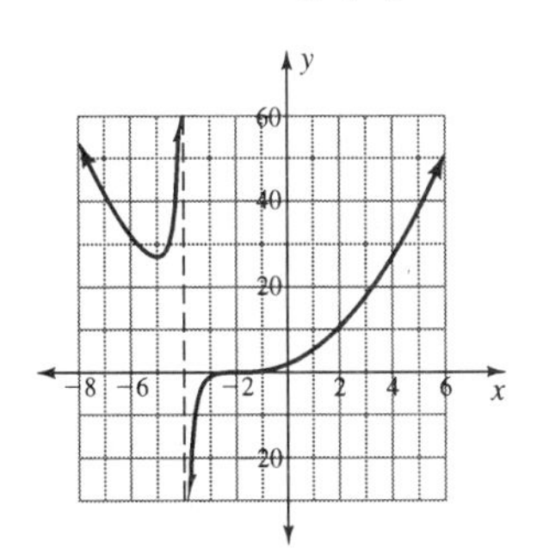

Section 6.0

1. a. $2x^{-1} = \frac{2}{x}$

b. $\left(\frac{y}{x}\right)^{-1} = \frac{x}{y}$

c. $\left(\frac{s}{t}\right)^{0} = 1$

d. $0.25^{-1} = \left(\frac{1}{4}\right)^{-1} = 4$

e. $\left(\frac{1}{2}\right)^{0} = 1$

3. a. $\left(\frac{m}{n}\right)^{-1} = \frac{n}{m}$

b. $\left(\frac{b}{c}\right)^{0} = 1$

c. $\left(\frac{c}{ab}\right)^{-1} = \frac{ab}{c}$

d. $\left(\frac{3}{4}\right)^{0} = 1$

e. $2.5^{-1} = \left(\frac{5}{2}\right)^{-1} = \frac{2}{5} = 0.4$

5. a. $(2y)^{-2} = \left(\frac{1}{2y}\right)^{2} = \frac{1^2}{2^2 y^2} = \frac{1}{4y^2}$

b. $2x^{-4} = \frac{2}{x^4}$

c. $\left(\frac{2x}{y}\right)^{-2} = \left(\frac{y}{2x}\right)^{2} = \frac{y^2}{4x^2}$

d. $\left(\frac{a}{2c}\right)^{-3} = \left(\frac{2c}{a}\right)^{3} = \frac{8c^3}{a^3}$

e. $\frac{2x^3y^{-2}}{6x^{-1}y^2} = \frac{2x^3x^1}{6y^2y^2} = \frac{2x^4}{6y^4} = \frac{x^4}{3y^4}$

f. $b^1 \cdot b^n = b^{1+n}$

7. a. $\left(\frac{3a^2}{c}\right)^{-3} = \left(\frac{c}{3a^2}\right)^{3} = \frac{c^3}{3^3 a^{2\cdot 3}} = \frac{c^3}{27a^6}$

b. $\left(\frac{a}{c^2}\right)^{-2} = \left(\frac{c^2}{a}\right)^{2} = \frac{c^{2\cdot 2}}{a^2} = \frac{c^4}{a^2}$

c. $\frac{1}{c^{-3}} = c^3$

d. $\frac{2}{a^{-2}} = 2a^2$

e. $\frac{12a^3b^{-2}}{4a^{-1}b^{-5}} = \frac{12a^3a^1b^5}{4b^2} = 3a^{3+1}b^{5-2}$

$= 3a^4b^3$

f. $3 \cdot 3^x = 3^1 \cdot 3^x = 3^{x+1}$

Section 6.0 (con't)

9. The exponent does not affect the 3.

$$3x^{-2} = \frac{3}{x^2}$$

11. $5\left(\frac{5}{6}\right)^3 = \frac{5}{1}\cdot\frac{5^3}{6^3} = \frac{5^4}{6^3}$;

$6\left(\frac{5}{6}\right)^4 = \frac{6}{1}\cdot\frac{5^4}{6^4} = \frac{6\cdot 5^4}{6\cdot 6^3} = \frac{5^4}{6^3}$;

expressions are equal.

13. If $2^x = 2^y$ then x = y; however we can not come to the same conclusion for $x^2 = y^2$.

$(3)^2 = (-3)^2$ but $3 \neq -3$.

15. Use a guess and check table to find pairs that will satisfy $x^3 + y^3 = 1729$, or solve the equation for y and use a graphing calculator to find integer pairs. Solutions are $1^3 + 12^3$ and $9^3 + 10^3$.

17. a. $\frac{1}{t} = \frac{1}{t_1} + \frac{1}{t_2}$

b. $g = \frac{9.81 \text{ m}}{\text{sec}^2}$

c. $P = \frac{A}{e^{rt}}$

d. $d = \frac{1 \text{ g}}{\text{cm}^3}$

e. $p = \frac{31 \text{ lb}}{\text{in}^2}$

19. $\frac{\frac{1}{x+h} - \frac{1}{x}}{h} = \frac{\frac{x}{x(x+h)} - \frac{x+h}{x(x+h)}}{h}$

$= \frac{x-(x+h)}{x(x+h)} \div h = \frac{x-x-h}{x(x+h)}\cdot\frac{1}{h}$

$= \frac{-h}{x(x+h)h} = \frac{-1}{x(x+h)}$

21. $x^2(-1)(x+1)^{-2} + \frac{2x}{x+1} = \frac{-x^2}{(x+1)^2} + \frac{2x}{x+1}$

$= \frac{-x^2}{(x+1)^2} + \frac{2x(x+1)}{(x+1)^2} = \frac{-x^2+2x^2+2x}{(x+1)^2}$

$= \frac{x^2+2x}{(x+1)^2}$

23. a. $0.001 = 10^{-3}$

b. $\frac{1}{100{,}000} = \frac{1}{10^5} = 10^{-5}$

c. $10{,}000 = 10^4$

d. 1 hundredth = 0.01 = 10^{-2}

e. 1 million = 1,000,000 = 10^6

f. 1 millionth = 0.000001 = 10^{-6}

25. a. 3000 = 3 x 10^3

b. 350 = 3.5 x 10^2

c. $350{,}\bar{0}00 = 3.500 \text{ x } 10^5$

d. 0.00350 = 3.50 x 10^{-3}

27. Significant digits: 25a 1; 25b 2; 25c 4; 25d 3.

Section 6.0 (con't)

29. a. 2.9979×10^{10} cm/sec =

29,979,000,000 cm/sec

b. 1.6726×10^{-24} g =

0.0000000000000000000000016726 g

c. -1.6022×10^{-19} C =

- 0.00000000000000000016022 C

d. 1.6022×10^{-19} J =

0.00000000000000000016022 J

31. $(4.0 \times 10^{-2})(1.5 \times 10^{6})$

$= (4.0)(1.5)(10^{6-2}) = 6 \times 10^{4}$

33. $\dfrac{2.4 \times 10^{6}}{0.3 \times 10^{-2}} = \dfrac{2.4}{0.3} \times 10^{6+2} = 8 \times 10^{8}$

35. a. 1.25 - 3 = 1.25×10^{-3} = 0.00125

b. 2.06 2 = 2.06×10^{2} = 206

c. 2.13 10 = 2.13×10^{10}

= 21,300,000,000

d. 4.23 - 15 = 4.23×10^{-15}

= 0.00000000000000423

37. a. $\dfrac{186{,}000\text{ mi}}{1\text{ sec}} \cdot \dfrac{60\text{ sec}}{1\text{ min}} \cdot \dfrac{60\text{ min}}{1\text{ hr}}$

669,600,000 mph or 6.70×10^{8} mph

b. $2^{3}(6.70 \times 10^{8}\text{ mph}) = 53.6 \times 10^{8}$ mph

$= 5.36 \times 10^{9}$ mph

c. $3^{3}(6.70 \times 10^{8}\text{ mph}) = 180.9 \times 10^{8}$

$= 1.81 \times 10^{10}$ mph

d. $14.1^{3}(6.70 \times 10^{8}\text{ mph}) \approx 18{,}782 \times 10^{8}$

$\approx 1.88 \times 10^{12}$ mph

Section 6.1

1. a. $x = \frac{2}{3}$

b. $x = \frac{3}{2}$

c. $x = \frac{3}{4}$

3. $3^4 = 3 \cdot 3 \cdot 3 \cdot 3 = 81$

$9^2 = (3 \cdot 3) \cdot (3 \cdot 3) = 81$

5. a. 4^x

$= (2 \cdot 2) \cdot (2 \cdot 2) \cdot (2 \cdot 2) \cdot (2 \cdot 2) \cdot (2 \cdot 2)$

$= 1024;\ x = 5$

b. 8^x

$= (2 \cdot 2 \cdot 2) \cdot (2 \cdot 2 \cdot 2) \cdot (2 \cdot 2 \cdot 2) \cdot 2$

$= 1024;\ x = 3\frac{1}{3} = \frac{10}{3}$

c. 16^x

$= (2 \cdot 2 \cdot 2 \cdot 2) \cdot (2 \cdot 2 \cdot 2 \cdot 2) \cdot 2 \cdot 2$

$= 1024;\ x = 2\frac{1}{2} = \frac{5}{2}$

d. 32^x

$= (2 \cdot 2 \cdot 2 \cdot 2 \cdot 2) \cdot (2 \cdot 2 \cdot 2 \cdot 2 \cdot 2)$

$= 1024;\ x = 2$

e. 64^x

$= (2 \cdot 2 \cdot 2 \cdot 2 \cdot 2 \cdot 2) \cdot 2 \cdot 2 \cdot 2 \cdot 2$

$= 1024;\ x = 1\frac{2}{3} = \frac{5}{3}$

f. 128^x

$= (2 \cdot 2 \cdot 2 \cdot 2 \cdot 2 \cdot 2 \cdot 2) \cdot 2 \cdot 2 \cdot 2$

$= 1024;\ x = 1\frac{3}{7} = \frac{10}{7}$

5. g. 256^x

$= (2 \cdot 2 \cdot 2 \cdot 2 \cdot 2 \cdot 2 \cdot 2 \cdot 2) \cdot 2 \cdot 2$

$= 1024;\ x = 1\frac{1}{4} = \frac{5}{4}$

7. a. $64 = (2 \cdot 2 \cdot 2) \cdot (2 \cdot 2 \cdot 2)$,

$64^{1/2} = (2 \cdot 2 \cdot 2) = 8$

b. $16 = 2 \cdot 2 \cdot 2 \cdot 2,\ 16^{1/4} = 2$

c. $32 = 2 \cdot 2 \cdot 2 \cdot 2 \cdot 2,\ 32^{1/5} = 2$

9. a. $32 = 2 \cdot 2 \cdot 2 \cdot 2 \cdot 2,\ 32^{2/5} = 2 \cdot 2 = 4$

b. $16 = (2 \cdot 2) \cdot (2 \cdot 2)$,

$16^{3/2} = (2 \cdot 2)^3 = 4^3 = 64$

c. $36 = (2 \cdot 3) \cdot (2 \cdot 3)$,

$36^{3/2} = (2 \cdot 3)^3 = 6^3 = 216$

d. $32 = 2 \cdot 2 \cdot 2 \cdot 2 \cdot 2,\ 32^{6/5} = 2^6 = 64$

11. a. $16 = 2 \cdot 2 \cdot 2 \cdot 2,\ 16^{5/4} = 2^5 = 32$

b. $64 = (2 \cdot 2 \cdot 2) \cdot (2 \cdot 2 \cdot 2)$,

$64^{3/2} = (2 \cdot 2 \cdot 2)^3 = 8^3 = 512$

c. $25 = 5 \cdot 5,\ 25^{3/2} = 5^3 = 125$

d. $81 = 3 \cdot 3 \cdot 3 \cdot 3,\ 81^{3/4} = 3^3 = 27$

13. Some possible answers are; $a = b = 1$;

$a = b = 2$; $a = 2, b = 4$

15. This is true for all values of $b > 0$ except $b = 1$.

Section 6.1 (con't)

17. To solve the equations from the graph, trace the y-value horizontally until it intersects the graph, trace vertically to the x-axis to locate the x value.

a. $x \approx 4.5$

b. $x \approx 4.7$

c. $x \approx 5$

19. a. $25^{1.5} = (5 \cdot 5) \cdot 5 = 5^3 = 125$

b. $4^{2.5} = (2 \cdot 2) \cdot (2 \cdot 2) \cdot 2 = 2^5 = 32$

c. $16^{2.5} = (2 \cdot 2 \cdot 2 \cdot 2) \cdot (2 \cdot 2 \cdot 2 \cdot 2) \cdot 2 \cdot 2$

$= 2^{10} = 1024$

d. $0.09 = 0.3 \cdot 0.3, \quad 0.09^{0.5} = 0.3$

21. a. Once a year is *annual.*

b. Twice a year is *semiannual.*

c. 4 times a year is *quarterly.*

d. 12 times a year is *monthly.*

e. 52 times a year is *weekly.*

f. 365 times a year is *daily.*

23. The compound interest formula is:

$S = P\left(1 + \frac{r}{n}\right)^{nt}$, the base is $\left(1 + \frac{r}{n}\right)$

25. a. $10{,}000\left(1 + \frac{0.04}{12}\right)^{12 \cdot 3} = 11{,}272.72$

b. $10{,}000\left(1 + \frac{0.04}{4}\right)^{4 \cdot 3} = 11{,}268.25$

27. a. $10{,}000\left(1 + \frac{0.08}{12}\right)^{12 \cdot 3} = 12{,}702.37$

b. $10{,}000\left(1 + \frac{0.08}{4}\right)^{4 \cdot 3} = 12{,}682.42$

29. 25 a) \$10,000 earning 4% interest compounded monthly for 3 years.

25 b) \$10,000 earning 4 interest compounded quarterly for 3 years.

31. $\$6.00\left(1 + \frac{0.05}{1}\right)^{1 \cdot 10} = \$9.77$

33. $\$1000\left(1 + \frac{0.07}{4}\right)^{4 \cdot 5} = \$1414.77$

35. year 3 $\approx$ \$6.55; year 4 $\approx$ \$6.94;

year 5 $\approx$ \$7.36, errors would be due to rounding.

37. The total value of the base would be smaller.

39. $\$1000\left(1 + \frac{0.08}{12}\right)^{12 \cdot 1.75} \approx \$1149.73$

41. $\$1000\left(1 + \frac{0.08}{12}\right)^{12 \cdot 2.25} \approx \$1196.50$

Section 6.1 (con't)

43. $\$1000\left(1+\frac{0.07}{4}\right)^{4\cdot 5.33} \approx \$1447.88$

45. $\$1000\left(1+\frac{0.07}{4}\right)^{4\cdot 4.75} \approx \$1390.44$

47. $t \approx 29.73$ years, viewing windows will vary, a possible window would be;

Xmin = 29.5, Xmax = 29.9,

Ymin = 7990, Ymax = 8010

49. Convert to metric units:

5 ft 1 in = (61 in.)(2.54 cm per in)

= 154.94 cm

100 lb ÷ 2.205 lb per kg ≈ 45.351 kg

130 lb ÷ 2.205 lb per kg ≈ 58.957 kg

Percent body fat is:

$$\frac{45.351^{1.2}}{154.94^{3.3}} \cdot 4{,}000{,}000 \approx 23.04$$

$$\frac{58.957^{1.2}}{154.94^{3.3}} \cdot 4{,}000{,}000 \approx 31.56$$

51. Convert to metric units:

6 ft 2 in = (74 in)(2.54 cm per in)

= 187.96 cm

153 lb ÷ 2.205 lb per kg ≈ 69.388 kg

193 lb ÷ 2.205 lb per kg ≈ 87.528 kg

Percent body fat is:

$$\frac{69.388^{1.2}}{187.96^{3.3}} \cdot 3{,}000{,}000 \approx 15.21$$

$$\frac{87.528^{1.2}}{187.96^{3.3}} \cdot 3{,}000{,}000 \approx 20.10$$

53. Convert to metric units:

5 ft 4 in = (64 in)(2.54 cm per in)

= 162.56 cm

118 lb ÷ 2.205 lb per kg ≈ 53.515 kg

149 lb ÷ 2.205 lb per kg ≈ 67.574 kg

Percent body fat is:

$$\frac{53.515^{1.2}}{162.56^{3.3}} \cdot 3{,}000{,}000 \approx 17.99$$

$$\frac{67.574^{1.2}}{162.56^{3.3}} \cdot 3{,}000{,}000 \approx 23.80$$

Section 6.2

1. a. $\sqrt{0.4}=\sqrt{0.2^2\cdot 10}=0.2\sqrt{10}$

b. $\sqrt{4}=\sqrt{2^2}=2$

c. $\sqrt{40}=\sqrt{2^2\cdot 10}=2\sqrt{10}$

d. $\sqrt{400}=\sqrt{2^2\cdot 10^2}=2\cdot 10=20$

e. $\sqrt{4000}=\sqrt{2^2\cdot 10^3}=2\cdot 10\sqrt{10}$

$=20\sqrt{10}$

The patterns reflect whether the exponent on 10 is even or odd.

3. a. $\sqrt{400{,}000}=\sqrt{2^2\cdot 10^5}=2\cdot 10^2\sqrt{10}$

$=200\sqrt{10}$

b. $\sqrt{4{,}000{,}000}=\sqrt{2^2\cdot 10^6}=2\cdot 10^3$

$=2000$

c. $\sqrt{0.04}=\sqrt{0.2^2}=0.2$

d. $\sqrt{0.000004}=\sqrt{0.2^2\cdot 10^{-4}}=0.2\cdot 10^{-2}$

$=0.002$

5. $\sqrt{100n}=\sqrt{10^2 n}=10\sqrt{n}$

7. $\sqrt{0.01n}=\sqrt{0.1^2 n}=0.1\sqrt{n}$

9. $\sqrt{\dfrac{n}{100}}=\dfrac{\sqrt{n}}{\sqrt{10^2}}=\dfrac{\sqrt{n}}{10}$

11. a. $\sqrt[3]{64}=\sqrt[3]{4^3}=4$

b. $\sqrt[5]{32}=\sqrt[5]{2^5}=2$

13. a. $\sqrt[3]{-27}=\sqrt[3]{(-3)^3}=-3$

b. $\sqrt[4]{625}=\sqrt[4]{5^4}=5$

15. a. $\sqrt{-4}$ This is not a real number

b. $\sqrt[4]{-16}$ This is not a real number

17. a. $-\sqrt[4]{10000}=-\sqrt[4]{10^4}=-10$

b. $\sqrt[3]{-1000}=\sqrt[3]{(-10)^3}=-10$

19. a. $x^{1/6}=\sqrt[6]{x}$

b. $\sqrt[3]{x}=x^{1/3}$

c. $x^{0.5}=x^{1/2}=\sqrt{x}$

21. a. $x^{3/2}=\sqrt{x^3}$ *or* $\left(\sqrt{x}\right)^3$

b. $\sqrt[4]{x^3}=x^{3/4}$

c. $x^{0.8}=x^{4/5}=\sqrt[5]{x^4}$ *or* $\left(\sqrt[5]{x}\right)^4$

23. a. $\left(\sqrt[3]{-8}\right)^5=\left(\sqrt[3]{(-2)^3}\right)^5=(-2)^5=-32$

b. $\sqrt[3]{27^2}=\left(\sqrt[3]{27}\right)^2=\left(\sqrt[3]{3^3}\right)^2=3^2=9$

25. a. $\sqrt[4]{16^2}=\sqrt[4]{(4^2)^2}=\sqrt[4]{4^4}=4$

b. $\left(\sqrt[4]{-64}\right)^2$ This is not a real number.

Section 6.2 (con't)

27. a. $\sqrt{y^2} = |y|$

b. $\sqrt{z^8} = z^{8/2} = z^4$

c. $\sqrt{x^6} = x^{6/2} = |x^3|$

29. a. $(x^4)^{1/2} = x^{4/2} = x^2$

b. $(x^{2/3})^3 = x^{6/3} = x^2$

c. $(y^6)^{1/6} = y^{6/6} = |y|$

31. a. $\sqrt{2}\sqrt{18} = \sqrt{2 \cdot 18} = \sqrt{36} = 6$

b. $-\sqrt{3}\sqrt{27} = -\sqrt{3 \cdot 27} = -\sqrt{81} = -9$

c. $\sqrt{2}\sqrt{8} = \sqrt{2 \cdot 8} = \sqrt{16} = 4$

d. $\sqrt[2]{\sqrt[3]{64}} = \sqrt[2]{\sqrt[3]{4^3}} = \sqrt[2]{4} = \sqrt[2]{2^2} = 2$

33. a. $\dfrac{\sqrt[4]{243}}{\sqrt[4]{3}} = \sqrt[4]{\dfrac{243}{3}} = \sqrt[4]{81} = \sqrt[4]{3^4} = 3$

b. $\sqrt[2]{\dfrac{27}{3}} = \sqrt{9} = 3$

c. $\sqrt[3]{16} \cdot \sqrt[3]{4} = \sqrt[3]{16 \cdot 4} = \sqrt[3]{4^3} = 4$

d. $\dfrac{\sqrt[4]{64}}{\sqrt[4]{4}} = \sqrt[4]{\dfrac{64}{4}} = \sqrt[4]{16} = \sqrt[4]{2^4} = 2$

35. a. $\sqrt{a}\sqrt{a^3} = \sqrt{a^4} = a^2, \quad a \ge 0$

b. $\sqrt[3]{b}\sqrt[3]{b^2} = \sqrt[3]{b^3} = b$

37. a. $x^{3/4}x^{1/3}$, add the exponents:

$\frac{3}{4} + \frac{1}{3} = \frac{9}{12} + \frac{4}{12} = \frac{13}{12}, \quad x^{3/4 + 1/3} = x^{13/12}$

b. $\frac{3}{4} + \frac{2}{3} = \frac{9}{12} + \frac{8}{12} = \frac{17}{12}, \quad x^{3/4 + 2/3} = x^{17/12}$

c. $x^{-1/3} = \dfrac{1}{x^{1/3}}$

d. $\sqrt[3]{x^2}\sqrt[4]{x} = x^{2/3}x^{1/4} = x^{2/3 + 1/4} = x^{8/12 + 3/12}$

$= x^{11/12}$

39. a. $\sqrt[3]{x}\sqrt[2]{x^3} = x^{1/3}x^{3/2} = x^{1/3 + 3/2} = x^{11/6}$

b. $\dfrac{x^{3/4}}{x^{1/3}} = x^{3/4 - 1/3} = x^{5/12}$

c. $\left(x^{\frac{3}{4}}\right)^{\frac{2}{3}} = x^{\frac{3}{4}\cdot\frac{2}{3}} = x^{\frac{1}{2}}$

41. a. $\sqrt[2]{\sqrt[3]{x^2}} = \left(x^{\frac{2}{3}}\right)^{\frac{1}{2}} = x^{\frac{2}{3}\cdot\frac{1}{2}} = x^{\frac{1}{3}}$

b. $\left(\dfrac{16x^4}{y^8}\right)^{\frac{3}{4}} = \dfrac{16^{\frac{3}{4}}x^{4\cdot\frac{3}{4}}}{y^{8\cdot\frac{3}{4}}} = \dfrac{2^3x^3}{y^6} = \dfrac{8x^3}{y^6}$

c. $\dfrac{x^{-\frac{1}{2}}}{x^{\frac{3}{2}}} = x^{-\frac{1}{2}-\frac{3}{2}} = x^{-2} = \dfrac{1}{x^2}$

43. a. $\sqrt[3]{16} = \sqrt[3]{2^3 \cdot 2} = \sqrt[3]{2^3}\sqrt[3]{2} = 2\sqrt[3]{2}$

b. $\sqrt{75a^5} = \sqrt{25 \cdot 3 \cdot a^4 \cdot a}$

$= \sqrt{25}\sqrt{3}\sqrt{a^4}\sqrt{a} = 5a^2\sqrt{3}\sqrt{a} = 5a^2\sqrt{3a}$

Section 6.2 (con't)

43. c. $\sqrt[4]{32x^9} = \sqrt[4]{2^4 \cdot 2 \cdot x^8 \cdot x}$

$= \sqrt[4]{2^4}\sqrt[4]{x^8}\sqrt[4]{2x} = 2x^2\sqrt[4]{2x}$

d. $\sqrt[5]{-32a^4b^5c^6} = \sqrt[5]{(-2)^5}\sqrt[5]{b^5}\sqrt[5]{c^5}\sqrt[5]{a^4c}$

$= -2bc\sqrt[5]{a^4c}$

e. $\sqrt[3]{x^4y^8z^9} = \sqrt[3]{x^3}\sqrt[3]{y^6}\sqrt[3]{z^9}\sqrt[3]{xy^2}$

$= xy^2z^3\sqrt[3]{xy^2}$

45. Let's start by working out the units:

Units on G =

$$N \cdot \frac{m^2}{kg^2} = \frac{kg \cdot m}{sec^2} \cdot \frac{m^2}{kg^2} = \frac{m^3}{kg \cdot sec^2}$$

$$\text{Units on } G \cdot M = \frac{m^3}{kg \cdot sec^2} \cdot kg = \frac{m^3}{sec^2}$$

Finally, units on

$$\frac{GM}{R} = \frac{\frac{m^3}{sec^2}}{m} = \frac{m^3}{sec^2} \cdot \frac{1}{m} = \frac{m^2}{sec^2}$$

a. $V_{circ} = \sqrt{\frac{6.67 \times 10^{-11} \cdot 3.30 \times 10^{23}}{3.24 \times 10^6} \frac{m^2}{sec^2}}$

$V_{circ} = \sqrt{\frac{6.67 \cdot 3.30}{3.24} \times 10^6 \frac{m^2}{sec^2}},$

$V_{circ} \approx 2.61 \times 10^3 \ m/sec$

45. b. $V_{circ} = \sqrt{\frac{6.67 \times 10^{-11} \cdot 1.90 \times 10^{27}}{8.14 \times 10^7} \frac{m}{sec}}$

$V_{circ} = \sqrt{\frac{6.67 \cdot 1.90}{8.14} \times 10^9 \frac{m}{sec}},$

$V_{circ} \approx 3.95 \times 10^4 \ m/sec$

c. $V_{exc} = \sqrt{2}\sqrt{\frac{6.67 \times 10^{-11} \cdot 5.98 \times 10^{24}}{7.18 \times 10^6} \frac{m}{sec}}$

$V_{esc} \approx \sqrt{2} \cdot 7.45 \times 10^3 \ m/sec$

$V_{esc} \approx 1.05 \times 10^4 \ m/sec$

d. $V_{exc} = \sqrt{2}\sqrt{\frac{6.67 \times 10^{-11} \cdot 6.44 \times 10^{23}}{4.19 \times 10^6} \frac{m}{sec}}$

$V_{esc} \approx \sqrt{2} \cdot 3.20 \times 10^3 \ m/sec$

$V_{esc} \approx 4.53 \times 10^3 \ m/sec$

47. $\sqrt{(-x)}$, $x < 0$ gives a real number because $-x > 0$ if $x < 0$.

49. $5 \cdot 5^n = 5^1 \cdot 5^n = 5^{1+n}$, When the bases are the same we add the exponents.

51. No absolute value symbol is needed because x^2 is always positive.

Mid-Chapter 6 Test

1. $x^{1/2} = \sqrt{x}, x \geq 0$

2. $\frac{1}{x} = x^{-1}, x \neq 0$

3. $\frac{a^{-1}b^2}{c^0} = \frac{b^2}{a^1 \cdot 1} = \frac{b^2}{a}$

4. $\frac{a^2b^{-1}}{c^{-2}} = \frac{a^2c^2}{b^1} = \frac{a^2c^2}{b}$

5. $a \cdot a^x = a^1 \cdot a^x = a^{x+1}$

6. a. $\left(\frac{2x}{3}\right)^{-2} = \left(\frac{3}{2x}\right)^2 = \frac{3^2}{2^2x^2} = \frac{9}{4x^2}$

b. $\frac{x}{y^{-2}} = xy^2$

7. a. $4.3 \text{ x } 10^{-3} = 0.0043$

b. $0.000123 = 1.23 \text{ x } 10^{-4}$

c. $\frac{3.6 \text{ x } 10^{-2}}{1.2 \text{ x } 10^3} = 3 \text{ x } 10^{-5} = 0.00003$

d. 0.060 has 2 significant digits

e. Possible answers are; the zero after the decimal point in d above, the zeros after 1 in 1000.

8. a. $\$1000\left(1+\frac{0.07}{52}\right)^{52 \cdot 5} \approx \$1418.73$

b. $\$1000\left(1+\frac{0.07}{52}\right)^{52 \cdot 2.75} \approx \$1212.12$

9. a. $4^x = 64, 64 = 4^3, \ x = 3$

b. $3^x = 27, 27 = 3^3, \ x = 3$

c. $\left(\frac{1}{2}\right)^x = \frac{1}{8}, \frac{1}{8} = \frac{1}{2^3} = \left(\frac{1}{2}\right)^3, x = 3$

d. $\pi^x = 1, \pi^0 = 1, x = 0$

10. $(1+0.02)(1+0.02) = 1.0404$ a 4.04% annual increase for 2 semiannual raises.

11. a. $\sqrt{4n} = 2\sqrt{n}, n \geq 0$

b. $\sqrt{40n^2} = \sqrt{2^2n^2 \cdot 10} = 2|n|\sqrt{10}$

c. $\sqrt{400n} = 20\sqrt{n}, n \geq 0$

d. $\sqrt{4000n} = 20\sqrt{10n}, n \geq 0$

12. a. $8^{2/3} = \left(\sqrt[3]{8}\right)^2 = (2)^2 = 4$

b. $125^{1/3} = \sqrt[3]{125} = 5$

13. a. $16^{3/4} = \left(\sqrt[4]{16}\right)^3 = (2)^3 = 8$

b. $(a^3)^{2/3} = \left(\sqrt[3]{a^3}\right)^2 = a^2$

14. a. $\sqrt[3]{-64} = \sqrt[3]{(-4)^3} = -4$

b. $\sqrt[4]{16y^4} = 2|y|$

15. a. $\sqrt{9x^5y^2} = \sqrt{9x^4xy^2} = 3x^2y\sqrt{x}$

b. $\sqrt[3]{24x^5y^6} = \sqrt[3]{2^3 \cdot 3x^3x^2y^6} = 2xy^2\sqrt[3]{3x^2}$

Section 6.3

1. $3\sqrt{2}-4\sqrt{2}+7\sqrt{2}=(3-4+7)\sqrt{2}$

$=6\sqrt{2}$

3. $\sqrt{20}+\sqrt{45}=2\sqrt{5}+3\sqrt{5}$

$=(2+3)\sqrt{5}=5\sqrt{5}$

5. $\sqrt{75}+\sqrt{27}=5\sqrt{3}+3\sqrt{3}$

$=(5+3)\sqrt{3}=8\sqrt{3}$

7. $\sqrt{9x}+\sqrt{16x}-\sqrt{x}=3\sqrt{x}+4\sqrt{x}-\sqrt{x}$

$=(3+4-1)\sqrt{x}=6\sqrt{x}$

9. $\sqrt{0.01x}+\sqrt{49x}-\sqrt{16x}$

$=0.1\sqrt{x}+7\sqrt{x}-4\sqrt{x}=(0.1+7-4)\sqrt{x}$

$=3.1\sqrt{x}$

11. $a\sqrt{b}+a\sqrt{c}$, not like terms

13. $\sqrt{ab}+2\sqrt{ab}=(1+2)\sqrt{ab}=3\sqrt{ab}$

15. $3\sqrt[3]{x}-\sqrt[3]{x}=(3-1)\sqrt[3]{x}=2\sqrt[3]{x}$

17. $3\sqrt[4]{x}-3\sqrt[2]{x}$, not like terms

19. $\sqrt[3]{8x}+\sqrt[3]{27x}=2\sqrt[3]{x}+3\sqrt[3]{x}$

$=(2+3)\sqrt[3]{x}=5\sqrt[3]{x}$

21. $a\sqrt[4]{81ab}-\sqrt[4]{a^5b}=3a\sqrt[4]{ab}-a\sqrt[4]{ab}$

$=(3-1)a\sqrt[4]{ab}=2a\sqrt[4]{ab}$

23. $\sqrt[4]{16x^4y}+x\sqrt[4]{y}=2x\sqrt[4]{y}+x\sqrt[4]{y}$

$(2+1)x\sqrt[4]{y}=3x\sqrt[4]{y}$

25. a. $(2-\sqrt{3})(2+\sqrt{3})=2^2-(\sqrt{3})^2$

$=4-3=1$

b. $(5-\sqrt{2})(5-\sqrt{2})=5^2-10\sqrt{2}+(\sqrt{2})^2$

$25-10\sqrt{2}+2=27-10\sqrt{2}$

27. a. $(x-\sqrt{5})(x+\sqrt{5})=x^2-(\sqrt{5})^2$

$=x^2-5$

b. $(x-\sqrt{7})(x-\sqrt{7})=x^2-2\sqrt{7}x+(\sqrt{7})^2$

$=x^2-2\sqrt{7}x+7$

29. a. $(\sqrt{x}+3)^2=(\sqrt{x})^2+6\sqrt{x}+3^2$

$=x+6\sqrt{x}+9$

b. $(\sqrt{a}-3)^2=(\sqrt{a})^2-6\sqrt{a}+(-3)^2$

$=a-6\sqrt{a}+9$

31. a. $(1-\sqrt{a})^2=1^2-2\sqrt{a}+(-\sqrt{a})^2$

$=1-2\sqrt{a}+a$

b. $(\sqrt{a}-\sqrt{b})^2=(\sqrt{a})^2-2\sqrt{a}\sqrt{b}+(\sqrt{b})^2$

$=a-2\sqrt{a}\sqrt{b}+b$

Section 6.3 (con't)

33. $(-2-\sqrt{2})^2+4(-2-\sqrt{2})+2=0$

$(-2)^2+2(-2)(-\sqrt{2})+(-\sqrt{2})^2-8-4\sqrt{2}+2$

$=4+4\sqrt{2}+2-8-4\sqrt{2}+2=0$

35. $(1+\sqrt{3})^2-2(1+\sqrt{3})-2=0$

$1^2+2\sqrt{3}+(\sqrt{3})^2-2-2\sqrt{3}-2$

$=1+2\sqrt{3}+3-2-2\sqrt{3}-2=0$

37. a. $\frac{1+\sqrt{5}}{2}\cdot\frac{1-\sqrt{5}}{2}=\frac{1-5}{4}=\frac{-4}{4}=-1$

b. $\frac{1+\sqrt{5}}{2}\approx 1.618,\ \frac{1-\sqrt{5}}{2}\approx -0.618$

c. Reciprocal of $\left(\frac{1+\sqrt{5}}{2}\right)^{-1}=\frac{2}{1+\sqrt{5}}$

$=\frac{2(1-\sqrt{5})}{(1+\sqrt{5})(1-\sqrt{5})}=\frac{2(1-\sqrt{5})}{1-5}=-\frac{1-\sqrt{5}}{2}$

Reciprocal of $\left(\frac{1-\sqrt{5}}{2}\right)^{-1}=\frac{2}{1-\sqrt{5}}$

$\frac{2(1+\sqrt{5})}{(1-\sqrt{5})(1+\sqrt{5})}=\frac{2(1+\sqrt{5})}{1-5}=-\frac{1+\sqrt{5}}{2}$

39. Observe this is a sum of cubes:

$(x+\sqrt[3]{2})(x^2-x\sqrt[3]{2}+\sqrt[3]{4})=x^3+(\sqrt[3]{2})^3$

$=x^3+2$

41. a. Conjugate is $3-\sqrt{2}$

b. Conjugate is $3+\sqrt{a}$

c. Conjugate is $a+\sqrt{b}$

d. Conjugate is $\sqrt{2}+\sqrt{3}$

43. a. $\frac{4}{\sqrt{5}}\cdot\frac{\sqrt{5}}{\sqrt{5}}=\frac{4\sqrt{5}}{5}$

b. $\frac{8}{\sqrt{6}}\cdot\frac{\sqrt{6}}{\sqrt{6}}=\frac{8\sqrt{6}}{6}=\frac{4\sqrt{6}}{3}$

45. a. $\frac{a}{\sqrt{c}}\cdot\frac{\sqrt{c}}{\sqrt{c}}=\frac{a\sqrt{c}}{c}$

b. $\frac{a}{\sqrt{a}}\cdot\frac{\sqrt{a}}{\sqrt{a}}=\frac{a\sqrt{a}}{a}=\sqrt{a}$

47. $\frac{4}{7-\sqrt{5}}\cdot\frac{7+\sqrt{5}}{7+\sqrt{5}}=\frac{4(7+\sqrt{5})}{49-5}$

$=\frac{4(7+\sqrt{5})}{44}=\frac{7+\sqrt{5}}{11}$

49. $\frac{x}{x-\sqrt{y}}\cdot\frac{x+\sqrt{y}}{x+\sqrt{y}}=\frac{x^2+x\sqrt{y}}{x^2-y}$

51. a. $\frac{1}{\sqrt[3]{2}}\cdot\frac{\sqrt[3]{4}}{\sqrt[3]{4}}=\frac{\sqrt[3]{4}}{2}$

b. $\frac{2}{\sqrt[3]{4}}\cdot\frac{\sqrt[3]{2}}{\sqrt[3]{2}}=\frac{2\sqrt[3]{2}}{2}=\sqrt[3]{2}$

c. $\frac{3}{\sqrt[3]{3}}\cdot\frac{\sqrt[3]{9}}{\sqrt[3]{9}}=\frac{3\sqrt[3]{9}}{3}=\sqrt[3]{9}$

Section 6.3 (con't)

53. $\dfrac{y}{x+r}=\dfrac{\sqrt{r^2-x^2}}{x+r}=\dfrac{\sqrt{r^2-x^2}}{x+r}\cdot\dfrac{\sqrt{r^2-x^2}}{\sqrt{r^2-x^2}}$

$$=\frac{r^2-x^2}{(x+r)\sqrt{r^2-x^2}}=\frac{(r-x)(r+x)}{(x+r)\sqrt{r^2-x^2}}$$

$$=\frac{r-x}{\sqrt{r^2-x^2}}=\frac{r-x}{y}=\frac{(-1)(x-r)}{y}$$

55. $d^2=(x_2-x_1)^2+(y_2-y_1)^2$

$$d=\sqrt{(x_2-x_1)^2+(y_2-y_1)^2}$$

57. $slope=\dfrac{-3-3}{4-(-6)}=\dfrac{-6}{10}=-\dfrac{3}{5}$

$$d=\sqrt{10^2+(-6)^2},\ d=\sqrt{100+36},$$

$$d=\sqrt{136},\ d=\sqrt{4\cdot 34},\ d=2\sqrt{34}$$

59. $slope=\dfrac{-4-7}{-3-6}=\dfrac{-11}{-9}=\dfrac{11}{9}$

$$d=\sqrt{9^2+11^2},\ d=\sqrt{81+121},\ d=\sqrt{202}$$

61. $slope=\dfrac{b-\frac{b}{2}}{a-\frac{a}{2}}=\dfrac{\frac{b}{2}}{\frac{a}{2}}=\dfrac{b}{2}\cdot\dfrac{2}{a}=\dfrac{b}{a}$

$$d=\sqrt{\left(\tfrac{a}{2}\right)^2+\left(\tfrac{b}{2}\right)^2},\ d=\sqrt{\frac{a^2+b^2}{4}},$$

$$d=\frac{\sqrt{a^2+b^2}}{2}$$

Section 6.4

1. $(\frac{1}{2},1), (\frac{1}{4},2), (\frac{1}{8},3), (\frac{1}{16},4)$

3. $(2, 1), (5, 2), (8, 3), (11, 4)$

5.

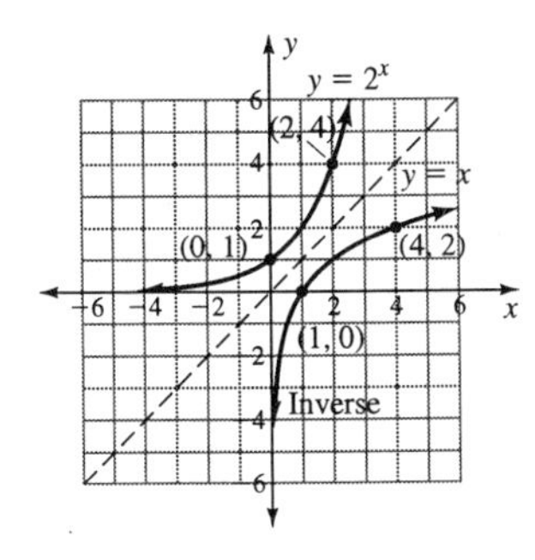

7.

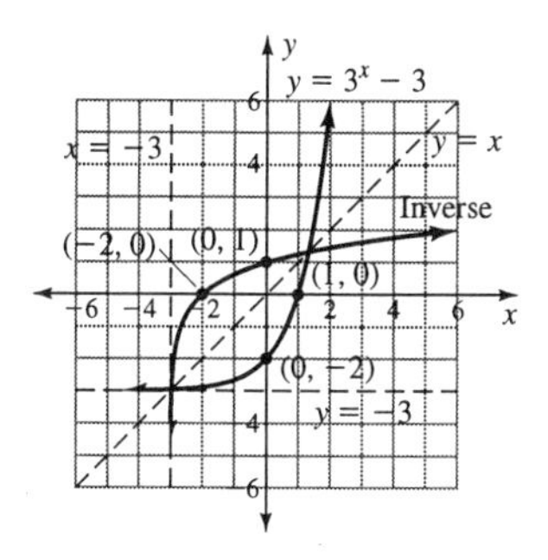

9. $x = 3y,\ yx = 3,\ y = \dfrac{3}{x}$

11. $x = 4y - 2,\ x + 2 = 4y,\ y = \dfrac{x+2}{4}$

$y = \dfrac{x}{4} + \dfrac{1}{2}$

13. $x = y + 3,\ y = x - 3$

15. $x = \sqrt{y},\ x^2 = (\sqrt{y})^2,\ y = x^2$

17. $x = \sqrt[5]{y},\ x^5 = \left(\sqrt[5]{y}\right)^5,\ y = x^5$

19. $x = y^4,\ y = \sqrt[4]{x}$

21. $x = y,\ y = x$

23. $x = -\dfrac{1}{y},\ xy = -1,\ y = -\dfrac{1}{x}$

25. $x = 1 - y,\ y + x = 1,\ y = 1 - x$

27. The inverse function is the same as the original function.

29. To solve for x in a power equation, $y = x^n$, take the nth root of both sides.

31. $A = \dfrac{\pi(5^2)(44)}{360},\ A \approx 9.6\,\text{in}^2$

33. $22 = \dfrac{\pi r^2(20)}{360},\ 7920 = 20\pi r^2,$

$396 = \pi r^2,\ r^2 = \dfrac{396}{\pi},\ r = \sqrt{\dfrac{396}{\pi}},$

$r \approx 11.2\ \text{in}$

35. $V = \frac{4}{3}\pi r^3,\ r^3 = \dfrac{3V}{4\pi},\ r = \sqrt[3]{\dfrac{3V}{4\pi}}$

37. $v = \dfrac{\pi p r^4}{8Ln},\ r^4 = \dfrac{8vLn}{\pi p},\ r = \sqrt[4]{\dfrac{8vLn}{\pi p}}$

39. $119{,}000 = 55{,}000(1+r)^{13},$

$\dfrac{119}{55} = (1+r)^{13},\ \sqrt[13]{\dfrac{119}{55}} = 1 + r,$

$r = \sqrt[13]{\dfrac{119}{55}} - 1,\ r \approx 0.061 \text{ or } 6.1\%$

$55{,}000(1+0.061)^{25} \approx \$242{,}000$

Section 6.4 (con't)

41. $87 = 55(1+r)^1$, $\frac{87}{55} = 1+r$

$r = \frac{87}{55} - 1$, $r \approx 0.5818$ *or* 58.18%

$55(1+0.5818)^{16} \approx \$84,487$

43. $5 = 96(1+r)^{21}$, $\frac{5}{96} = (1+r)^{21}$

$\sqrt[21]{\frac{5}{96}} = 1+r$, $r = \sqrt[21]{\frac{5}{96}} - 1$, $r \approx -0.131$

or $r \approx -13.1\%$,

45. $3000 = 40{,}000(1+r)^{10}$,

$\frac{3}{40} = (1+r)^{10}$, $\sqrt[10]{\frac{3}{40}} = 1+r$,

$r = \sqrt[10]{\frac{3}{40}} - 1$, $r \approx -0.228$

depreciation rate $\approx 22.8\%$

47. $x^3 = 10$, $x = \sqrt[3]{10}$, $x = 10^{1/3}$, $x \approx 2.154$

$x^4 = 10$, $x = \sqrt[4]{10}$, $x = 10^{1/4}$, $x \approx 1.778$

$x^5 = 10$, $x = \sqrt[5]{10}$, $x = 10^{1/5}$, $x \approx 1.585$

$x^6 = 10$, $x = \sqrt[6]{10}$, $x = 10^{1/6}$, $x \approx 1.468$

49. The base x approaches 1 as the exponent gets larger.

51. $a \geq 0$

53. The set of inputs is all real numbers.

55. There are limitations on the domain because the square root of a negative number is not a real number.

57. The extra number is an extraneous root and will give a false statement when substituted back into the original equation.

59. $\sqrt{x+2} = 4$, $x = 14$

61. $\sqrt{x+2} = -1$, never, { }

63. Defined when $2 - x \geq 0$, $x \leq 2$;

$\sqrt{2-x} = 2$, $(\sqrt{2-x})^2 = 2^2$, $2 - x = 4$,
$x = -2$

65. Defined when $x - 5 \geq 0$, $x \geq 5$;

$\sqrt{x-5} = 2$, $(\sqrt{x-5})^2 = 2^2$
$x - 5 = 4$, $x = 9$

67. Defined when $x - 5 \geq 0$, $x \geq 5$;

$\sqrt{x-5} - 6 = -2$, $\sqrt{x-5} = 4$,
$(\sqrt{x-5})^2 = 4^2$, $x - 5 = 16$, $x = 21$

69. Defined when $2x + 14 \geq 0$, $2x \geq -14$,

$x \geq -7$;

$\sqrt{2x+14} = 0$, $(\sqrt{2x+14})^2 = 0$,
$2x + 14 = 0$, $2x = -14$, $x = -7$

Section 6.4 (con't)

71. Defined when x ≥ -7 (see # 69);

$\sqrt{2x+14} = 6$, $(\sqrt{2x+14})^2 = 6^2$,
$2x+14 = 36$, $2x = 22$, $x = 11$

73. Defined when 3x - 5 ≥ 0, 3x ≥ 5,

$x \geq \frac{5}{3}$;

$\sqrt{3x-5} = 2$, $3x-5 = 4$,
$3x = 9$, $x = 3$

75. See # 73 for defined values,

$\sqrt{3x-5} = -1$, not possible for the square root to equal a negative, { }

77. Defined when 5x - 1 ≥ 0, 5x ≥ 1,

$x \geq \frac{1}{5}$;

$\sqrt{5x-1} = 3$, $5x-1 = 9$,
$5x = 10$, $x = 2$

79. Defined when 2x + 7 ≥ 0 **and** x ≥ 0,

2x ≥ -7, $x \geq \frac{-7}{2}$, $\frac{-7}{2}$ not ≥ 0 so x ≥ 0 defines domain.

$\sqrt{2x+7} = \sqrt{x}+2$,
$\left(\sqrt{2x+7}\right)^2 = \left(\sqrt{x}+2\right)^2$,
$2x+7 = x+4\sqrt{x}+4$, $x+3 = 4\sqrt{x}$,
$(x+3)^2 = (4\sqrt{x})^2$, $x^2+6x+9 = 16x$,
$x^2-10x+9 = 0$, $(x-1)(x-9) = 0$
$x = 1$ or $x = 9$
$\sqrt{2(1)+7} = \sqrt{1}+2$, $\sqrt{9} = 1+2$, $3 = 3$
$\sqrt{2(9)+7} = \sqrt{9}+2$, $\sqrt{25} = 3+2, 5 = 5$

Both values give true statements, {1, 9}

81. Defined when x + 4 ≥ 0 **and** x ≥ 0, since the solution to x + 4 ≥ 0 is x ≥ -4, x ≥ 0 defines the domain.

$\sqrt{3x}-2 = \sqrt{x+4}$, $\left(\sqrt{3x}-2\right)^2 = \left(\sqrt{x+4}\right)^2$
$3x-4\sqrt{3x}+4 = x+4$, $2x = 4\sqrt{3x}$,
$(2x)^2 = \left(4\sqrt{3x}\right)^2$, $4x^2 = 16(3x)$,
$4x^2-48x = 0$, $4x(x-12) = 0$,
$x = 0$ or $x = 12$,
$\sqrt{0}-2 = \sqrt{0+4}$, $-2 = 2$, false
$\sqrt{3(12)}-2 = \sqrt{12+4}$, $6-2 = 4$, $4 = 4$,
the solution is $x = 12$

Section 6.4 (con't)

83. Defined when 3x - 3 ≥ 0 **and** x ≥ 0,

3x ≥ 3, x ≥ 1, this is the domain.

$$\sqrt{3x-3} = \sqrt{4x}-1,$$
$$\left(\sqrt{3x-3}\right)^2 = \left(\sqrt{4x}-1\right)^2,$$
$$3x-3 = 4x-2\sqrt{4x}+1,\ 2\sqrt{4x} = x+4,$$
$$\left(2\sqrt{4x}\right)^2 = (x+4)^2,\ 4(4x) = x^2+8x+16,$$
$$x^2-8x+16 = 0,\ (x-4)(x-4) = 0,$$
$$x = 4,$$
$$\sqrt{3(4)-3} = \sqrt{4(4)}-1,\ \sqrt{9} = 4-1,\ 3 = 3$$

the solution is $x = 4$

85. $r = \sqrt{24L},\ r^2 = 24L,\ L = \dfrac{r^2}{24}$

87. $r_{dry} = \sqrt{24(50)},\ r_{dry} \approx 34.64$ mph

$r_{wet} = \sqrt{12(50)},\ r_{wet} \approx 24.49$ mph

$$\frac{r_{wet}}{r_{dry}} = \frac{\sqrt{12(50)}}{\sqrt{24(50)}},\ \frac{r_{wet}}{r_{dry}} = \sqrt{\frac{12(50)}{24(50)}}$$

$$\frac{r_{wet}}{r_{dry}} = \sqrt{\frac{1}{2}},\ \frac{r_{wet}}{r_{dry}} = \frac{1}{\sqrt{2}},\ \frac{r_{wet}}{r_{dry}} = \frac{1\cdot\sqrt{2}}{\sqrt{2}\sqrt{2}},$$
$$\frac{r_{wet}}{r_{dry}} = \frac{\sqrt{2}}{2};$$

You might conclude that it is best to slow down when the pavement is wet.

89. a. $29 = \sqrt{\dfrac{3h}{2}},\ 29^2 = \dfrac{3h}{2},$

$$\frac{29^2\cdot 2}{3} = h,\quad h \approx 560 \text{ ft}$$

b. $35.8 = \sqrt{\dfrac{3h}{2}},\ 35.8^2 = \dfrac{3h}{2},$

$$\frac{35.8^2\cdot 2}{3} = h,\quad h \approx 850 \text{ ft}$$

c. $29 = \sqrt{\dfrac{3h}{8}},\ 29^2 = \dfrac{3h}{8}$

$$\frac{29^2\cdot 8}{3} = h,\quad h \approx 2240 \text{ ft}$$

$$35.8 = \sqrt{\frac{3h}{8}},\ 35.8^2 = \frac{3h}{8},$$

$$\frac{35.8^2\cdot 8}{3} = h,\quad h \approx 3420 \text{ ft}$$

When solving for h the distance, d, is being multiplied by a factor of 8 instead of 2. As for why you can see farther on the moon, there is no atmosphere.

Chapter 6 Review

1. $\dfrac{a^0b^{-1}}{c^2} = \dfrac{1}{bc^2}$

3. $\left(\dfrac{2x}{y^2}\right)^4 = \dfrac{2^4x^4}{y^{2\cdot 4}} = \dfrac{16x^4}{y^8}$

5. $\left(\dfrac{2x^2}{y}\right)^3 = \dfrac{2^3x^{2\cdot 3}}{y^3} = \dfrac{8x^6}{y^3}$

7. **a.** 0.00000000000000345

   b. 0.006400

   c. 400,500

   d. 4780.0

9. **a.** $(2.5 \text{ x } 10^6)(37 \text{ x } 10^9) = 92.5 \text{ x } 10^{6+9}$

   $= 92.5 \text{ x } 10^{15} = 9.25 \text{ x } 10^{16}$
   $= 92,500,000,000,000,000$

   b. $(6.3 \text{ x } 10^6)(3.7 \text{ x } 10^{10}) = 23.31 \text{ x } 10^{16}$

   $= 2.331 \text{ x } 10^{17}$
   $= 233,100,000,000,000,000$

   c. $(2.3 \text{ x } 10^5)(3.7 \text{ x } 10^{10}) = 8.51 \text{ x } 10^{15}$

   $= 8,510,000,000,000,000$

   d. $(1.2 \text{ x } 10^7)(3.7 \text{ x } 10^{10}) = 4.44 \text{ x } 10^{17}$

   $= 444,000,000,000,000,000$

   e. $(4.9 \text{ x } 10^5)(3.7 \text{ x } 10^{10}) = 18.13 \text{ x } 10^{15}$

   $= 1.813 \text{ x } 10^{16}$
   $= 18,130,000,000,000,000$

11. $\$1200\left(1+\dfrac{0.06}{365}\right)^{365\cdot 7} \approx \$1826.29$

13. $\$1200\left(1+\dfrac{0.06}{365}\right)^{365\cdot 2.5} \approx \$1394.18$

15. **a.** $64^{1/2} = \sqrt{64} = 8$

   b. $64^{1/6} = \sqrt[6]{64} = \sqrt[6]{2^6} = 2$

   c. $64^{5/6} = (\sqrt[6]{64})^5 = (2)^5 = 32$

17. **a.** $16^{0.75} = 16^{3/4} = (\sqrt[4]{16})^3 = (\sqrt[4]{2^4})^3 = 2^3 = 8$

   b. $25^{3/2} = (\sqrt{25})^3 = 5^3 = 125$

   c. $27^{4/3} = \left(\sqrt[3]{27}\right)^4 = \left(\sqrt[3]{3^3}\right)^4 = 3^4 = 81$

19. **a.** $(b^4)^{3/4} = b^{4\cdot\frac{3}{4}} = b^3$

   b. $a^{2/3} \cdot a^{1/3} = a^{\frac{2}{3}+\frac{1}{3}} = a^{3/3} = a$

   c. $x^{3/4} \cdot x^{1/2} = x^{\frac{3}{4}+\frac{1}{2}} = x^{\frac{3}{4}+\frac{2}{4}} = x^{5/4}$

   d. $\dfrac{b^{3/4}}{b^{1/2}} = b^{\frac{3}{4}-\frac{1}{2}} = b^{\frac{3}{4}-\frac{2}{4}} = b^{1/4}$

   e. $b \cdot b^x = b^{x+1}$

Chapter 6 Review (con't)

21. a. $x^{1/2} \cdot x^{1/4} = x^{\frac{1}{2}+\frac{1}{4}} = x^{3/4}$

b. $\dfrac{x^n}{x} = x^{n-1}$

c. $\left(x^{1/4}\right)^{2/3} = x^{\frac{1}{4}\cdot\frac{2}{3}} = x^{\frac{1}{6}}$

d. $\dfrac{a^{2/3}}{a^{4/3}} = a^{\frac{2}{3}-\frac{4}{3}} = a^{-2/3} = \dfrac{1}{a^{2/3}}$

e. $\dfrac{a}{a^x} = a^{1-x}$

23. a. $\sqrt[3]{-8} = (-8)^{1/3} = -2$

b. $\sqrt[4]{\dfrac{625}{16}} = \left(\dfrac{625}{16}\right)^{1/4} = \dfrac{625^{1/4}}{16^{1/4}} = \dfrac{5}{2}$

c. $\sqrt[3]{8^2} = 8^{2/3} = 2^2 = 4$

25. a. $125^{2/3} = \sqrt[3]{125^2} = \left(\sqrt[3]{125}\right)^2 = 5^2 = 25$

b. $64^{1/3} = \sqrt[3]{64} = 4$

c. $32^{2/5} = \sqrt[5]{32^2} = \left(\sqrt[5]{32}\right)^2 = 2^2 = 4$

27. a. $(-64)^{1/3} = \sqrt[3]{-64} = -4$

b. $-64^{1/2} = -\sqrt{64} = -8$

c. $(-16)^{1/4}$, not a real number

29. a. $x^{2/3} = \sqrt[3]{x^2} = \left(\sqrt[3]{x}\right)^2$

b. $x^{1.5} = x^{3/2} = \sqrt{x^3} = \left(\sqrt{x}\right)^3$

c. $a^{0.75} = a^{3/4} = \sqrt[4]{a^3} = \left(\sqrt[4]{a}\right)^3$

31. a. $\sqrt[3]{8000} = \sqrt[3]{2^3 \cdot 10^3} = 2 \cdot 10 = 20$

b. $\sqrt[3]{800} = \sqrt[3]{2^3 \cdot 10^2} = 2 \cdot 10^{\frac{2}{3}} \approx 9.283$

c. $\sqrt[3]{80} = \sqrt[3]{2^3 \cdot 10^1} = 2 \cdot 10^{\frac{1}{3}} \approx 4.309$

d. $\sqrt[3]{8} = \sqrt[3]{2^3 \cdot 10^0} = 2$

e. $\sqrt[3]{0.8} = \sqrt[3]{2^3 \cdot 10^{-1}} = 2 \cdot 10^{-\frac{1}{3}} \approx 0.9283$

f. $\sqrt[3]{0.08} = \sqrt[3]{2^3 \cdot 10^{-2}} = 2 \cdot 10^{-\frac{2}{3}} \approx 0.4309$

g. $\sqrt[3]{0.008} = \sqrt[3]{2^3 \cdot 10^{-3}} = 2 \cdot 10^{-1} = 0.2$

33. a. $\sqrt{9n} = 3\sqrt{n},\ n \geq 0$

b. $\sqrt{90n^2} = \sqrt{9n^2 \cdot 10} = 3|n|\sqrt{10}$

c. $\sqrt{0.09x} = \sqrt{9 \cdot 10^{-2}x} = 3 \cdot 10^{-1}\sqrt{x}$

$= 0.3\sqrt{x},\ x \geq 0$

d. $\sqrt{900x^4} = 30x^2$

35. a. $\sqrt{z}\sqrt{z^5} = \sqrt{z^6} = z^3,\ z \geq 0$

b. $\sqrt[3]{x^2} \cdot \sqrt[3]{x} = \sqrt[3]{x^3} = x$

c. $\sqrt[3]{x^5} \cdot \sqrt[3]{x^1} = \sqrt[3]{x^6} = x^2$

37. a. $\sqrt[3]{x} \cdot \sqrt{x} = x^{\frac{1}{3}} \cdot x^{\frac{1}{2}} = x^{\frac{2}{6}+\frac{3}{6}} = x^{\frac{5}{6}},\ x \geq 0$

b. $\sqrt{x^3} \cdot \sqrt{x^2} = \sqrt{x^5} = x^{\frac{5}{2}},\ x \geq 0$

c. $\sqrt{\sqrt[3]{x}} = \left(x^{\frac{1}{3}}\right)^{\frac{1}{2}} = x^{\frac{1}{3}\cdot\frac{1}{2}} = x^{\frac{1}{6}},\ x \geq 0$

Chapter 6 Review (con't)

39. a. $\sqrt{3}+2\sqrt{12}=\sqrt{3}+2\sqrt{2^2\cdot 3}$

$=\sqrt{3}+4\sqrt{3}=(1+4)\sqrt{3}=5\sqrt{3}$

b. $\sqrt{9x}+\sqrt{x}=3\sqrt{x}+\sqrt{x}=4\sqrt{x},\ x\geq 0$

41. a. $(3+\sqrt{6})(3-\sqrt{6})=3^2-(\sqrt{6})^2$

$=9-6=3$

b. $(3-\sqrt{x})(3-\sqrt{x})=3^2-6\sqrt{x}+(-\sqrt{x})^2$

$=9-6\sqrt{x}+x,\ x\geq 0$

43. a. $\frac{1}{b-\sqrt{a}}\cdot\frac{b+\sqrt{a}}{b+\sqrt{a}}=\frac{b+\sqrt{a}}{b^2-a}$

b. $\frac{1}{\sqrt{a}+\sqrt{b}}\cdot\frac{\sqrt{a}-\sqrt{b}}{\sqrt{a}-\sqrt{b}}=\frac{\sqrt{a}-\sqrt{b}}{a-b}$

c. $\frac{1}{\sqrt[3]{9x}}\cdot\frac{\sqrt[3]{3x^2}}{\sqrt[3]{3x^2}}=\frac{\sqrt[3]{3x^2}}{3x}$

45. $(1-\sqrt{2})^2-2(1-\sqrt{2})-1=0$

$1-2\sqrt{2}+2-2+2\sqrt{2}-1=0$
$0=0$, answer checks

$$x=\frac{-(-2)\pm\sqrt{(-2)^2-4(1)(-1)}}{2(1)}$$

$x=\frac{2\pm\sqrt{8}}{2}, x=\frac{2\pm 2\sqrt{2}}{2},$
$x=1\pm\sqrt{2}$, other solution is $1+\sqrt{2}$

47. a. (3, 3), (5, 4), (7, 5) one output for each input, this is a function

b. (-1, -2), (-2, 0), (-3, 2) one output for each input, this is a function

49. a. $x=2y-1,\ 2y=x+1,\ y=\frac{x+1}{2}$

$y=\frac{x}{2}+\frac{1}{2}$

b. $x=3y,\ y=\frac{x}{3}$

c. $x=y^3,\ y=\sqrt[3]{x}$

d. $x=\sqrt{y},\ y=x^2$

e. $x=-y,\ y=-x$

f. $x=\frac{1}{y},\ y=\frac{1}{x},\ x\neq 0,\ y\neq 0$

51. $9.00=5.50(1+r)^{10},\ \frac{9.00}{5.50}=(1+r)^{10}$

$\left(\frac{9.00}{5.50}\right)^{\frac{1}{10}}=1+r,\left(\frac{9.00}{5.50}\right)^{\frac{1}{10}}-1=r$

$r\approx 0.0505$ or 5.05%

53. $15{,}000=10{,}000(1+r)^5,\ 1.5=(1+r)^5$

$1.5^{\frac{1}{5}}=1+r,\ 1.5^{\frac{1}{5}}-1=r,$
$r\approx 0.0845$ or 8.45%

Chapter 6 Review (con't)

55. $T = 2\pi\sqrt{\frac{L}{g}}, \frac{T}{2\pi} = \sqrt{\frac{L}{g}}, \left(\frac{T}{2\pi}\right)^2 = \frac{L}{g}$

$\frac{T^2}{4\pi^2} = \frac{L}{g}, gT^2 = 4\pi^2 L, g = \frac{4\pi^2 L}{T^2}$

57. $F = \frac{Q_1Q_2}{kr^2}, Fr^2 = \frac{Q_1Q_2}{k}, r^2 = \frac{Q_1Q_2}{kF},$

$r = \sqrt{\frac{Q_1Q_2}{kF}}$

59. $r^3 = a^2p, r = \sqrt[3]{a^2p}$

61. a. From the graph $\approx \{0.25, 2\}$

b. $2 = \sqrt{5x-1}, 2^2 = \left(\sqrt{5x-1}\right)^2,$

$4 = 5x-1, 5x = 5, x = 1$

c. $\sqrt{8x} - 1 = 1, \sqrt{8x} = 2, \left(\sqrt{8x}\right)^2 = 2^2$

$8x = 4, x = \frac{1}{2}$

d. $\sqrt{8x} - 1 = \sqrt{5x-1},$

$\left(\sqrt{8x} - 1\right)^2 = \left(\sqrt{5x-1}\right)^2,$

$(\sqrt{8x})^2 - 2\sqrt{8x} + 1 = 5x - 1,$

$8x - 2\sqrt{8x} + 1 = 5x - 1, 3x + 2 = 2\sqrt{8x},$

$(3x+2)^2 = \left(2\sqrt{8x}\right)^2,$

$9x^2 + 12x + 4 = 4(8x),$

$9x^2 + 12x + 4 = 32x, 9x^2 - 20x + 4 = 0,$

$(9x-2)(x-2) = 0, x = 2, 9x = 2, x = \frac{2}{9},$

$\{\frac{2}{9}, 2\}$

63. a. From the graph $x \approx 12$

b. $\sqrt{6x-8} = 4, \left(\sqrt{6x-8}\right)^2 = 4^2$

$6x - 8 = 16, 6x = 24, x = 4$

c. $\sqrt{3x} + 2 = 2, \sqrt{3x} = 0, x = 0$

d. $\sqrt{3x} + 2 = \sqrt{6x-8},$

$\left(\sqrt{3x}+2\right)^2 = \left(\sqrt{6x-8}\right)^2,$

$(\sqrt{3x})^2 + 4\sqrt{3x} + 4 = 6x - 8,$

$3x + 4\sqrt{3x} + 4 = 6x - 8,$

$4\sqrt{3x} = 3x - 12, \left(4\sqrt{3x}\right)^2 = (3x-12)^2$

$16(3x) = 9x^2 - 72x + 144,$

$9x^2 - 72x + 144 = 48x,$

$9x^2 - 120x + 144 = 0, (9x-12)(x-12) = 0,$

$x = 12,$ or $9x = 12, x = \frac{4}{3}$

Graph shows only one intersection, at $x = 12$, other solution is extraneous

Chapter 6 Test

1. $\dfrac{a^0b^2}{b^{-1}} = 1\cdot b^{2-(-1)} = b^3$

2. $\dfrac{a^{-2}b^3}{(bc)^2} = \dfrac{a^{-2}b^3}{b^2c^2} = \dfrac{b^{3-2}}{a^2c^2} = \dfrac{b}{a^2c^2}$

3. $\left(\dfrac{x^2y}{3x}\right)^{-2} = \left(\dfrac{3x}{x^2y}\right)^2 = \left(\dfrac{3}{xy}\right)^2 = \dfrac{9}{x^2y^2}$

4. $(2x)^2 = 2^2 \cdot x^2 = 4x^2$

5. $\sqrt{4x^2} = \sqrt{4}\cdot\sqrt{x^2} = 2|x|$

   Principal square root is always positive,

   2x could be negative.

6. **a.** 0.00003450

   b. $\dfrac{6.3 \text{ x } 10^{-2}}{7 \text{ x } 10^4} = 0.9 \text{ x } 10^{-2-4}$

   $= 0.9 \text{ x } 10^{-6} = 9 \text{ x } 10^{-7}$

7. **a.** $\sqrt{36x} = \sqrt{36}\cdot\sqrt{x} = 6\sqrt{x},\ x \ge 0$

   b. $\sqrt{0.36x^2} = \sqrt{0.36}\cdot\sqrt{x^2} = 0.6|x|$

   c. $\sqrt{360x^4} = \sqrt{36}\cdot\sqrt{10}\cdot\sqrt{x^4} = 6x^2\sqrt{10}$

8. $a\left(\dfrac{b}{a}\right)^4 = a\dfrac{b^4}{a^4} = \dfrac{b^4}{a^3} = b\dfrac{b^3}{a^3} = b\left(\dfrac{b}{a}\right)^3$

9. $2000\left(1+\dfrac{0.08}{4}\right)^{4\cdot 3} \approx \$2536.48$

10. $18{,}000 = 12{,}000(1+r)^5,\ 1.5 = (1+r)^5$

    $1.5^{\frac{1}{5}} = 1+r,\ \sqrt[5]{1.5}-1 = r,$
    $r \approx 0.0845$ or 8.45%

11. $M = \dfrac{mgl}{\pi r^2 s},\ r^2M = \dfrac{mgl}{\pi s},\ r^2 = \dfrac{mgl}{\pi sM},$

    $r = \sqrt{\dfrac{mgl}{\pi sM}}$

12. $F = \dfrac{mv^2}{r},\ Fr = mv^2,\ \dfrac{Fr}{m} = v^2,$

    $v = \sqrt{\dfrac{Fr}{m}}$

13. $64^{2/3} = \sqrt[3]{64^2} = \left(\sqrt[3]{64}\right)^2 = 4^2 = 16$

14. $32^{4/5} = \sqrt[5]{32^4} = \left(\sqrt[5]{32}\right)^4 = 2^4 = 16$

15. $81^{3/4} = \sqrt[4]{81^3} = \left(\sqrt[4]{81}\right)^3 = 3^3 = 27$

16. **a.** $\sqrt[4]{\dfrac{16}{81}} = \dfrac{\sqrt[4]{16}}{\sqrt[4]{81}} = \dfrac{2}{3}$

    b. $\sqrt[3]{-27} = -3$

    c. $-\sqrt[3]{-64} = -(-4) = 4$

17. **a.** $(b^2)^{3/2} = b^{2\cdot\frac{3}{2}} = b^3$

    b. $b^{2/3}b^{1/3} = b^{\frac{2}{3}+\frac{1}{3}} = b$

    c. $3\cdot 3^n = 3^{n+1}$

Chapter 6 Test (con't)

18. a. $(2-\sqrt{x})(2+\sqrt{x}) = 2^2 - (\sqrt{x})^2$

$= 4 - x$

b. $(8-\sqrt{3})(3+\sqrt{8})$

$= 24 - 3\sqrt{3} + 8\sqrt{8} - \sqrt{3}\sqrt{8}$

$= 24 - 3\sqrt{3} + 8\sqrt{4}\sqrt{2} - \sqrt{3}\sqrt{4}\sqrt{2}$

$= 24 - 3\sqrt{3} + 16\sqrt{2} - 2\sqrt{6}$

19. a. $\frac{1}{x+\sqrt{y}} \cdot \frac{x-\sqrt{y}}{x-\sqrt{y}} = \frac{x-\sqrt{y}}{x^2-y}$

b. $\frac{1}{\sqrt[3]{x}} \cdot \frac{\sqrt[3]{x^2}}{\sqrt[3]{x^2}} = \frac{\sqrt[3]{x^2}}{x}$

20. $(4-\sqrt{10})^2 - 8(4-\sqrt{10}) + 6 = 0$

$16 - 8\sqrt{10} + 10 - 32 + 8\sqrt{10} + 6 = 0,$

$32 - 32 - 8\sqrt{10} + 8\sqrt{10} = 0, 0 = 0$

21. (-1, -1), (-3, -2), (-5, -3), one output for each input, this is a function

22.

x	$y = \left(\frac{1}{2}\right)^x$
-2	$\left(\frac{1}{2}\right)^{-2} = 2^2 = 4$
-1	$\left(\frac{1}{2}\right)^{-1} = 2$
0	$\left(\frac{1}{2}\right)^0 = 1$
1	$\left(\frac{1}{2}\right)^1 = \frac{1}{2}$
2	$\left(\frac{1}{2}\right)^2 = \frac{1}{4}$

y	$x = \left(\frac{1}{2}\right)^y$
-2	4
-1	2
0	1
1	$\frac{1}{2}$
2	$\frac{1}{4}$

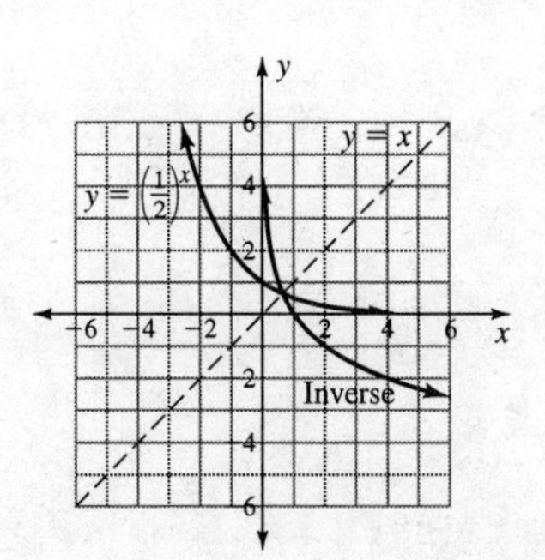

Chapter 6 Test (con't)

23. $3y - 2x = 6,\ 3y = 2x + 6,\ y = \frac{2}{3}x + 2$

24. a. From the graph x = 8

b. $\sqrt{x+1} = 2,\ \left(\sqrt{x+1}\right)^2 = 2^2$

$x + 1 = 4,\ x = 3$

c. $\sqrt{2x} - 1 = 1,\ \sqrt{2x} = 2,\ \left(\sqrt{2x}\right)^2 = 2^2$

$2x = 4,\ x = 2$

d. $\sqrt{x+1} = \sqrt{2x} - 1,\ \left(\sqrt{x+1}\right)^2 = \left(\sqrt{2x} - 1\right)^2$

$x + 1 = (\sqrt{2x})^2 - 2\sqrt{2x} + 1,$

$x + 1 = 2x - 2\sqrt{2x} + 1,\ 2\sqrt{2x} = x,$

$\left(2\sqrt{2x}\right)^2 = x^2,\ 4(2x) = x^2,\ 8x = x^2$

$x = 8$

Cumulative Review 1 to 6

1. $5 - 3(x+2) = 32,\ -3(x+2) = 27,$

$x + 2 = -9,\ x = -11$

3. $A = \dfrac{a+b+c}{3},\ 3A = a+b+c,$

$b = 3A - a - c$

5. Let y = total cost, x = kilowatt hours; assume a linear equation

$m = \dfrac{83.33 - 68.46}{1410 - 1120},\ m \approx 0.051$

$b = 83.33 - 0.051(1410),\ b \approx 11.42$

$y \approx 0.051x + 11.42$

Note: value of b will vary depending on which data point is used and how many decimal places are used for m.

7. a. $m = \dfrac{1-(-2)}{-1-4},\ m = -\dfrac{3}{5}$

b. Perpendicular line has negative reciprocal slope, in this case $\dfrac{5}{3}$

9. Domain is all real numbers, range is 4

11. a. $f(-2) = 2(-2)^2 - 5(-2) + 6$

$f(-2) = 24$

$f(0) = 2(0)^2 - 5(0) + 6,$

$f(0) = 6$

11. b. f(0) is the y-intercept

c. $x = \dfrac{-(-5)}{2(2)},\ x = \frac{5}{4} = 1\frac{1}{4}$

$y = 2(\frac{5}{4})^2 - 5(\frac{5}{4}) + 6,$

$y = 2(\frac{25}{16}) - \frac{25}{4} + 6,\ y = \frac{25}{8} - \frac{50}{8} + \frac{48}{8}$

$y = \frac{23}{8} = 2\frac{7}{8}$

vertex is at $\left(1\frac{1}{4},\ 2\frac{7}{8}\right)$

d. $x = \dfrac{-b \pm \sqrt{b^2 - 4ac}}{2a};$

$x = \dfrac{-(-5) \pm \sqrt{(-5)^2 - 4(2)(6)}}{2(2)},$

$x = \dfrac{5 \pm \sqrt{-23}}{4}$, no real x-intercepts

13. 12a. $x^2 + 3x = 4,\ x^2 + 3x - 4 = 0,$

$(x+4)(x-1) = 0,$

$x + 4 = 0,\ x = -4$ or

$x - 1 = 0,\ x = 1,\ \{-4, 1\}$

12b. $x^2 + 3x = 0,\ x(x+3) = 0,$

$x = 0,$ or $x + 3 = 0,\ x = -3\ \ \{-3, 0\}$

15. $\left(\dfrac{7}{2}\right)^2 = \dfrac{7^2}{4^2} = \dfrac{49}{4}$

Cumulative Review 1 to 6 (con't)

17. a. First differences: 3, 5, 7, 9

Second differences: 2, 2, 2

Next number: 26 + (9 + 2) = 37

Quadratic function

2a = 2, a = 1

c = 2 - (3 - 2), c = 1

$2 = 1(1)^2 + b(1) + 1,$

2 = 2 + b, b = 0

$y = x^2 + 1$

b. First differences: 3, 3, 3, 3

Next number: 7 + 3 = 10

Linear function

y = -5 + (x - 1)3, y=-5 + 3x - 3,

y = 3x - 8

c. First differences: 4, 6, 8, 10

Second differences: 2, 2, 2

Next number: 25 + (10 + 2) = 37

Quadratic function

2a = 2, a = 1

c = -3 - (4 - 2) = -5

$-3 = 1(1)^2 + b(1) - 5,$ b = 1

$y = x^2 + x - 5$

17. d. First differences: 1, 2, 4, 8

Second differences: 1, 2, 4

Observe first differences increase by a factor of 2, next first difference will be 16, next number is 13 + 16 = 29

Function is neither linear or quadratic

19. $x^3 - 1 = (x-1)(x^2 + x + 1),$

$(x-1) = 0,\ x = 1\ or$

$x^2 + x + 1 = 0$

$$x = \frac{-1 \pm \sqrt{1^2 - 4(1)(1)}}{2(1)},$$

$$x = \frac{-1 \pm \sqrt{-3}}{2},\ x = -\frac{1}{2} \pm \frac{i\sqrt{3}}{2}$$

$\{1, -\frac{1}{2} \pm \frac{\sqrt{3}}{2} i\}$

21. a. $x = \dfrac{-0.09}{2(-0.16)},\ x \approx 0.28$

$y = 0.09(0.28) - 0.16(0.28)^2,\ y \approx 0.01$

vertex is ≈ (0.28, 0.01)

b. $-115 = 0.09x - 0.16x^2$

$0.16x^2 - 0.09x - 115 = 0$

$$x = \frac{-(-0.09) \pm \sqrt{(-0.09)^2 - 4(0.16)(-115)}}{2(0.16)}$$

$$x = \frac{0.09 \pm \sqrt{73.6081}}{0.32},\ x \approx \frac{0.09 \pm 8.58}{0.32}$$

$x \approx 27.1\,\text{ft}$, (negative solution was discarded)

23. $\dfrac{x+y}{y} = \dfrac{x}{y} + \dfrac{y}{y} = \dfrac{x}{y} + 1,$ **c**

25. $\dfrac{x}{x+y} = \dfrac{x}{y+x},$ **a**

Cumulative Review 1 to 6 (con't)

27. $\frac{a}{a-b}=\frac{a-b+b}{a-b}=\frac{a-b}{a-b}+\frac{b}{a-b}$

$=1+\frac{b}{a-b}$, **b**

29. $\frac{x^2+3x}{x-1}\cdot\frac{x^2-1}{3x^2}=\frac{x(x+3)}{(x-1)}\cdot\frac{(x+1)(x-1)}{3x^2}$

$=\frac{(x+3)(x+1)}{3x}$

31. $19=\frac{(150)(704.5)}{x^2}$, $19x^2=105{,}675$,

$x^2\approx 5561.8$, $x\approx\sqrt{5561.8}$, $x\approx 75$

$\frac{105{,}675}{x^2}=24$, $105{,}675=24x^2$

$x^2\approx 4403.1$, $x\approx\sqrt{4403.1}$, $x\approx 66$

66 in $\le$ x $\le$ 75 in

33. 1 billion = 10^9,

(3)0.825 x 10^9 = 2.475 x 10^9

= 2,475,000,000

35. a. $x+2=3\sqrt{x}$, $(x+2)^2=(3\sqrt{x})^2$

$x^2+4x+4=9x$, $x^2-5x+4=0$,

$(x-4)(x-1)=0$,

$x-4=0$, $x=4$

$x-1=0$, $x=1$

$\{1, 4\}$

b. $x=\sqrt{6-x}$, $x^2=\left(\sqrt{6-x}\right)^2$,

$x^2=6-x$, $x^2+x-6=0$,

$(x+3)(x-2)=0$,

$x+3=0$, $x=-3$, principal square root can not be negative, extraneous solution

$x-2=0$, $x=2$

$\{2\}$

37. $W=nB^{1.6}$, $W=nB^{\frac{8}{5}}$, $\frac{W}{n}=B^{\frac{8}{5}}$,

$\left(B^{\frac{8}{5}}\right)^{\frac{5}{8}}=\left(\frac{W}{n}\right)^{\frac{5}{8}}$, $B=\left(\frac{W}{n}\right)^{\frac{5}{8}}$

Section 7.0

1. a. $\frac{1.5}{0.5} = 3, \frac{4.5}{1.5} = 3, \frac{13.5}{4.5} = 3$

Common ratio is 3

b. $a_n = 0.5 \cdot 3^{n-1}$

c. $y = 0.167 \cdot 3^x$

d. $0.5 \cdot 3^{n-1} = 0.5 \cdot 3^n \cdot 3^{-1}$

$= 0.5 \cdot \frac{1}{3} \cdot 3^n = \frac{0.5}{3} \cdot 3^n = 0.167 \cdot 3^n$

3. a. $\frac{20}{10} = 2, \frac{40}{20} = 2, \frac{80}{40} = 2$

Common ratio is 2

b. $a_n = 10 \cdot 2^{n-1}$

c. $y = 5 \cdot 2^x$

d. $10 \cdot 2^{n-1} = 10 \cdot 2^n \cdot 2^{-1}$

$= 10 \cdot \frac{1}{2} \cdot 2^n = \frac{10}{2} \cdot 2^n = 5 \cdot 2^n$

5. a. $\frac{\frac{1}{2}}{\frac{1}{4}} = \frac{1}{2} \cdot \frac{4}{1} = 2, \frac{1}{\frac{1}{2}} = 2, \frac{2}{1} = 2, \frac{4}{2} = 2$

Common ratio is 2

b. $a_n = \frac{1}{4} \cdot 2^{n-1}$

c. $y = 0.125 \cdot 2^x$

d. $\frac{1}{4} \cdot 2^{n-1} = \frac{1}{4} \cdot 2^n \cdot 2^{-1} = \frac{1}{4} \cdot \frac{1}{2} \cdot 2^n$

$= \frac{1}{8} \cdot 2^n = 0.125 \cdot 2^n$

7. a. $\frac{2}{1} = 2, \frac{4}{2} = 2, \frac{8}{4} = 2, \frac{16}{8} = 2$

Common ratio is 2

b. $a_n = 1 \cdot 2^{n-1}$

c. $y = 0.5 \cdot 2^x$

d. $1 \cdot 2^{n-1} = 1 \cdot 2^n \cdot 2^{-1} = 1 \cdot \frac{1}{2} \cdot 2^n$

$= 0.5 \cdot 2^n$

9. a. $\frac{16}{32} = \frac{1}{2}, \frac{8}{16} = \frac{1}{2}, \frac{4}{8} = \frac{1}{2}, \frac{2}{4} = \frac{1}{2}$

Common ratio is $\frac{1}{2}$

b. $a_n = 32 \cdot (\frac{1}{2})^{n-1}$

c. $y = 64 \cdot 0.5^x$

d. $32 \cdot (\frac{1}{2})^{n-1} = 32 \cdot (\frac{1}{2})^n \cdot (\frac{1}{2})^{-1}$

$= 32 \cdot 2 \cdot (\frac{1}{2})^n = 64 \cdot (\frac{1}{2})^n = 64 \cdot 0.5^n$

11 .a. $\frac{1}{3} = \frac{1}{3}, \frac{\frac{1}{3}}{1} = \frac{1}{3}, \frac{\frac{1}{9}}{\frac{1}{3}} = \frac{1}{9} \cdot \frac{3}{1} = \frac{1}{3},$

$\frac{\frac{1}{27}}{\frac{1}{9}} = \frac{1}{27} \cdot \frac{9}{1} = \frac{1}{3},$ Common ratio is $\frac{1}{3}$

b. $a_n = 3 \cdot (\frac{1}{3})^{n-1}$

c. $y \approx 9 \cdot 0.333^x$

d. $3 \cdot (\frac{1}{3})^{n-1} = 3 \cdot (\frac{1}{3})^n \cdot (\frac{1}{3})^{-1}$

$= 3 \cdot 3 \cdot (\frac{1}{3})^n = 9 \cdot (\frac{1}{3})^n \approx 9 \cdot 0.333^n$

Section 7.0 (con't)

13. No common ratio,

First differences: 6, 10, 14

Second differences: 4, 4

Quadratic function

$4 = 2a, \ a = 2$

$c = 2 - (6\text{-}4), \ c = 0$

$2 = 2(1^2) + b(1) + 0, \ b = 0$

$y = 2x^2$

15. Common ratio is 3, (see #1 above)

$y = 0.5 \cdot 3^{x-1}$ or $y = 0.167 \cdot 3^x$

17. No common ratio,

First differences: 4, 5, 6

Second differences: 1, 1

Quadratic function

$1 = 2a, \ a = 0.5$

$c = 3 - (4 - 1), \ c = 0$

$3 = 0.5(1^2) + b(1) + 0, \ b = 2.5$

$y = 0.5x^2 = 2.5x$

19. No common ratio,

First differences: 8, 8, 8, 8

Linear function

$y = 30 + (x - 1)8, \ y = 8x + 22$

21. Common ratio is 2

$y = 5 \cdot 2^{x-1}$ or $y = \frac{5}{2} \cdot 2^x, \ y = 2.5 \cdot 2^x$

23. $2^{x+2} = 2^x \cdot 2^2 = 2^x \cdot 4 = 4 \cdot 2^x$

25. $2^{x-1} = 2^x \cdot 2^{-1} = 2^x \cdot \frac{1}{2} = \dfrac{2^x}{2}$

27. $\left(\frac{1}{2}\right)^{-x} = 2^x$

29. $\dfrac{2^n}{2} = 2^n \cdot 2^{-1} = 2^{n-1}$

31. $4 \cdot 2^n = 2^2 \cdot 2^n = 2^{2+n} = 2^{n+2}$

33. $\dfrac{3^n}{3} = 3^n \cdot 3^{-1} = 3^{n-1}$

35. $9 \cdot 3^n = 3^2 \cdot 3^n = 3^{2+n} = 3^{n+2}$

37. a. Common ratio is 2,

$f(x) = \frac{1}{8} \cdot 2^{x-1}, \ \frac{1}{8} = 2^{-3},$
$f(x) = 2^{-3} \cdot 2^{x-1}, \ f(x) = 2^{x-1-3},$
$f(x) = 2^{x-4}$

b. Common ratio is 2,

$f(x) = 4 \cdot 2^{x-1}, \ f(x) = 2^2 \cdot 2^{x-1},$
$f(x) = 2^{x-1+2}, \ f(x) = 2^{x+1}$

Section 7.0 (con't)

39. a. Common ratio is $\frac{1}{2}$,

$f(x) = 4 \cdot (\frac{1}{2})^{x-1}$, $f(x) = 2^2 \cdot (\frac{1}{2})^{x-1}$,
$f(x) = (\frac{1}{2})^{-2} \cdot (\frac{1}{2})^{x-1}$, $f(x) = (\frac{1}{2})^{x-1-2}$,
$f(x) = (\frac{1}{2})^{x-3}$

b. Common ratio is $\frac{1}{2}$,

$f(x) = 8 \cdot (\frac{1}{2})^{x-1}$, $f(x) = 2^3 \cdot (\frac{1}{2})^{x-1}$,
$f(x) = (\frac{1}{2})^{-3} \cdot (\frac{1}{2})^{x-1}$, $f(x) = (\frac{1}{2})^{x-1-3}$,
$f(x) = (\frac{1}{2})^{x-4}$

41. Common ratio is -3,

$a_n = 2 \cdot (-3)^{n-1}$, Base is negative, this is not an exponential function

43. Common ratio is $-\frac{1}{2}$,

$a_n = 81 \cdot (-\frac{1}{2})^{n-1}$, Base is negative, this is not an exponential function

45. Common ratio is 2,

$a_n = 27.5 \cdot 2^{n-1}$

Section 7.1

1. a. In a^x the base is a, the exponent is x.

b. In $4x^3$ the base is x, the exponent is 3.

c. In 2^{-x} the base is 2, the exponent is -x.

d. In x^a the base is x, the exponent is a.

3. a. In π^x the base is π, the exponent is x.

b. In $2x^e$ the base is x, the exponent is *e*.

c. In a_1r^{n-1} the base is r, the exponent is n - 1.

d. In $100(1.06)^t$ the base is 1.06, the exponent is t.

5. a. As x gets larger $y = 5^x$ will get larger, the graph will be increasing.

b. As x gets larger $y = (\frac{3}{4})^x$ will get smaller, (*observe that the denominator will get increasing larger than the numerator)* the graph will be decreasing.

7. a. As x gets larger $y = (\frac{3}{2})^x$ will get larger, (*observe that the numerator will get increasing larger than the denominator*) the graph will be increasing.

b. As x gets larger $y = 0.5^x$ will get smaller (*base is between 0 and* 1), the graph will be decreasing.

9. a. As x gets larger, $y = (1 + 0.05)^x$ will get larger (*base is* > *1*), the graph will be increasing.

b. As x gets larger, $y = (\frac{4}{3})^x$ will get larger, the graph will be increasing.

11. Remember the equation with the largest base will increase the fastest. $a = 10^x$, $b = 2.72^x$, $c = 2^x$.

13. The graph of $y = 2^x$ is never below the x-axis. { }

15. There is no input that will result in a negative output. { }

17. $3 \cdot 2^0 = 3 \cdot 1 = 3$; $4 \cdot 2^0 = 4 \cdot 1 = 4$;

$\frac{1}{2} \cdot 2^0 = \frac{1}{2} \cdot 1 = \frac{1}{2}$;

the y-intercept of $y = a \cdot 2^x$ is a.

Section 7.1 (con't)

19. a. y-intercept is 16;

$y = 16 \cdot 2^x, y = 2^4 \cdot 2^x, y = 2^{x+4}$

b. y-intercept is 8;

$y = 8 \cdot 2^x, y = 2^3 \cdot 2^x, y = 2^{x+3}$

21. a. y-intercept is 27;

$y = 27 \cdot 3^x, y = 3^3 \cdot 3^x, y = 3^{x+3}$

b. y-intercept is $\frac{1}{27}$;

$y = \dfrac{3^x}{27}, y = 3^{-3} \cdot 3^x, y = 3^{x-3}$

23. a. $y = 2^{x-4}, y = 2^{-4} \cdot 2^x, y = \dfrac{2^x}{16}$;

y-intercept is $\frac{1}{16}$

b. $y = 2^{x+1}, y = 2^1 \cdot 2^x, y = 2 \cdot 2^x$;

y-intercept is 2

25. The coefficient a is the y-intercept value.

27. The graph passes through (1, b) because when $x = 1$, $b^x = b$.

29. One possible answer is $y = x^{-2}$

31.

x	$y = 2^{x-1}$	$y = 2^x$	$y = 2^{x+1}$
-2	$2^{-2-1} = \frac{1}{8}$	$2^{-2} = \frac{1}{4}$	$2^{-2+1} = \frac{1}{2}$
-1	$2^{-1-1} = \frac{1}{4}$	$2^{-1} = \frac{1}{2}$	$2^{-1+1} = 1$
0	$2^{0-1} = \frac{1}{2}$	$2^0 = 1$	$2^{0+1} = 2$
1	$2^{1-1} = 1$	$2^1 = 2$	$2^{1+1} = 4$
2	$2^{2-1} = 2$	$2^2 = 4$	$2^{2+1} = 8$
	graph c	graph b	graph a

Adding 1 in the exponent shifts the graph one unit left, subtracting 1 in the exponent shifts the graph one unit right.

33.

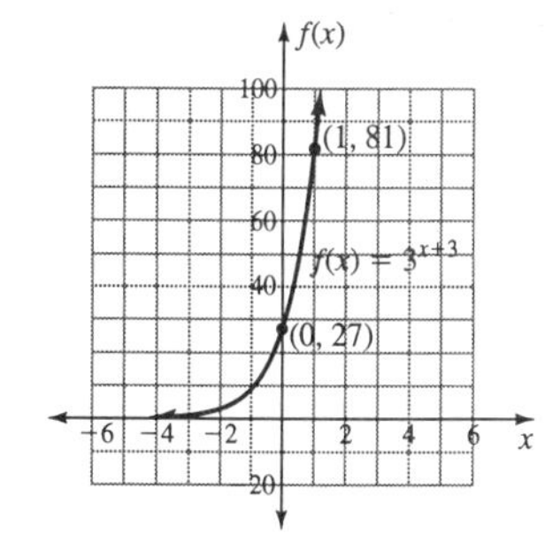

35.

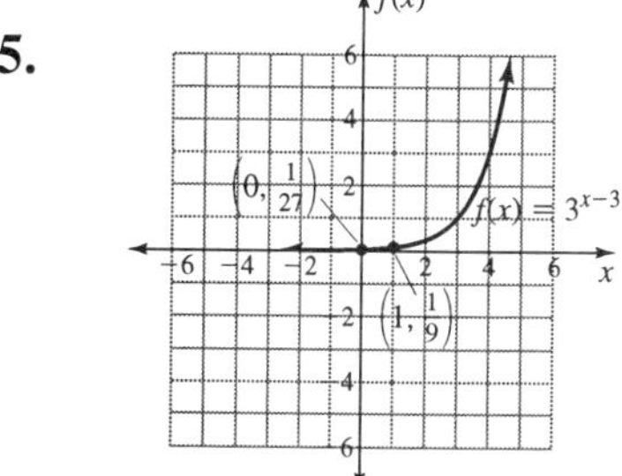

37. The graphs of 3^x and 3^{x+1} never intersect. There is no value of x that will make 3^x equal 3^{x+1}.

Section 7.1 (con't)

39. a. When $a^x = 2$, $x \approx 2.4$

b. When $a^x = 4$, $x \approx 4.8$

c. $a^2 \approx 1.8$

d. $a^5 \approx 4.3$

e. $a \approx 1.34$

41. a. Graph should be a horizontal line at $y = 1$.

c. Graph is a horizontal line at $y = 1$

d. The function is defined for all inputs.

e. The range is $y = 1$

43. b. Graph is a series of points

c. The equation is not defined for all inputs.

d. The equation is not an exponential function, the base is negative.

Section 7.2

1. a. $2^x = 256,\ 2^x = 2^8,\ x = 8$

b. $16^x = 256,\ (2^4)^x = 2^8,\ 2^{4x} = 2^8,$

$4x = 8,\ x = 2$

3. a. $4^n = 8,\ (2^2)^n = 2^3,\ 2^{2n} = 2^3,$

$2n = 3,\ n = \frac{3}{2}$

b. $8^n = 16,\ (2^3)^n = 2^4,\ 2^{3n} = 2^4,$

$3n = 4,\ n = \frac{4}{3}$

5. a. $16^n = 4,\ (2^4)^n = 2^2,\ 2^{4n} = 2^2,$

$4n = 2,\ n = \frac{1}{2}$

b. $100^x = 1000,\ (10^2)^x = 10^3,\ 10^{2x} = 10^3,$

$2x = 3,\ x = \frac{3}{2}$

7. a. $64^x = \frac{1}{8},\ (2^6)^x = 2^{-3},\ 2^{6x} = 2^{-3},$

$6x = -3,\ x = -\frac{1}{2}$

b. $125^n = \frac{1}{5},\ (5^3)^n = 5^{-1},\ 5^{3n} = 5^{-1}$

$3n = -1,\ n = -\frac{1}{3}$

9. a. $8^n = \frac{1}{2},\ (2^3)^n = 2^{-1},\ 2^{3n} = 2^{-1},$

$3n = -1,\ n = -\frac{1}{3}$

b. $81^x = \frac{1}{3},\ (3^4)^x = 3^{-1},\ 3^{4x} = 3^{-1},$

$4x = -1,\ x = -\frac{1}{4}$

11. a. $\left(\frac{1}{2}\right)^{x+5} = 32,\ (2^{-1})^{x+5} = 2^5,\ 2^{-x-5} = 2^5$

$-x - 5 = 5,\ -x = 10,\ x = -10$

b. $\left(\frac{1}{8}\right)^{x+3} = 8,\ (8^{-1})^{x+3} = 8^1,\ 8^{-x-3} = 8^1,$

$-x - 3 = 1,\ -x = 4,\ x = -4$

13. a. $2^{x-1} = 2^{2x+5},\ x - 1 = 2x + 5,\ x = -6$

b. $3^{2x-1} = 3^{x+3},\ 2x - 1 = x + 3,\ x = 4$

15. a. $2^{x+5} = 4^{x-1},\ 2^{x+5} = (2^2)^{x-1},$

$2^{x+5} = 2^{2x-2},\ x + 5 = 2x - 2,\ x = 7$

b. $5^{x+1} = 125^{x-3},\ 5^{x+1} = (5^3)^{x-3},$

$5^{x+1} = 5^{3x-9},\ x + 1 = 3x - 9,\ 2x = 10,$

$x = 5$

Section 7.2 (con't)

17. a.

x	$y = 3^x$
-2	$3^{-2} = \frac{1}{9}$
-1	$3^{-1} = \frac{1}{3}$
0	$3^0 = 1$
1	$3^1 = 3$
2	$3^2 = 9$
3	$3^3 = 27$
4	$3^4 = 81$

b.

Input, Inverse	Output, Inverse
$\frac{1}{9}$	-2
$\frac{1}{3}$	-1
1	0
3	1
9	2
27	3
81	4

17. c.

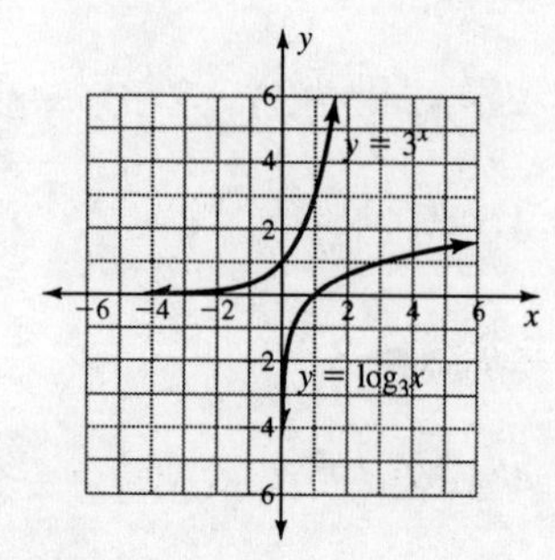

d. In $y = 3^x$, the base is 3, the function can be written $x = \log_3 y$, the inverse would be $y = \log_3 x$

19. The set of outputs from a function is its *range.*

21. The domain of a function is the *range* of its inverse.

23. If the ordered pair (a, b) belongs to a function, then the ordered pair *(b, a)* belongs to its inverse.

25.

Exponential Equation	Logarithmic Equation
$3^3 = 27$	$\log_3 27 = 3$
$2^3 = 8$	$\log_2 8 = 3$
$10^1 = 10$	$\log_{10} 10 = 1$
$5^3 = 125$	$\log_5 125 = 3$
$3^{-2} = \frac{1}{9}$	$\log_3 (\frac{1}{9}) = -2$

Section 7.2 (con't)

27. a. $\log_{10} 1000 = 3$ is $10^3 = 1000$

b. $\log_{10} 1 = 0$ is $10^0 = 1$

c. $\log_3 81 = 4$ is $3^4 = 81$

d. $\log_{10} 100 = 2$ is $10^2 = 100$

29. a. $\log_{10} 0.001 = -3$ is $10^{-3} = 0.001$

b. $\log_4 1 = 0$ is $4^0 = 1$

c. $\log_m n = k$ is $m^k = n$

d. $\log_5(\frac{1}{25}) = -2$ is $5^{-2} = \frac{1}{25}$

31. a. $2^5 = 32$ is $\log_2 32 = 5$

b. $2^1 = 2$ is $\log_2 2 = 1$

c. $2^0 = 1$ is $\log_2 1 = 0$

d. $10^2 = 100$ is $\log_{10} 100 = 2$

33. a. $10^{-3} = 0.001$ is $\log_{10} 0.001 = -3$

b. $f^d = g$ is $\log_f g = d$

c. $3^{-2} = \frac{1}{9}$ is $\log_3 (\frac{1}{9}) = -2$

d. $4^0 = 1$ is $\log_4 1 = 0$

35. a. $\log_{10} 1 = 0$

b. $\log_{10} 6 \approx 0.77815$

c. $\log_{10} 10 = 1$

d. $\log_{10} 60 \approx 1.77815$

e. $\log_{10} 100 = 2$

f. $\log_{10} 600 \approx 2.77815$

37. a. $\log_{10} 0.01 = -2$

b. $\log_{10} 0.1 = -1$

c. $\log_{10} 0.6 \approx -0.22185$

d. $\log_{10} 0.06 \approx -1.22185$

e. $\log_{10} 0.006 \approx -3.22185$

f. $\log_{10} 0.000006 \approx -5.22185$

39. a. The logs of 6, 60 and 6000 have the same numbers to the right of the decimal place and they are all positive numbers

b. The logs of 0.6, 0.006 and 0.006 have the same numbers to the right of the decimal place and they are all negative numbers.

41. a. $17 = 10^x$, $\log_{10} 17 = x$, $x \approx 1.23045$

b. $125 = 10^x$, $\log_{10} 125 = x$, $x \approx 2.09691$

c. $10^x = 400$, $\log_{10} 400 = x$, $x \approx 2.60206$

d. $10^x = 0.05$, $\log_{10} 0.05 = x$,

$x \approx -1.30102$

43.

Exponential Equation	Logarithmic Equation	Solve for x
$5^0 = x$	$\log_5 x = 0$	$x = 1$
$x^3 = 64$	$\log_x 64 = 3$	$x = 4$
$10^x = -10$	$\log_{10} (-10) = x$	{ }

Section 7.2 (con't)

45. a. $\log_7 x = 2,\ 7^2 = x,\ x = 49$

b. $\log_3 3 = x,\ 3^x = 3,\ x = 1$

c. $\log_{10} 0.01 = x,\ 10^x = 0.01,$

$10^x = 10^{-2},\ x = -2$

d. $\log_{10} x = 0,\ 10^0 = x,\ x = 1$

e. $\log_x 64 = 3,\ x^3 = 64,\ x^3 = 4^3,\ x = 4$

f. $\log_a x = 1,\ a^1 = x,\ x = a$

47. a. $\log_2 x = 8,\ 2^8 = x,\ x = 256$

b. $\log_{10} x = -2,\ 10^{-2} = x,\ x = 0.01$

c. $\log_2 (\frac{1}{4}) = x,\ 2^x = \frac{1}{4},\ 2^x = 2^{-2},\ x = -2$

d. $\log_{10} 1 = x,\ 10^x = 1,\ x = 0$

e. $\log_x 100 = 2,\ x^2 = 100,\ x^2 = 10^2,$

$x = 10$

f. $\log_a a = x,\ a^x = a,\ x = 1$

49. The line of symmetry for the graph of a function and its inverse is $y = x$.

51. The logarithm is the *exponent* to which we raise a base to obtain a given number.

53. The *base* of a logarithmic equation must be a positive number not equal to 1.

55. The set of outputs (range) to a logarithmic equation, $y = \log_b x$, is □ .

57. The set of outputs (range) to an exponential equation , $y = b^x$, is $y > 0$.

59. In the equation the denominator is missing the input value for the logarithm operation.

61. Not necessarily, in $x^2 = 3^2$, x could be -3.

63. $4^{6t} = (2^2)^{6t} = 2^{12t},\ 8^{4t} = (2^3)^{4t} = 2^{12t}$

The rules are the same.

Mid Chapter 7 Test

1. a. Common ratio of 3,

Geometric sequence,

$a_n = 3 \cdot 3^{n-1}$, $f(x) = 3^x$,

$3 \cdot 3^{n-1} = 3^1 \cdot 3^{n-1} = 3^{1+n-1} = 3^n$

b. First differences: 6, 6, 6, 6

Arithmetic sequence,

$y = -1 + (x - 1)6$, $y = 6x - 7$

c. Common ratio of 2,

Geometric sequence,

$a_n = 3 \cdot 2^{n-1}$, $f(x) = 1.5 \cdot 2^x$,

$3 \cdot 2^{n-1} = 3 \cdot 2^{-1} \cdot 2^n = \frac{3}{2} \cdot 2^n = 1.5 \cdot 2^n$

d. Common ratio of $\frac{1}{3}$,

Geometric sequence,

$a_n = 81 \cdot (\frac{1}{3})^{n-1}$, $f(x) = 243 \cdot (\frac{1}{3})^x$,

$81 \cdot (\frac{1}{3})^{n-1} = 81 \cdot (\frac{1}{3})^{-1} \cdot (\frac{1}{3})^n$

$= 81 \cdot 3 \cdot (\frac{1}{3})^n = 243 \cdot (\frac{1}{3})^n$

e. First differences: 5, 9, 13, 17

Second differences: 4, 4, 4

Quadratic sequence,

$2a = 4$, $a = 2$; $c = 1 - (5 - 4)$, $c = 0$;

$1 = 2(1)^2 + b(1) + 0$, $b = -1$;

$y = 2x^2 - x$

1. f. Common ratio of 2,

Geometric sequence,

$a_n = \frac{1}{8} \cdot 2^{n-1}$, $f(x) = 0.0625 \cdot 2^x$,

$\frac{1}{8} \cdot 2^{n-1} = \frac{1}{8} \cdot 2^{-1} \cdot 2^n = \frac{1}{8} \cdot \frac{1}{2} \cdot 2^n$

$= \frac{1}{16} \cdot 2^n = 0.0625 \cdot 2^n$

2.

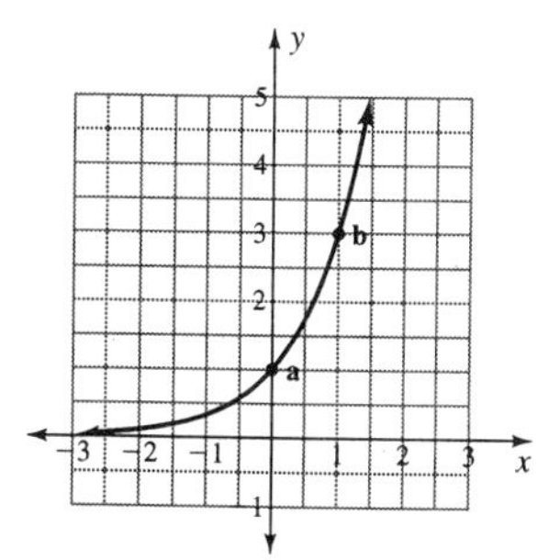

3. a. The graphs both approach the x-axis on the left and rise rapidly as we move to the right.

b. The graph of $f(x) = 2^{x-1}$ is $f(x) = 2^x$ shifted 1 unit to the right.

c. The graphs do not intersect, $2^x \neq 2^{x-1}$

4. a. $x \approx 3.3$

b. $2^x = 10$ or $\log_2 10 = x$

5. $y = \frac{1}{2} \cdot 3^x$ grows rapidly as x gets larger,

$y = 2 \cdot 3^{-x}$ approaches zero as x gets larger.

a. The constants $\frac{1}{2}$ and 2 show the y-intercepts.

b. The negative exponent indicates a decreasing function when $b > 0$.

Mid Chapter 7 Test (con't)

6. a. $2^x = 32,\ 2^x = 2^5, x = 5$

b. $(\frac{1}{2})^x = 4, (2^{-1})^x = 2^2, 2^{-x} = 2^2,$

$-x = 2,\ x = -2$

c. $(\frac{1}{2})^x = \frac{1}{16}, (2^{-1})^x = 2^{-4}, 2^{-x} = 2^{-4},$

$-x = -4,\ x = 4$

d. $5^x = \frac{1}{125}, 5^x = 5^{-3}, x = -3$

e. $10^x = -10$, { }

f. $10^x = 0.01, 10^x = 10^{-2}, x = -2$

7. a. $\log_4 x = -1, 4^{-1} = x,\ x = \frac{1}{4}$

b. $\log_{10} 0.0001 = x,\ 10^x = 0.0001,$

$10^x = 10^{-4}, x = -4$

c. $\log_3 x = 3,\ 3^3 = x, x = 27$

d. $\log_2 x = 0,\ 2^0 = x,\ x = 1$

e. $\log_4 2 = x,\ 4^x = 2,\ (2^2)^x = 2^1,$

$2^{2x} = 2^1, 2x = 1, x = \frac{1}{2}$

f. $\log_4 -2 = x$, { }

8. a.

x	$y = 4^x$
-2	$4^{-2} = \frac{1}{16}$
-1	$4^{-1} = \frac{1}{4}$
0	$4^0 = 1$
1	$4^1 = 4$
2	$4^2 = 16$
3	$4^3 = 64$

b.

x, inverse	y, inverse
$\frac{1}{16}$	-2
$\frac{1}{4}$	-1
1	0
4	1
16	2
64	3

c.

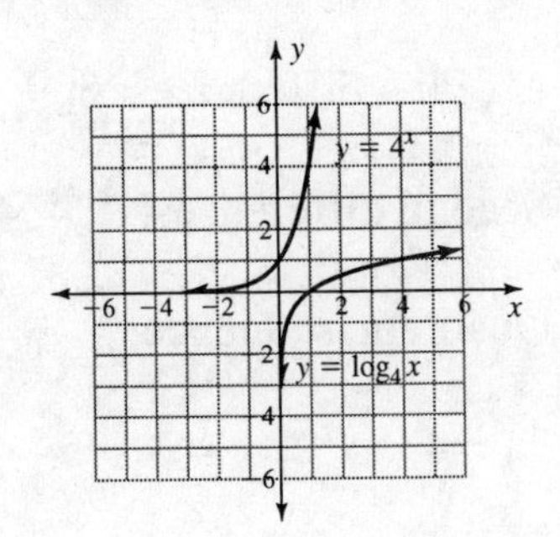

d. $y = 4^x$ is $\log_4 y = x$, inverse is $y = \log_4 x$

Mid Chapter 7 Test (con't)

9.

Exponential Equation	Logarithmic Equation	Solve for x
$10^x = 15$	$\log_{10} 15 = x$	$x \approx 1.17609$
$10^x = 13$	$\log_{10} 13 = x$	$x \approx 1.11394$
$3^x = 24$ *graph this*	$\log_3 24 = x$	$x \approx 2.893$ *trace graph*
$3^9 = x$	$\log_3 x = 9$	$x = 19{,}683$
4^{x+1} *graph this*	$\log_4 50 = x+1$	$x \approx 1.822$ *trace graph*

Section 7.3

1. a. The base is 3.

b. The base is 10.

c. The base is 4.

d. The base is 5.

e. The base is 10.

f. The base is m.

3. a. $pH = -\log(1.0 \times 10^{-14})$, $pH = 14$

b. $pH = -\log(0.20)$, $pH \approx 0.7$

c. $pH = -\log(5.01 \times 10^{-3})$, $pH \approx 2.3$

d. $pH = -\log(1.00 \times 10^{-10})$, $pH = 10$

e. $pH = -\log(1.26 \times 10^{-3})$, $pH \approx 2.9$

5. a. base

b. acid

c. acid

d. base

e. acid

7. a. $4.1 = -\log[H^+]$, $-4.1 = \log[H^+]$,

$10^{-4.1} = [H^+]$, $[H^+] \approx 7.94 \times 10^{-5}$ M

b. $-0.5 = -\log[H^+]$, $0.5 = \log[H^+]$,

$10^{0.5} = [H^+]$, $[H^+] \approx 3.16$ M

c. $12.0 = -\log[H^+]$, $-12.0 = \log[H^+]$,

$10^{-12} = [H^+]$, $[H^+] = 1.0 \times 10^{-12}$ M

7. d. $3.5 = -\log[H^+]$, $-3.5 = \log[H^+]$,

$10^{-3.5} = [H^+]$, $[H^+] \approx 3.16 \times 10^{-4}$ M

9. pH is negative when $[H^+] > 1$; pH is positive when $[H^+] < 1$.

11. a. $I = 3.2623 \times 10^7$,

$R = \log(3.2623 \times 10^7)$, $R \approx 7.5$

b. $I = 1.2589 \times 10^8$,

$R = \log(1.2589 \times 10^8)$, $R \approx 8.1$

13. a. $8.6 = \log I$, $10^{8.6} = I$,

$I \approx 398{,}107{,}000$

b. $6.8 = \log I$, $10^{6.8} = I$,

$I \approx 6{,}310{,}000$

15. $6.9 = \log I$, $10^{6.9} = I$, $I \approx 7{,}943{,}000$

$6.8 = \log I$, $10^{6.8} = I$, $I \approx 6{,}310{,}000$

$$\frac{7{,}943{,}000}{6{,}310{,}000} \approx 1.3 \text{ times as strong}$$

17. a. $2^x = 1$, $x = 0$

b. $2^x = 3$, $x = \log_2 3$,

$$\log_2 3 = \frac{\log 3}{\log 2},\ x \approx 1.58$$

c. $2^x = 4$, $2^x = 2^2$, $x = 2$

d. $2^x = 5$, $x = \log_2 5$,

$$\log_2 5 = \frac{\log 5}{\log 2},\ x \approx 2.32$$

Section 7.3 (con't)

17. e. $2^x = 6$, $x = \log_2 6$,

$$\log_2 6 = \frac{\log 6}{\log 2},\ x \approx 2.58$$

f. $2^x = 10$, $x = \log_2 10$,

$$\log_2 10 = \frac{\log 10}{\log 2},\ x \approx 3.32$$

g. $2^x = 12$, $x = \log_2 12$,

$$\log_2 12 = \frac{\log 12}{\log 2},\ x \approx 3.58$$

h. $2^x = 20$, $x = \log_2 20$,

$$\log_2 20 = \frac{\log 20}{\log 2},\ x \approx 4.32$$

i. $2^x = 24$, $x = \log_2 24$,

$$\log_2 24 = \frac{\log 24}{\log 2},\ x \approx 4.58$$

19. a. $5^x = 2$, $x = \log_5 2$,

$$\log_5 2 = \frac{\log 2}{\log 5},\ x \approx 0.43$$

b. $5^x = 5$, $x = 1$

c. $5^x = 6$, $x = \log_5 6$,

$$\log_5 6 = \frac{\log 6}{\log 5},\ x \approx 1.11$$

d. $5^x = 10$, $x = \log_5 10$,

$$\log_5 10 = \frac{\log 10}{\log 5},\ x \approx 1.43$$

e. $5^x = 25$, $5^x = 5^2$, $x = 2$

19. f. $5^x = 30$, $x = \log_5 30$,

$$\log_5 30 = \frac{\log 30}{\log 5},\ x \approx 2.11$$

21. a. $3^x = 10$, $x = \log_3 10$,

$$\log_3 10 = \frac{\log 10}{\log 3},\ x \approx 2.09590$$

b. $x = \log_4 6$, $\log_4 6 = \frac{\log 6}{\log 4}$,

$x \approx 1.29248$

c. $20 = 10(1.07)^x$, $2 = 1.07^x$,

$x = \log_{1.07} 2$, $\log_{1.07} 2 = \frac{\log 2}{\log 1.07}$,

$x \approx 10.24477$

d. $15 = 30(0.95)^x$, $0.5 = 0.95^x$,

$x = \log_{0.95} 0.5$, $\log_{0.95} 0.5 = \frac{\log 0.5}{\log 0.95}$,

$x \approx 13.51341$

e. $30 = 60(0.5)^x$, $0.5 = 0.5^x$, $x = 1$

23. x-intercept is 1, $f(1) = 0$

25. The x-intercept on both functions is 1,

$y = \log_6 x$ is below $y = \log_3 x$ to the right of the x-intercept and above to the left of the x-intercept.

27. False

29. False, $b \neq \log_b 1$

Section 7.3 (con't)

31. Graphs y_2 and y_3 mirror each other across $y = x$. y_2 and y_3 are inverse functions. $1 = 2^0$ is the same fact as $0 = \log_2 1$.

33. $1{,}000{,}000 = 1000(1 + 0.02)^t$

$1{,}000 = 1.02^t$, $t = \log_{1.02} 1000$,

$\log_{1.02} 1000 = \dfrac{\log 1000}{\log 1.02}$, $t \approx 349$ yrs

35. $1{,}000{,}000 = 1000(1 + 0.10)^t$,

$1000 = 1.10^t$, $t = \log_{1.10} 1000$

$\log_{1.10} 1000 = \dfrac{\log 1000}{\log 1.10}$, $t \approx 73$ yrs

37. a. Rate is 8%

b. Rate is 6%

39. a. Rate is -3%

b. $1 - r = 0.95$, $r = -0.05$

Rate is -5%

41. $2c = c(1.06)^t$, $2 = 1.06^t$, $t = \log_{1.06} 2$,

$\log_{1.06} 2 = \dfrac{\log 2}{\log 1.06}$, $t \approx 12$ yrs

43. $72 \div 1 = 72$ yrs, $72 \div 2 = 36$ yrs,

$72 \div 5 = 14.4$ yrs, $72 \div 8 = 9$ yrs,

$72 \div 10 = 7.2$ yrs, $72 \div 20 = 3.6$ yrs.

45. a. $72 \div 17 \approx 4.2\%$

b. $72 \div 34 \approx 2.1\%$

47. $15{,}000 = 30{,}000(1 - 0.10)^t$, $0.5 = 0.90^t$,

$t = \log_{0.90} 0.5$, $t = \dfrac{\log 0.5}{\log 0.90}$, $t \approx 6.58$ yrs

49. a. $0.5 = \left(\frac{2}{3}\right)^{n-1}$, $\log_{\frac{2}{3}} 0.5 = n - 1$,

$\dfrac{\log 0.5}{\log \frac{2}{3}} + 1 = n$, $n \approx 2.7$

b. $32 = 108\left(\frac{2}{3}\right)^{n-1}$, $\frac{32}{108} = \left(\frac{2}{3}\right)^{n-1}$,

$n - 1 = \log_{\frac{2}{3}} \frac{32}{108}$, $n = 1 + \dfrac{\log \frac{32}{108}}{\log \frac{2}{3}}$, $n = 4$

c. $1 = 108\left(\frac{2}{3}\right)^{n-1}$, $\frac{1}{108} = \left(\frac{2}{3}\right)^{n-1}$,

$n - 1 = \log_{\frac{2}{3}} \frac{1}{108}$, $n = 1 + \dfrac{\log \frac{1}{108}}{\log \frac{2}{3}}$, $n \approx 12.5$

51. $2P = P(1 + r)^t$, $2 = (1 + r)^t$, $t = \log_{(1+r)} 2$,

$t = \dfrac{\log 2}{\log(1+r)}$, P does not appear in the formula, doubling time is not dependent on the initial amount.

53. a. $\dfrac{1 \text{ half-life}}{26.3 \text{ hrs}} \cdot \dfrac{24 \text{ hrs}}{1 \text{ day}} \cdot \dfrac{365\text{days}}{1 \text{ yr}}$

≈ 333 half-lives in 1 year

b. $y \approx 2{,}500{,}000(0.974)^x$, x = hours

55. a. $\dfrac{1 \text{ half-life}}{14.3 \text{ days}} \cdot \dfrac{365 \text{ days}}{1 \text{ yr}}$

≈ 26 half-lives in 1 year

b. $y \approx 230{,}000\ (0.953)^x$, x = days

Section 7.3 (con't)

57. a. $\frac{1 \text{ half-life}}{15 \text{ hrs}} \cdot \frac{24 \text{ hr}}{1 \text{ day}} \cdot \frac{365 \text{ days}}{1 \text{ yr}}$

≈ 584 half-lives in 1 year

b. $y \approx 12{,}000{,}000(0.955)^x$, x = hours

59. a. Estimates will vary

b. Using x = years, y = cost and letting 1986 be x = 0 point; $y \approx 0.054x + 0.5$

Slope is the change in cost per year.

c. $y \approx 0.5(1.079)^x$, y-intercept is 0.5, the cost in 1986,

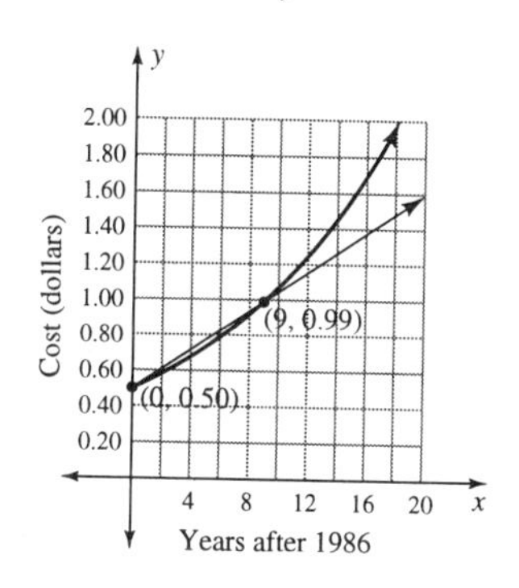

d. Rate of inflation is 7.9%

e. We assumed the change in price or percent increase each year was constant.

f. $y \approx 0.0544(19) + 0.5$, $y \approx \$1.53$

$y \approx 0.5(1.079)^{19}$, $y \approx \$2.12$

61. $y \approx 24{,}780(1.049)^x$, rate = 4.9%

$y \approx 24{,}780(1.049)^{70}$, $y \approx 705{,}000$

63. $y \approx 42{,}738(0.998)^x$, rate = -0.2%

$y \approx 42{,}738(0.998)^{70}$, $y \approx 37{,}000$

65. a. $y \approx 25{,}600(0.87)^x$

b. $y \approx 50(0.87)^x$

c. $y \approx 50(0.5)^x$

In part a and b the x values are in 5-unit increments.

Section 7.4

1. a. $\log 2 + \log x = \log 2x$

b. $\log x^2 = 2\log x$

c. $\log\left(\frac{2}{x}\right) = \log 2 - \log x$

d. $\log 2^x = x\log 2$

e. $\log x - \log 2 = \log\left(\frac{x}{2}\right)$

3. Property 1, $\log(2 \cdot 3) = \log 2 + \log 3$

5. Property 2, $\log 3^2 = 2\log 3$

7. $\log\sqrt{x} = \log x^{\frac{1}{2}} = \frac{1}{2}\log x$

9. $x = \log_b m$ and $y = \log_b n$

$m = b^x$ and $n = b^y$

$\frac{m}{n} = \frac{b^x}{b^y} = b^{x-y}$

$\log_b\left(\frac{m}{n}\right) = x - y = \log_b m - \log_b n$

11. a. $\log(x + 1) + \log(x - 1)$

$= \log[(x + 1)(x - 1)] = \log(x^2 - 1)$

b. $\log(x^2 - 1) - \log(x - 1) = \log\left(\frac{x^2 - 1}{x - 1}\right)$

$\log\left(\frac{(x+1)(x-1)}{(x-1)}\right) = \log(x+1)$

13. a. $\log(x^2 + 3x + 2) = \log[(x + 2)(x + 1)]$

$\log[(x + 2)(x + 1)] - \log(x + 1)$

$= \log\left(\frac{(x+2)(x+1)}{(x+1)}\right) = \log(x+2)$

13. b. $\log(x + 3) + \log(x - 2)$

$= \log[(x + 3)(x - 2)] = \log(x^2 + x - 6)$

15. $\log_2 x^2 + \log_2 x = \log_2(x^2 \cdot x) = \log_2 x^3$

$= 3\log_2 x;\ 3\log_2 x = 6, \log_2 x = 2,$

$2^2 = x, x = 4$

17. $\log x + \log x = \log x^2 = 2\log x;$

$2\log x = 2, \log x = 1, 10^1 = x, x = 10$

19. $\log(x^2 + x) - \log(x + 1)$

$= \log[x(x + 1)] - \log(x + 1)$

$= \log\left(\frac{x(x+1)}{(x+1)}\right) = \log x$

$\log x = 1, 10^1 = x, x = 10$

21. $\log_3 x^3 - \log_3 x = \log_3\left(\frac{x^3}{x}\right) = \log_3 x^2$

$= 2\log_3 x, 2\log_3 x = 4, \log_3 x = 2,$

$3^2 = x, x = 9$

23. a. $5^x = 20, \log 5^x = \log 20,$

$x\log 5 = \log 20, x = \frac{\log 20}{\log 5}, x \approx 1.8614$

b. $5^x = 20, \log_5 20 = x,$

$x = \frac{\log 20}{\log 5}, x \approx 1.8614$

Section 7.4 (con't)

25. a. $3^{x+2} = 48$, $\log 3^{x+2} = \log 48$,

$(x + 2)\log 3 = \log 48$, $x+2 = \dfrac{\log 48}{\log 3}$,

$x = \dfrac{\log 48}{\log 3} - 2$, $x \approx 1.5237$

b. $\log_3 48 = x + 2$, $\dfrac{\log 48}{\log 3} = x + 2$

$x = \dfrac{\log 48}{\log 3} - 2$, $x \approx 1.5237$

27. a. $4^{x-1} = 28$, $\log 4^{x-1} = \log 28$,

$(x - 1)\log 4 = \log 28$, $x - 1 = \dfrac{\log 28}{\log 4}$,

$x = \dfrac{\log 28}{\log 4} + 1$, $x \approx 3.4037$

b. $\log_4 28 = x - 1$, $\dfrac{\log 28}{\log 4} = x - 1$,

$x = \dfrac{\log 28}{\log 4} + 1$, $x \approx 3.4037$

29. a. $9^{x+1} = 42$, $\log 9^{x+1} = \log 42$,

$(x + 1)\log 9 = \log 42$, $x + 1 = \dfrac{\log 42}{\log 9}$

$x = \dfrac{\log 42}{\log 9} - 1$, $x \approx 0.7011$

b. $\log_9 42 = x + 1$, $\dfrac{\log 42}{\log 9} = x + 1$,

$x = \dfrac{\log 42}{\log 9} - 1$, $x \approx 0.7011$

31. a. $1000(1.08)^t = 2000$,

$\log[1000(1.08)^t] = \log 2000$,

$\log 1000 + \log(1.08)^t = \log 2000$,

$t\log 1.08 = \log 2000 - \log 1000$,

$t = \dfrac{\log\left(\frac{2000}{1000}\right)}{\log 1.08}$, $t \approx 9.0065$

b. $1.08^t = 2$, $\log_{1.08} 2 = t$,

$t = \dfrac{\log 2}{\log 1.08}$, $t \approx 9.0065$

33. a. *Although problem 31 demonstrates it is not necessary to divide by the constant before taking the log of both sides, it does make the solution easier.*

$15(1.07)^t = 45$, $1.07^t = 3$,

$\log (1.07)^t = \log 3$, $t\log 1.07 = \log 3$,

$t = \dfrac{\log 3}{\log 1.07}$, $t \approx 16.2376$

b. $\log_{1.07} 3 = t$, $t = \dfrac{\log 3}{\log 1.07}$, $t \approx 16.2376$

35. a. $1000(1 - 0.08)^t = 500$,

$1000(0.92)^t = 500$, $0.92^t = 0.5$,

$\log 0.92^t = \log 0.5$, $t\log 0.92 = \log 0.5$,

$t = \dfrac{\log 0.5}{\log 0.92}$, $t \approx 8.3130$

Section 7.4 (con't)

35. b. $\log_{0.92} 0.5 = t,\ t = \dfrac{\log 0.5}{\log 0.92},$

$t \approx 8.3130$

37. a. $45(1 - 0.07)^t = 15,\ 45(0.93)^t = 15,$

$0.93^t = \frac{1}{3},\ \log 0.93^t = \log \frac{1}{3},$

$t\log 0.93 = \log \frac{1}{3},\ t = \dfrac{\log \frac{1}{3}}{\log 0.93},$

$t \approx 15.1385$

b. $\log_{0.93} \frac{1}{3} = t,\ \ t = \dfrac{\log \frac{1}{3}}{\log 0.93},$

$t \approx 15.1385$

39. $\log \dfrac{1}{[H^+]} = \log[H^+]^{-1} = -\log[H^+]$

formulas are the same

41.

x	$y = 3^x$	x	$y = 3^x$
-1	$3^{-1} = \frac{1}{3}$	3	$3^3 = 27$
0	$3^0 = 1$	4	$3^4 = 81$
1	$3^1 = 3$	5	$3^5 = 243$
2	$3^2 = 9$	6	$3^6 = 729$

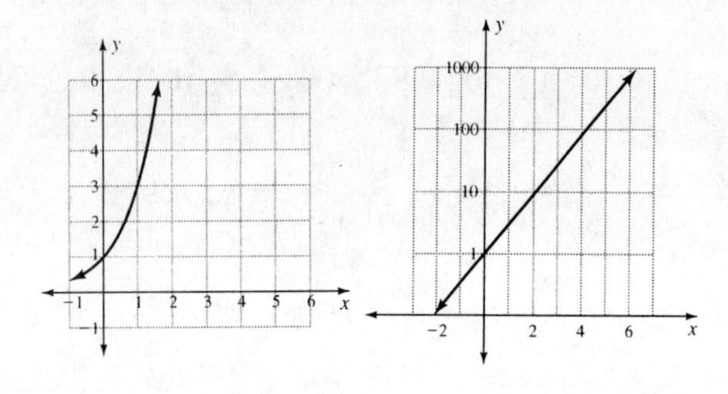

43. The exponential data is a straight line on semilog graph paper.

45. Approximately 22 feet

47. Approximately 70 years

49. Loss is more rapid between year 2 and year 6. Breakage due to wind is one explanation.

51.

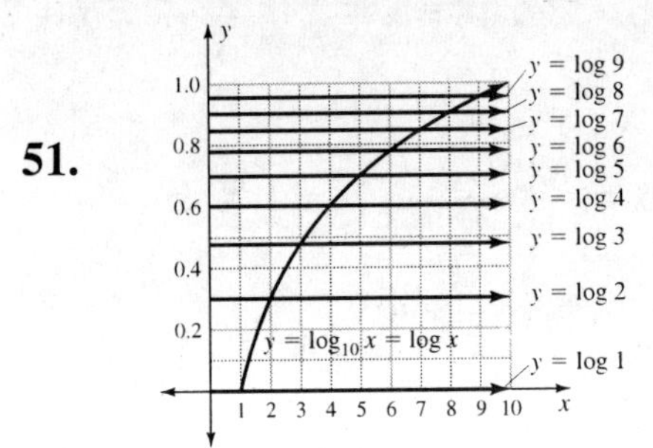

53. a. $y \approx 52.2(0.95)^x$

b. As the water gets deeper there is less light penetration.

c. $\log_{10} y \approx -0.02x + 1.72$

d. $y \approx 10^{-0.02x + 1.72},\ y \approx 10^{-0.02x} \cdot 10^{1.72},$

$y \approx (10^{-0.02})^x \cdot 10^{1.72},\ \ y \approx (0.95)^x \cdot 52.5,$

$y \approx 52.5(0.95)^x$, *error due to rounding*

55. a. $y = 0.0013(0.881)^x$

b. Air density decreases as altitude increases.

c. $\log y = -0.055x - 2.882$

d. $y \approx 10^{-0.055x - 2.882},$

$y \approx (10^{-0.055})^x(10^{-2.882}),$

$y \approx (0.881)^x(0.0013),$

Section 7.5

1. Annually: $1000(1+\frac{0.04}{1})^{3(1)} \approx \$1124.86$,

Monthly: $1000(1+\frac{0.04}{12})^{3(12)} \approx \$1127.27$,

Daily: $1000(1+\frac{0.04}{365})^{3(365)} \approx \$1127.49$

Monthly gains \$2.41 over annually,

daily gains \$2.63 over annually.

3. a. $e^2 \approx 7.39$

b. $e^{\pi} \approx 23.14$

c. $2\sqrt{e} \approx 3.30$

5. a. $3^e \approx 19.81$

b. $e^{0.4 \cdot 2} \approx 2.23$

c. $e^{\frac{\pi}{2}} \approx 4.81$

7. $e^{(e^1)}$

9. $1000e^{0.08 \cdot 1} \approx \$1083.29$

11. $1000e^{0.10 \cdot 1} \approx \$1105.17$

13. $1000e^{0.045 \cdot 1} \approx \$1046.03$

15. $500{,}000 = Pe^{0.10 \cdot 12}$, $P = \dfrac{500{,}000}{e^{1.2}}$,

$P \approx \$150{,}597.11$

17. $500{,}000 = Pe^{0.10 \cdot 20}$, $P = \dfrac{500{,}000}{e^{2}}$,

$P \approx \$67{,}667.64$

19.

$\ln y = 1$	$\log_e y = 1$	$y = e^1$	$y = e$
$\ln y = 0$	$\log_e y = 0$	$y = e^0$	$y = 1$
$y = \ln(-1)$	$y = \log_e(-1)$	$e^y = -1$	{ }
$y = \ln e^2$	$y = \log_e e^2$	$e^y = e^2$	$y = 2$
$y = \ln e^e$	$y = \log_e e^e$	$e^y = e^e$	$y = e$

21. a. $e^x = 4$, $x = \ln 4$, $x \approx 1.3863$

b. $\ln x = -2$, $e^{-2} = x$, $x \approx 0.1353$

23. a. $\ln x = 1.5$, $e^{1.5} = x$, $x \approx 4.4817$

b. $e^x = 2$, $x = \ln 2$, $x \approx 0.6931$

25. a. $69 \div 8 \approx 8.66\%$

b. $69 \div 11 \approx 6.27\%$

c. $r \approx \frac{69}{t}$

27. $2P = Pe^{11r}$, $2 = e^{11r}$, $\ln 2 = 11r$,

$r = \dfrac{\ln 2}{11}$, $r \approx 0.0630$ or 6.30%

29. $3P = Pe^{0.08t}$, $3 = e^{0.08t}$, $\ln 3 = 0.08t$,

$t = \dfrac{\ln 3}{0.08}$, $t \approx 13.7$ yrs

31. $3 = e^{rt}$, $\ln 3 = rt$, $t = \dfrac{\ln 3}{r}$

33. $0.5P = Pe^{-0.08t}$, $0.5 = e^{-0.08t}$,

$\ln 0.5 = -0.08t$, $t = \dfrac{\ln 0.5}{-0.08}$, $t \approx 8.7$ yrs

Section 7.5 (con't)

35. a. $\log_6 8 = \dfrac{\ln 8}{\ln 6} \approx 1.1606$

$\log_6 8 = \dfrac{\log 8}{\log 6} \approx 1.1606$

b. $\log 8 \approx 0.903$, $\log 6 \approx 0.778$

$\ln 8 \approx 2.079$, $\ln 6 \approx 1.792$

c. The ratios are equal the logarithms are not.

37. $K = e^{\left(-\frac{Ea}{RT}+C\right)}$

39. $\dfrac{-EnF}{RT} = \ln[H^+]$, $[H^+] = e^{\left(\frac{-EnF}{RT}\right)}$

41. a.

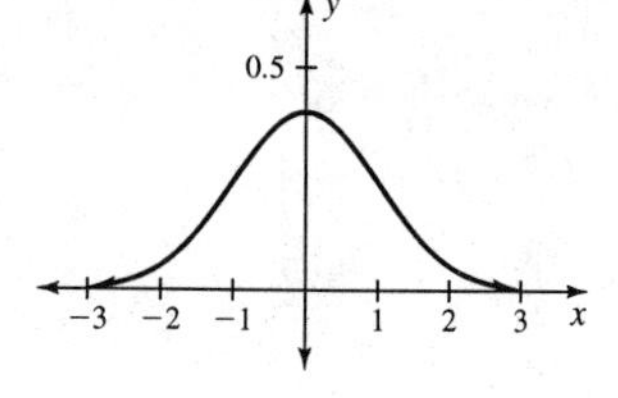

b. Highest point is $y \approx 0.4$ at $x = 0$

c. y-intercept is $\dfrac{1}{\sqrt{2\pi}} \approx 0.39894$

d. Estimates will vary, should be around 0.65.

43. a. In 1950 the population was 1,849,568 (the y-intercept point) or 1,850,000 rounded to 3 significant digits.

b. $1{,}849{,}568e^{-0.0147(20)} \approx 1{,}380{,}000$

c. $1{,}849{,}568e^{-0.0147(40)} \approx 1{,}030{,}000$

43. d. The sign on the exponent indicates the function is decreasing.

e. $0.5 = e^{-0.0147t}$, $\ln 0.5 = -0.0147t$,

$t = \dfrac{\ln 0.5}{-0.0147}$, $t \approx 47$ yrs

$1950 + 47 = 1997$

f. $1{,}849{,}568e^{-0.0147(50)} \approx 887{,}000$

45. The maximum value is attained when $x = e$, the point is $(e, e^{\frac{1}{e}})$

47. a. $4! = 4 \cdot 3 \cdot 2 \cdot 1 = 24$

b. $10! = 10 \cdot 9 \cdot 8 \cdot 7 \cdot 6 \cdot 5 \cdot 4 \cdot 3 \cdot 2 \cdot 1$
$= 3{,}628{,}800$

c. $6! = 6 \cdot 5 \cdot 4 \cdot 3 \cdot 2 \cdot 1 = 720$

d. $9! = 9 \cdot 8 \cdot 7 \cdot 6 \cdot 5 \cdot 4 \cdot 3 \cdot 2 \cdot 1$
$= 362{,}880$

e. 9 terms ≈ 7.3873, $e^2 \approx 7.3891$

f. 15 terms ≈ 0.72537, $\ln 2 \approx 0.69315$

Chapter 7 Review

1. Common ratio = 3,

Next number = $243 \cdot 3 = 729$

Geometric sequence

$y = 9 \cdot 3^{x-1}$, or $y = 3 \cdot 3^x$

3. First differences; 2, 4, 6, 8

Second differences; 2, 2, 2

Next number = $21 + (8 + 2) = 31$

Quadratic sequence

$2a = 2$, $a = 1$, $c = 1 - (2 - 2) = 1$

$1 = 1(1^2) + b(1) + 1$, $b = -1$

$y = x^2 - x + 1$

5. First differences; 3, 3, 3, 3

Next number = $17 + 3 = 20$

Arithmetic sequence

$y = 5 + (x - 1)3$, $y = 3x + 2$

7. $a_n = \frac{1}{4}(2)^{n-1}$, $f(x) = 0.125 \cdot 2^x$

$0.25(2)^{n-1} = 0.25 \cdot 2^{-1} \cdot 2^n = \frac{0.25}{2} \cdot 2^n$,

$= 0.125 \cdot 2^n$

9. $a_n = \frac{1}{16}(2)^{n-1}$, $f(x) = 0.03125 \cdot 2^n$

$\frac{1}{16}(2)^{n-1} = \frac{1}{16} \cdot \frac{1}{2} \cdot 2^n = \frac{1}{32} \cdot 2^n$

$= 0.03125 \cdot 2^n$

11. a. $y = 2^{x-3} = 2^{-3} \cdot 2^x = \frac{1}{8} \cdot 2^x$,

y-intercept = $\frac{1}{8}$

b. $y = 3^{x+1}$, $y = 3 \cdot 3^x$,

y-intercept = 3

c. $y = 3^{x-3}$, $y = 3^{-3} \cdot 3^x$, $y = \frac{1}{27} \cdot 3^x$

y-intercept = $\frac{1}{27}$

13. a. y-intercept = 2, $y = 2^1 \cdot 2^x$, $y = 2^{x+1}$

b. y-intercept = $\frac{1}{9}$, $y = 3^{-2} \cdot 3^x$, $y = 3^{x-2}$

15. a. $4^x = 64$, $4^x = 4^3$, $x = 3$

b. $2^x = 2$, $x = 1$

c. $a^x = \frac{1}{a}$, $a^x = a^{-1}$, $x = -1$

d. $b^n = 1$, $n = 0$

17. a. $25^n = 125$, $(5^2)^n = 5^3$, $5^{2n} = 5^3$,

$2n = 3$, $n = \frac{3}{2}$

b. $27^n = \frac{1}{3}$, $(3^3)^n = 3^{-1}$, $3^{3n} = 3^{-1}$,

$3n = -1$, $n = -\frac{1}{3}$

c. $(\frac{1}{25})^n = 125$, $(5^{-2})^n = 5^3$, $5^{-2n} = 5^3$,

$-2n = 3$, $n = -\frac{3}{2}$

d. $(\frac{1}{16})^n = 16$, $16^{-1n} = 16$, $-n = 1$, $n = -1$

Chapter 7 Review (con't)

19. a. $(\frac{1}{100})^n = 10$, $10^{-2n} = 10^1$, $-2n = 1$,

$n = -\frac{1}{2}$

b. $100^n = 10$, $10^{2n} = 10^1$, $2n = 1$, $n = \frac{1}{2}$

c. $4^{x-4} = 64$, $4^{x-4} = 4^3$, $x - 4 = 3$, $x = 7$

d. $27^{x+1} = 81$, $(3^3)^{x+1} = 3^4$, $3^{3x+3} = 3^4$,

$3x + 3 = 4$, $3x = 1$, $x = \frac{1}{3}$

21. $(\frac{1}{2})^x = (2^{-1})^x = 2^{-x}$

23.

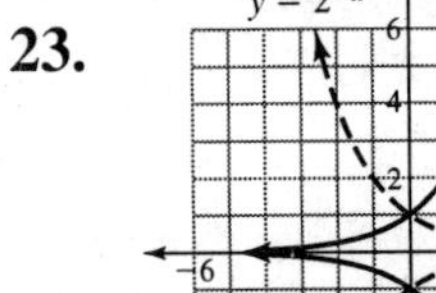

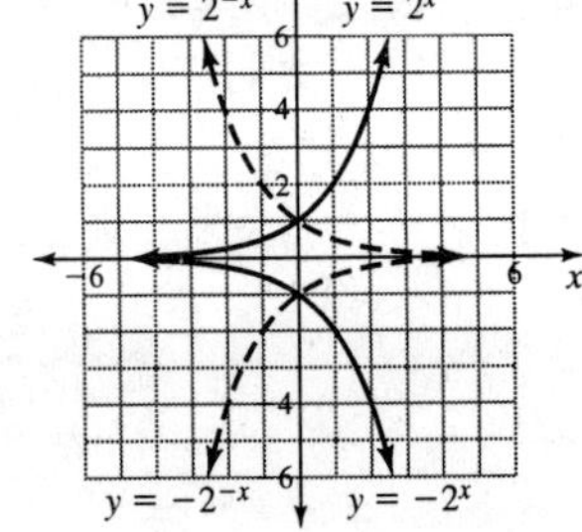

25.

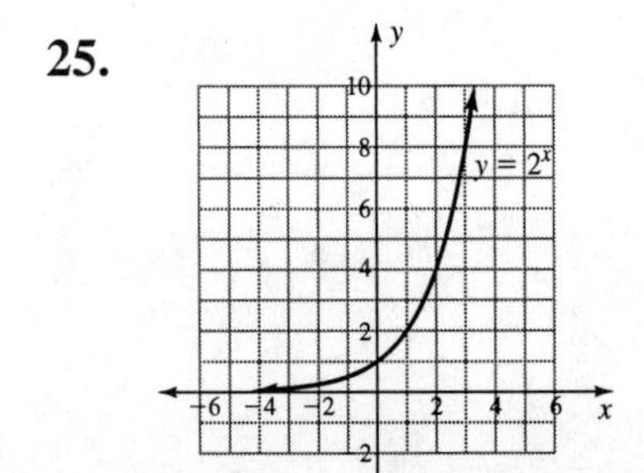

a. Shift 1 unit to the left.

b. Shift 2 units to the right.

27.

Exponential Equation	Logarithmic Equation	Solve for x
$2^x = 16$	$\log_2 16 = x$	$2^x = 2^4$ $x = 4$
$x^2 = 25$	$\log_x 25 = 2$	$x^2 = 5^2$ $x = 5$
$3^x = 81$	$\log_3 81 = x$	$3^x = 3^4$ $x = 4$
$10^{\frac{1}{2}} = x$	$\log_{10} x = \frac{1}{2}$	$x = \sqrt{10}$
$10^x = 19$	$\log_{10} 19 = x$	$x \approx 1.2788$
$4^0 = x$	$\log_4 x = 0$	$x = 1$

29. $10^x = 0.1$, $10^x = 10^{-1}$, $x = -1$

31. $36 = 10^x$, $\log 36 = x$, $x \approx 1.556$

33. $10^x = 0.75$, $\log 0.75 = x$, $x \approx -0.125$

35. $4^{x+1} = 32$, $(2^2)^{x+1} = 2^5$, $2^{2x+2} = 2^5$,

$2x + 2 = 5$, $2x = 3$, $x = \frac{3}{2}$

37. $3^{2x} = 6$, $\log 3^{2x} = \log 6$, $2x\log 3 = \log 6$,

$$2x = \frac{\log 6}{\log 3},\quad x = \frac{\log 6}{2\log 3},\quad x \approx 0.815$$

39. $\log_{10} x = -1$, $10^{-1} = x$, $x = 0.1$

41. $\log_2 x = 1$, $2^1 = x$, $x = 2$

43. $\log_2 2 = x$, $2^x = 2$, $x = 1$

45. $\log 10000 = x$, $x = 4$

47. $\log x = 2$, $10^2 = x$, $x = 100$

49. $\log_3 9 = x$, $3^x = 9$, $3^x = 3^2$, $x = 2$

Chapter 7 Review (con't)

51. $\log_{27} 9 = x,\ 27^x = 9,\ (3^3)^x = 3^2,\ 3^{3x} = 3^2,$

$3x = 2,\ x = \frac{2}{3}$

53. $y = \log_3 x$ is $3^y = x$ inverse of $y = 3^x$

55. a. $1 = 3^0$ and $\log_3 1 = 0$, also functions are inverses

b. $3 = 3^1$ and $\log_3 3 = 1$, also functions are inverses

c. Graphs mirror each other across y = x

57. $1000(1 + \frac{0.06}{4})^{4(2)} \approx \$1126.49$

$2000 = 1000(1 + \frac{0.06}{4})^{4t},\ 2 = (1 + \frac{0.06}{4})^{4t},$

$\log 2 = \log(1 + \frac{0.06}{4})^{4t},$

$\log 2 = 4t\log(1 + \frac{0.06}{4}),\ t = \dfrac{\log 2}{4\log(1+\frac{0.06}{4})}$

$t \approx 11.64$ yrs

59. $1000(1 + \frac{0.06}{365})^{365(2)} \approx \$1127.49$

$2 = (1 + \frac{0.06}{365})^{365t},\ \log 2 = \log(1 + \frac{0.06}{365})^{365t}$

$\log 2 = 365t\log(1 + \frac{0.06}{365}),$

$t = \dfrac{\log 2}{365\log(1+\frac{0.06}{365})},\ t \approx 11.55$ yrs

61. $\log(x - 1) + \log(x - 2) = \log(x - 1)(x - 2)$

$= \log(x^2 - 3x + 2)$

63. $\log(x^2 + x) - \log x = \log\left(\dfrac{x(x+1)}{x}\right)$

$= \log(x + 1)$

65. $\log(x^2 + 2x + 1) = \log(x + 1)^2$

$= 2\log(x + 1)$

67. $\log\sqrt{x} = \log x^{\frac{1}{2}} = \frac{1}{2}\log x$

69. $\log_3 x^2 - \log_3 x = \log_3\left(\frac{x^2}{x}\right) = \log_3 x$

$\log_3 x = 2,\ 3^2 = x,\ x = 9$

71. $\log 500 = \log(5 \cdot 100) = \log 5 + \log 100$

$= \log 5 + 2 = 0.69897 + 2 = 2.69897$

73. Refer to #71, they are all 2 + a decimal.

75. $\log(1000x)$ is $\log x + 3$

77.

1, 3, 8, 10, 30, 100, 300, 800, 1000, 3000, 10,000

10^0 10^1 10^2 10^3 10^4

79. a. $\ln 3 \approx 1.10$

b. $\ln \pi \approx 1.14$

c. $\ln e = 1.00$

81. a. $1000 = Pe^{0.06(5)},\ P = \dfrac{1000}{e^{0.3}},$

$P \approx \$740.82$

b. $1000 = Pe^{0.08(5)},\ P = \dfrac{1000}{e^{0.4}},$

$P \approx \$670.32$

Chapter 7 Review (con't)

81. c. $1000 = Pe^{0.12(5)}$, $P = \dfrac{1000}{e^{0.6}}$,

$P \approx \$548.81$

83. $5 = \dfrac{69}{100r}$, $r = \dfrac{69}{500}$, $r = 0.138$ or 13.8%

$e^{0.138(5)} \approx 1.994 \approx 2$

85. $e^x = 1$, $x = 0$

87. $e^x = e$, $x = 1$

89. $\ln x = -2$, $x = e^{-2}$, $x \approx 0.135$

91. $\ln x = e$, $x = e^e$, $x \approx 15.154$

93. $y \approx 49.3(0.626)^x$ million, with x = 0 for 1989; $49.3(0.626)^{16} \approx 0.027$ million or $27,000

95. $P \approx 13{,}426(1.0265)^x$,

rate of growth ≈ 2.65%,

$13{,}426(1.0265)^{70} \approx 84{,}000$, this does not seem reasonable because the population decreased between 1980 and 1990.

97. a.

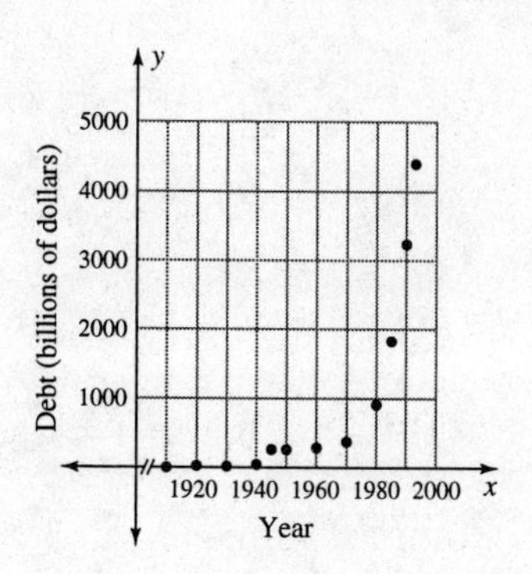

b. $y \approx 4.1986(1.0868)^x$ billion

c. Answers may vary, possible events are WWI, WWII and government spending in the 1980's.

e.

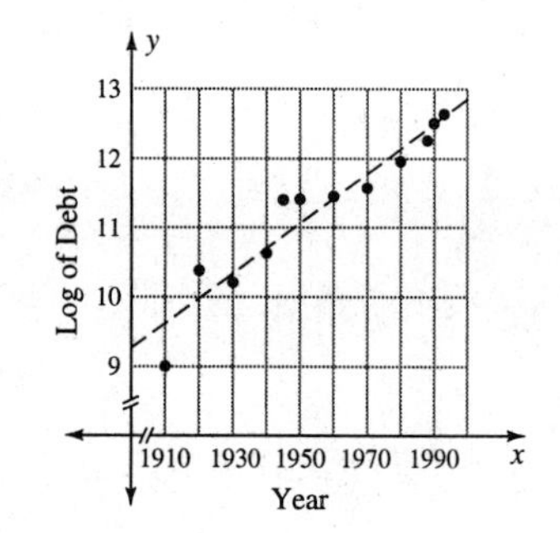

Chapter 7 Test

1. First differences: 3, 3, 3, 3

Arithmetic sequence,

$a_n = 3 + (n - 1)3$, $a_n = 3n$,

$f(x) = 3x$

2. Common ratio = 2,

Geometric sequence,

$a_n = \frac{1}{8} \cdot 2^{n-1}$,

$f(x) = 0.0625 \cdot 2^x$,

$\frac{1}{8} \cdot 2^{n-1} = \frac{1}{8} \cdot 2^{-1} \cdot 2^n = \frac{1}{8} \cdot \frac{1}{2} \cdot 2^n$

$= \frac{1}{16} \cdot 2^n = 0.0625 \cdot 2^n$

3. Common ratio = $\frac{1}{2}$,

Geometric sequence,

$a_n = \frac{1}{4}(\frac{1}{2})^{n-1}$,

$f(x) = 0.5 \cdot 0.5^x$,

$\frac{1}{4}(\frac{1}{2})^{n-1} = \frac{1}{4}(\frac{1}{2})^{-1} \cdot (\frac{1}{2})^n = \frac{1}{4} \cdot 2 \cdot (\frac{1}{2})^n$

$= \frac{1}{2}(\frac{1}{2})^n = 0.5 \cdot 0.5^n$

4. First differences: 9, 15, 21, 27

Second differences: 6, 6, 6

Quadratic sequence,

$f(x) = 3x^2$

5. $y = 40.4x + 16{,}093$,

$f(x) \approx 16{,}093(1.0025)^x$,

$40.4(60) + 16{,}093 = 18{,}517$

$16{,}093(1.0025)^{60} \approx 18{,}694$

both methods over estimated the actual population

6. a.

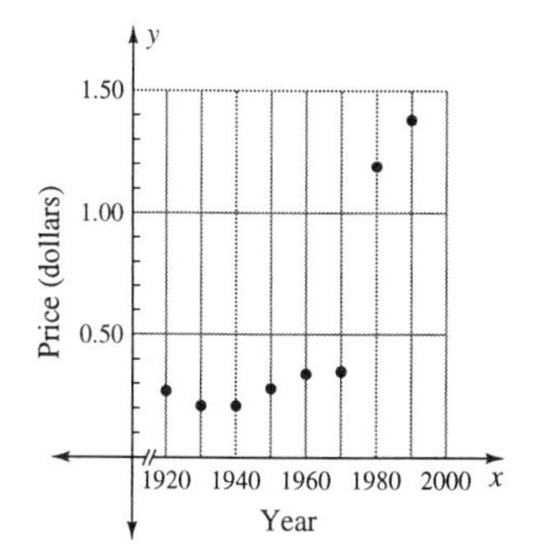

b. Data starts off slowly increasing then increases rapidly, and exponential equation would fit best.

c. $f(x) \approx 0.161(1.0263)^x$

7. a. $3^x = 27$, $3^x = 3^3$, $x = 3$

b. $(\frac{1}{16})^n = 64$, $(4^{-2})^n = 4^3$, $4^{-2n} = 4^3$

$-2n = 3$, $n = -\frac{3}{2}$

c. $\frac{1}{1000} = 0.1^x$, $10^{-3} = (10^{-1})^x$, $10^{-3} = 10^{-x}$,

$-3 = -x$, $x = 3$

d. $4^{0.5x} = 16$, $4^{0.5x} = 4^2$, $0.5x = 2$, $x = 4$

e. $4^x = \frac{1}{64}$, $4^x = 4^{-3}$, $x = -3$

f. $100^n = 10$, $10^{2n} = 10^1$, $2n = 1$, $n = \frac{1}{2}$

Chapter 7 Test (con't)

7. g. $27^x = 9$, $3^{3x} = 3^2$, $3x = 2$, $x = \frac{2}{3}$

h. $4^{x+1} = 8$, $(2^2)^{x+1} = 2^3$, $2^{2x+2} = 2^3$,

$2x + 2 = 3$, $2x = 1$, $x = \frac{1}{2}$

8. a. $10^x = 3$, $\log 3 = x$, $x \approx 0.477$

b. $5^{2x} = 0.25$, $\log 5^{2x} = \log 0.25$,

$2x\log 5 = \log 0.25$, $x = \dfrac{\log 0.25}{2\log 5}$,

$x \approx -0.431$

c. $10^x = -1$ { }

d. $e^x = \frac{1}{e}$, $e^x = e^{-1}$, $x = -1$

e. $5^{x-2} = \frac{1}{125}$, $5^{x-2} = 5^{-3}$, $x - 2 = -3$,

$x = -1$

f. $\log_2 x = -2$, $2^{-2} = x$, $x = \frac{1}{4}$

g. $\log_5 x = 3$, $5^3 = x$, $x = 125$

h. $\log_x 32 = 5$, $x^5 = 32$, $x^5 = 2^5$, $x = 2$

i. $\log x = 1$, $10^1 = x$, $x = 10$

j. $\ln x = 1.5$, $e^{1.5} = x$, $x \approx 4.482$

9. a. $\log_7 1 = 0$

b. $\log 1 = 0$

c. $\log_2 x + \log_2(x + 1) = \log_2[x(x + 1)]$

$= \log_2(x^2 + x)$

d. $\log_3(x^2 - 9) - \log_3(x + 3)$

$= \log\left(\dfrac{(x + 3)(x - 3)}{(x + 3)}\right) = \log(x - 3)$

10. $y = b^{x+2}$, $y = b^x b^2$, $y = b^2 b^x$

y-intercept $= b^2$

11.

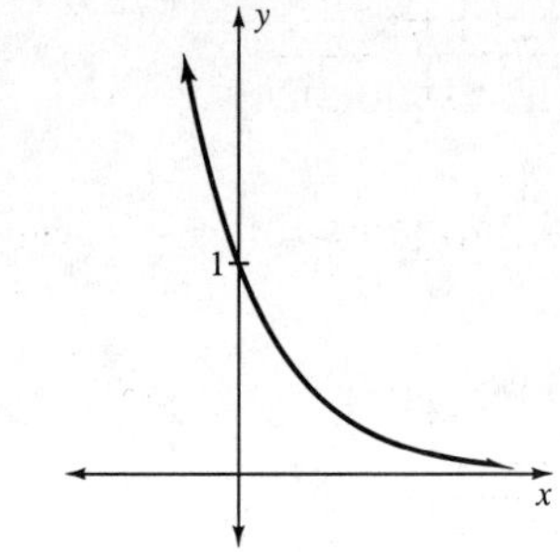

12. $1+r = 0.93$, $r = 0.93 - 1$, $r = -0.07$

13. $\log(100x) = \log x + \log 100 = \log x + 2$

14. Possible answers are (0, 1) and (1, 0); (1, 10) and (10, 1). $y = 10^x$ is $x = \log y$ which is the inverse of $y = \log x$.

15. a. $pH = -\log[5.01 \times 10^{-9}]$, $pH \approx 8.3$

b. $11.9 = -\log[H^+]$, $-11.9 = \log[H^+]$,

$10^{-11.9} = [H^+]$, $[H^+] \approx 1.26 \times 10^{-12}$ M

16.

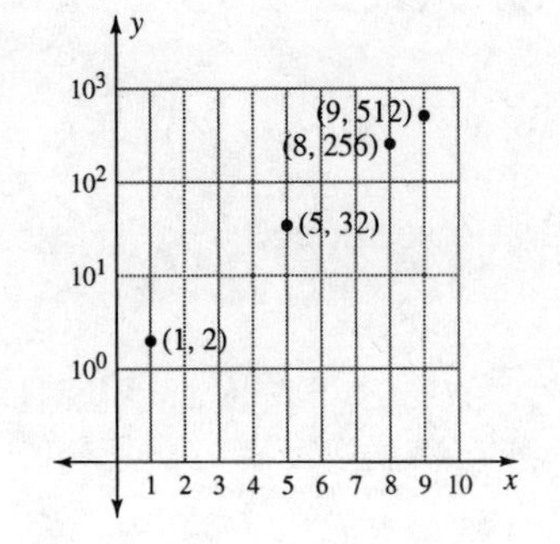

Chapter 7 Test (con't)

17. Equation appears to be $y = 3^x$

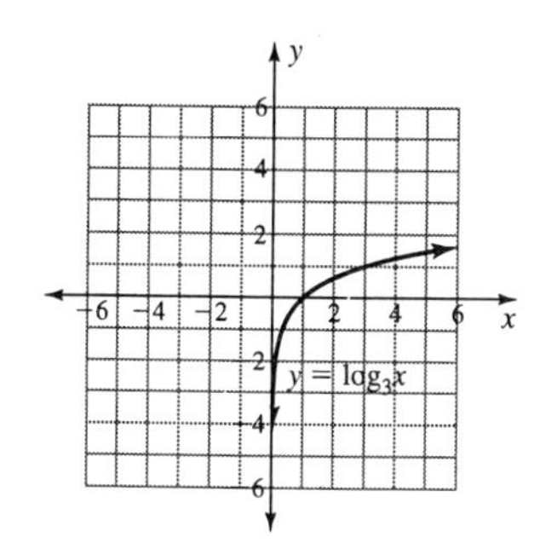

18. Equation appears to be $y = \log_2 x$

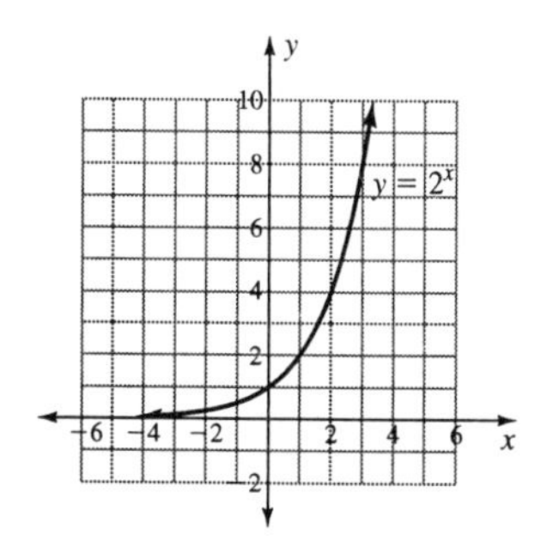

19. $1360 = 1000(1 + 0.075)^t$,

$1.36 = (1.075)^t$, $\log 1.36 = \log(1.075)^t$,

$\log 1.36 = t\log(1.075)$, $t = \dfrac{\log 1.36}{\log 1.075}$,

$t \approx 4.25$ yrs

20. $1906 = 1000e^{0.075t}$, $1.906 = e^{0.075t}$

$\ln 1.906 = 0.075t$, $t = \dfrac{\ln 1.906}{0.075}$,

$t \approx 8.6$ yrs

Section 8.0

1.

Work	Class	Study	Total
40	20	60	120(too high)
30	15	45	90 (too high)
20	10	30	60 (correct)

$w = 2c$, $s = 3c$, $w + c + s = 60$;

$2c + c + 3c = 60$, $6c = 60$, $c = 10$;

$w = 2(10)$, $w = 20$; $s = 3(10)$, $s = 30$

20 hrs work, 10 hrs class, 30 hrs study

3.

Carb.	Fat	Protein	Total
10 g 40 cal	13 g 117 cal	5 g 20 cal	28 g 177 cal
6 g 24 cal	19 g 171 cal	3 g 12 cal	28 g 207 cal
4 g 16 cal	22 g 198 cal	2 g 8 cal	28 g 222 cal

$c = 2p$, $c + f + p = 28$, $4c + 9f + 4p = 222$

$2p + f + p = 28$, $f = 28 - 3p$;

$4(2p) + 9(28 - 3p) + 4p = 222$,

$8p + 252 - 27p + 4p = 222$, $-15p = -30$,

$p = 2$; $c = 2(2)$, $c = 4$; $6 + f = 28$, $f = 22$

carbos. = 4g, fat = 22 g, protein = 2 g

5.

Carb.	Fat	Protein	Total
19 g 76 cal	5 g 45 cal	5 g 20 cal	29 g 141 cal
25 g 100 cal	2 g 18 cal	2 g 8 cal	29 g 126 cal
27 g 108 cal	1 g 9 cal	1 g 4 cal	29 g 121 cal

$f = p$, $c + f + p = 29$, $4c + 9f + 4p = 121$;

$c + p + p = 29$, $c = 29 - 2p$;

$4(29 - 2p) + 9p + 4p = 121$,

$116 - 8p + 13p = 121$, $5p = 5$, $p = 1$;

$f = 1$; $c + 1 + 1 = 29$, $c = 27$;

carbos = 27 g, fat = 1 g, protein = 1 g

7. $3x + y = 4$, $3x + y - 3x = 4 - 3x$,

$y = -3x + 4$

9. $x - 3y = 7$, $x - 3y + 3y = 7 + 3y$,

$x = 3y + 7$

11. $4y - 2x = 5$, $4y = 2x + 5$, $4y - 5 = 2x$,

$x = \dfrac{4y-5}{2}$, $x = 2y - \frac{5}{2}$

13. $4x - 2y = -3$, $4x = -3 + 2y$, $4x + 3 = 2y$,

$y = \dfrac{4x+3}{2}$, $y = 2x + \frac{3}{2}$

Section 8.0 (con't)

15. $y = x^2 + 5$, $y - 5 = x^2$, $x^2 = y - 5$

17. $x - 3 = -2y^2 + x^2$, $2y^2 + x - 3 = x^2$,

$2y^2 = x^2 - x + 3$, $y^2 = \dfrac{x^2 - x + 3}{2}$,

$y^2 = \frac{1}{2}x^2 - \frac{1}{2}x + \frac{3}{2}$

19. $x^2 + y^2 = 8$, $y^2 = 8 - x^2$, $y = \pm\sqrt{8 - x^2}$

21. $x^2 - y^2 = 10$, $x^2 = y^2 + 10$, $x = \pm\sqrt{y^2 + 10}$

23. $3(3x - 4) - 2x = 9$, $9x - 12 - 2x = 9$,

$7x = 21$, $x = 3$;

$y = 3(3) - 4$, $y = 5$

25. $(1.4x - 0.6x) + (y - y) + 3 = 0 - 3$

$0.8x = -6$, $x = -7.5$;

$0.6(-7.5) + y = 3$, $-4.5 + y = 3$, $y = 7.5$

27. $x = 9.4 - 2y$, $3(9.4 - 2y) - 5y = 4$,

$28.2 - 6y - 5y = 4$, $-11y = -24.2$, $y = 2.2$;

$x = 9.4 - 2(2.2)$, $x = 5$

29. $1.8x - 6.6 + 1.4x = 3$, $3.2x = 9.6$, $x = 3$;

$y = 1.8(3) - 6.6$, $y = -1.2$

31. $5x - 6y = 14.7$,

$(5x + 2x) + (-6y + 6y) = (14.7 - 4.2)$,

$7x = 10.5$, $x = 1.5$;

$5(1.5) = 14.7 + 6y$, $6y = -7.2$, $y = -1.2$

33. $2(3x - 2y) = 2(24)$, $6x - 4y = 48$,

$(2x + 6x) + (4y - 4y) = 0 + 48$,

$8x = 48$, $x = 6$;

$2(6) + 4y = 0$, $4y = -12$, $y = -3$

35. $7(5x - 2y) = 7(-9)$, $35x - 14y = -63$,

$2(6x + 7y) = 2(8)$, $12x + 14y = 16$,

$(35x + 12x) + (-14y + 14y) = (-63 + 16)$,

$47x = -47$, $x = -1$;

$5(-1) - 2y = -9$, $-2y = -4$, $y = 2$

37. $b = 1 - a$, $2c = 10 - 3a$, $c = 5 - \frac{3}{2}a$,

$2a + 3(1 - a) + 2(5 - \frac{3}{2}a) = 1$,

$2a + 3 - 3a + 10 - 3a = 1$, $-4a + 13 = 1$,

$-4a = -12$, $a = 3$

$b = 1 - 3$, $b = -2$

$c = 5 - \frac{3}{2}(3)$, $c = \frac{10}{2} - \frac{9}{3}$, $c = \frac{1}{2}$

Section 8.0 (con't)

39. $5a = 27 + 2b,\ a = \frac{27}{5} + \frac{2}{5}b,$

$3c = -b - 7,\ c = \frac{-1}{3}b - \frac{7}{3},$

$2(\frac{27}{5} + \frac{2}{5}b) + b - 4(\frac{-1}{3}b - \frac{7}{3}) = 17,$

$\frac{54}{5} + \frac{4}{5}b + b + \frac{4}{3}b + \frac{28}{3} = 17,$
$15(\frac{54}{5} + \frac{4}{5}b + b + \frac{4}{3}b + \frac{28}{3}) = (17)15,$
$162 + 12b + 15b + 20b + 140 = 255,$
$47b = -47,\ b = -1$

$a = \frac{27}{5} + \frac{2}{5}(-1),\ a = 5,$

$c = \frac{-1}{3}(-1) - \frac{7}{3},\ c = -2$

41. $x + y + z = 2,\ x + y - 2z = -7;$

$(x - x) + (y - y) + [z - (-2z)] = 2 - (-7)$

$3z = 9,\ z = 3$

$x + 3(3) = 0,\ x = -9$

$-9 + y + 3 = 2,\ y - 6 = 2,\ y = 8$

43. $3(2x + 3y - z) = (11)3,\ 6x + 9y - 3z = 33,$

$2(3x + y + 2z) = 2(13),$

$6x + 2y + 4z = 26,$

$(6x - 6x) + (9y - 2y) + (-3z - 4z) = 33 - 26$

$7y - 7z = 7;\ 2y = z + 7,\ z = 2y - 7,$

$7y - 7(2y - 7) = 7,\ 7y - 14y + 49 = 7,$

$-7y = -42,\ y = 6,$

$z = 2(6) - 7,\ z = 5$

$2x + 3(6) - 5 = 11,\ 2x = -2,\ x = -1$

45. No, for example $x = 4$ can not be written in $y = mx + b$ form because m is undefined.

47. $x = 2$ is $x + 0y = 2$ where $a = 1,\ b = 0$ and $c = 2$

49. $acx = ce - bcy,\ acx = af - ady,$

$ce - bcy = af - ady,\ ady - bcy = af - ce,$

$(ad - bc)y = af - ce,\ y = \dfrac{af - ce}{ab - bc}$

51. When the coefficient on a variable is 1 and it is easy to solve for that variable then substitution is more convenient.

53. At point K, $700 = a(0)^2 + b(0) + c,$

$c = 700,\ -0.04 = 2a(0) + b,\ b = -0.04;$

At point L, $0.03 = 2a(2000) - 0.04,$

$0.07 = 4000a,\ a = 1.75 \times 10^{-5}$ or

$a = \frac{7}{400{,}000};\ b = -\frac{1}{25}$

$y = \frac{7}{400{,}000}x^2 - \frac{1}{25}x + 700$

55. At point M, $-0.05 = 2a(0) + b,\ b = -0.05$

At point N, $0.06 = 2a(2400) - 0.05,$

$0.11 = 4800a,\ a = \frac{11}{480{,}000},\ b = \frac{1}{20},$

$1200 = \frac{11}{480{,}000}(2400)^2 - \frac{1}{20}(2400) + c,$

$1200 = 132 - 120 + c,\ c = 1188;$

$y = \frac{11}{480{,}000}x^2 - \frac{1}{20}x + 1188$

Section 8.1

1.

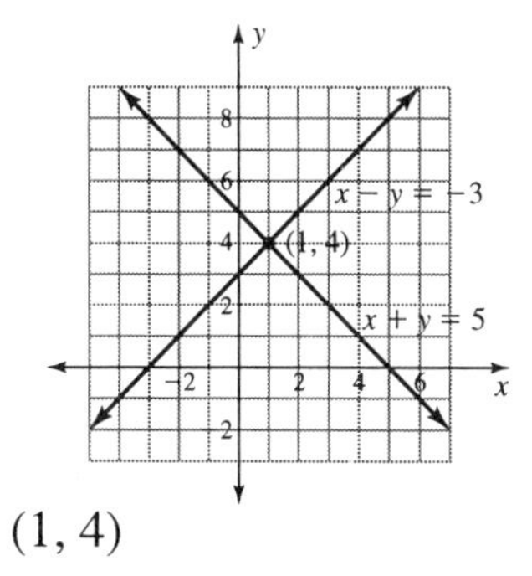

(1, 4)

3.

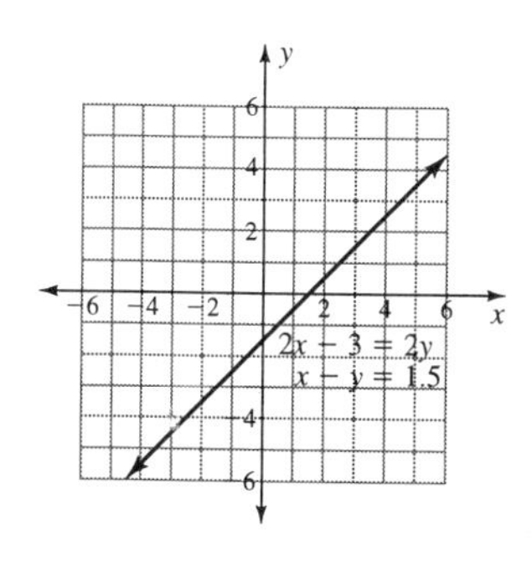

Coincident lines, if we solve using substitution or elimination the variables would drop out and we would get a true result.

5.

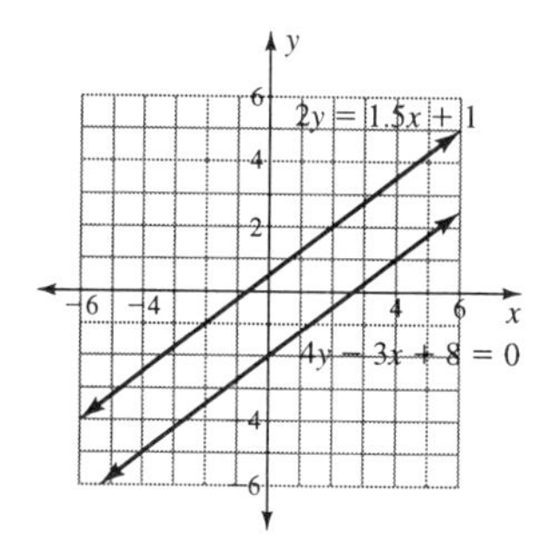

Parallel lines, if we solve using substitution or elimination the variables would drop out and we would get a false result.

7.

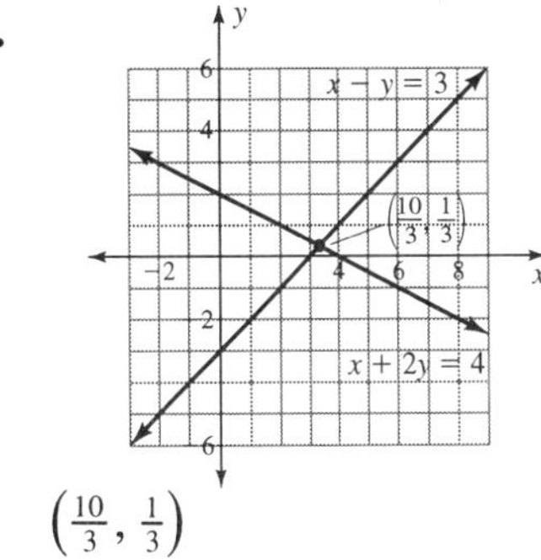

$\left(\frac{10}{3}, \frac{1}{3}\right)$

9.

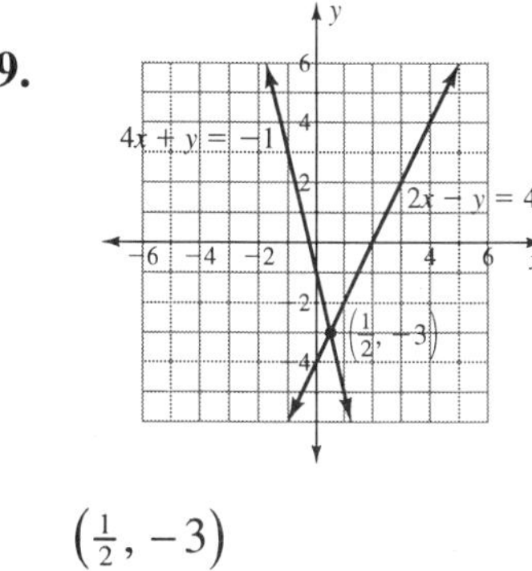

$\left(\frac{1}{2}, -3\right)$

11.

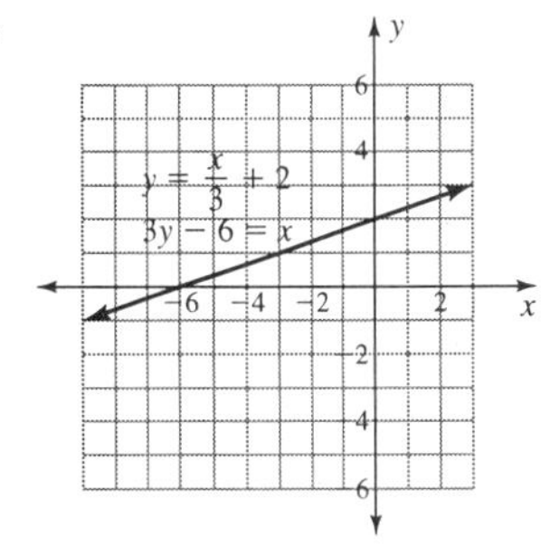

Coincident lines, if we solve using substitution or elimination the variables would drop out and we would get a true result.

13. Quantities are \$1600 and \$12,000; values 5% and 8% annual interest

15. Quantities are 20 hours and 15 hours; values \$6.50 and \$7.25 per hour

Section 8.1 (con't)

17. Quantities are 100 mL and 1000 mL;

values 8 and 0% solution

19. We might ask what is the total value. We will add that dimes are worth \$0.10 and quarters are worth \$0.25.

21. We might ask what is the measure of each angle. We will add the sum of all interior angles in a triangle is 180°.

23. GPA = total point ÷ total credit hours

Assuming that an A is worth 4 points, a B worth 3 points and a C worth 2 points:

$4(A) + 3(B) + 3(C) + 1(B) = \text{total points}$

$4(4) + 3(3) + 3(2) + 1(3) = 34,$

$4 + 3 + 3 + 1 = 11$ credit hours

$GPA = 34 \div 11, \ GPA \approx 3.09$

25.

Q, credits	V, points	Q · V
75	3.08	75(3.08) = 231
A	4.0	4A
75 + A	3.25	(75 + A)3.25

$3.25(75 + A) = 231 + 4A,$

$243.75 + 3.25A = 231 + 4A,$

$0.75A = 12.75, \ A = 17$ credit hours

27.

Q, lbs	V, % protein	Q · V
A	0.15	0.15A
B	0.10	0.12B
1000	0.12	1000(0.12)

$0.15A + 0.10B = 0.12(1000),$

$A + B = 1000; \ B = 1000 - A,$

$0.15A + 0.10(1000 - A) = 120,$

$0.15A + 100 - 0.10A = 120$

$0.05A = 20, \ A = 400$ pounds 15%

$B = 1000 - 400, \ B = 600$ pounds 10%

29.

Q, \$	V, interest	Q · V
x	0.06	0.06x
y	0.09	0.09y
75,000	0.08	75,000(0.08)

$0.06x + 0.09y = 75{,}000(0.08)$

$x + y = 75{,}000, \ y = 75{,}000 - x,$

$0.06x + 0.09(75{,}000 - x) = 6000,$

$0.06 + 6750 - 0.09x = 6000,$

$-0.03x = -750, \ x = 25{,}000$

$y = 75{,}000 - 25{,}000, \ y = 50{,}000$

\$25,000 at 6% and \$50,000 at 9%

Section 8.1 (con't)

31.

Q, $	V, interest	Q · V
x	0.21	0.21x
y	0.11	0.11y
10,000	0.13	10,000(0.13)

$x + y = 10{,}000,\ \ y = 10{,}000 - x;$

$0.21x + 0.11y = 10{,}000(0.13),$

$0.21x + 0.11(10{,}000 - x) = 1300,$

$0.21x + 1100 - 0.11x = 1300,$

$0.10x = 200,\ x = 2000,$

$y = 10{,}000 - 2000,\ \ y = 8000$

$2,000 credit cards, $8,000 bank

33. *Note: 5% glucose solution is 95% water*

Q, L	V, % water	Q · V
x	1.00	x
y	0.50	0.50y
0.5	0.95	0.5(0.95)

$x + y = 0.5,\ y = 0.5 - x,$

$x + 0.50y = 0.5(0.95),$

$x + 0.5(0.5 - x) = 0.475,$

$x + 0.25 - 0.5x - 0.475,\ 0.5x = 0.225,$

$x = 0.45\ ,\ \ y = 0.5 - 0.45,\ y = 0.05$

0.45 L distilled water, 0.05 L glucose

35. $250 + 0.10s = 350 + 0.08s,$

$0.02s = 100,\ s = \$5000$

37. $550 + 0.04s = 750 + 0.04s,$

$550 = 750$, false statement, these compensation packages will never be equal, package #4 will always be better.

39. $250 + 0.10s = 550 + 0.06s,$

$0.04s = 300,\ s = \$7500$

41. Increasing the percent of sales, increases the slope, the graph becomes steeper.

43. Parallel lines have no solution, make the slopes equal when you write the equations.

Section 8.2

1. Matrix A has one row and two columns, it is a 1 x 2 matrix

3. Matrix F has 3 rows and 1 column, it is a 3 x 1 matrix.

5. $[2\ \ 3] + [2\ \ 4] = [2+2\ \ \ 3+4] = [4\ \ 7]$

7. $\begin{bmatrix} 1 \\ 2 \\ 3 \end{bmatrix} + \begin{bmatrix} 4 \\ 5 \\ 6 \end{bmatrix} = \begin{bmatrix} 1+4 \\ 2+5 \\ 3+6 \end{bmatrix} = \begin{bmatrix} 5 \\ 7 \\ 9 \end{bmatrix}$

9. $\begin{bmatrix} 6 & 7 & 8 \\ 9 & 10 & 11 \\ 12 & 13 & 14 \end{bmatrix} - \begin{bmatrix} 1 & 2 & 3 \\ 4 & 5 & 6 \\ 7 & 8 & 9 \end{bmatrix}$

$= \begin{bmatrix} 6-1 & 7-2 & 8-3 \\ 9-4 & 10-5 & 11-6 \\ 12-7 & 13-8 & 14-9 \end{bmatrix} = \begin{bmatrix} 5 & 5 & 5 \\ 5 & 5 & 5 \\ 5 & 5 & 5 \end{bmatrix}$

11. $\begin{bmatrix} 1 & 3 \\ -1 & -2 \end{bmatrix}\begin{bmatrix} x \\ y \end{bmatrix} = \begin{bmatrix} x+3y \\ -x-2y \end{bmatrix}$

13. $\begin{bmatrix} 1 & 3 \\ -1 & -2 \end{bmatrix}\begin{bmatrix} 2 \\ -1 \end{bmatrix} = \begin{bmatrix} -1 \\ 0 \end{bmatrix}$

15. $\begin{bmatrix} 4 & -2 \\ 1 & 2 \end{bmatrix}\begin{bmatrix} 2 \\ -1 \end{bmatrix} = \begin{bmatrix} 10 \\ 0 \end{bmatrix}$

17. $\begin{bmatrix} 1 & 3 \\ -1 & -2 \end{bmatrix}\begin{bmatrix} 2 & 3 \\ 4 & 5 \end{bmatrix} = \begin{bmatrix} 2+12 & 3+15 \\ -2-8 & -3-10 \end{bmatrix}$

$= \begin{bmatrix} 14 & 18 \\ -10 & -13 \end{bmatrix}$

19. $\begin{bmatrix} 2 & 3 \\ 4 & 5 \end{bmatrix}\begin{bmatrix} 1 & 3 \\ -1 & -2 \end{bmatrix} = \begin{bmatrix} 2-3 & 6-6 \\ 4-5 & 12-10 \end{bmatrix}$

$= \begin{bmatrix} -1 & 0 \\ -1 & 2 \end{bmatrix}$

21. $\begin{bmatrix} 4 & -2 \\ 1 & 2 \end{bmatrix}\begin{bmatrix} -2 & -1 \\ 1 & 2 \end{bmatrix} = \begin{bmatrix} -8-2 & -4-4 \\ -2+2 & -1+4 \end{bmatrix}$

$= \begin{bmatrix} -10 & -9 \\ 0 & 3 \end{bmatrix}$

23. $\begin{bmatrix} -2 & -1 \\ 1 & 2 \end{bmatrix}\begin{bmatrix} 4 & -2 \\ 1 & 2 \end{bmatrix} = \begin{bmatrix} -8-1 & 4-2 \\ 4+2 & -2+4 \end{bmatrix}$

$= \begin{bmatrix} -9 & 2 \\ 6 & 2 \end{bmatrix}$

25. det[A] = -2 - (-3) = 1

27. det[C] = 8 - (-2) = 10

29. det[A] = 1 (from exercise 25)

$[A]^{-1} = \begin{bmatrix} \frac{-2}{1} & \frac{-3}{1} \\ \frac{-(-1)}{1} & \frac{1}{1} \end{bmatrix} = \begin{bmatrix} -2 & -3 \\ 1 & 1 \end{bmatrix}$

31. det[C] = 10 (from exercise 27)

$[C]^{-1} = \begin{bmatrix} \frac{2}{10} & \frac{-(-2)}{10} \\ \frac{-1}{10} & \frac{4}{10} \end{bmatrix} = \begin{bmatrix} 0.2 & 0.2 \\ -0.1 & 0.4 \end{bmatrix}$

33. a. $[A][A]^{-1} = \begin{bmatrix} 1 & 0 \\ 0 & 1 \end{bmatrix}$

b. $[A]^{-1}[A] = \begin{bmatrix} 1 & 0 \\ 0 & 1 \end{bmatrix}$

Section 8.2 (con't)

35. $\begin{bmatrix} 1 & 2 \\ 4 & 4 \end{bmatrix}\begin{bmatrix} x \\ y \end{bmatrix} = \begin{bmatrix} 5 \\ 6 \end{bmatrix}$, $x = -2$ $y = 3.5$

37. $\begin{bmatrix} 1 & 2 \\ 2 & 4 \end{bmatrix}\begin{bmatrix} x \\ y \end{bmatrix} = \begin{bmatrix} 5 \\ 6 \end{bmatrix}$, det[A] = 0,

parallel lines.

39. $\begin{bmatrix} 1 & 2 \\ 0 & 4 \end{bmatrix}\begin{bmatrix} x \\ y \end{bmatrix} = \begin{bmatrix} 5 \\ 6 \end{bmatrix}$, $x = 2$, $y = 1.5$

41. $\begin{bmatrix} 1 & 2 \\ -2 & 4 \end{bmatrix}\begin{bmatrix} x \\ y \end{bmatrix} = \begin{bmatrix} 5 \\ 6 \end{bmatrix}$, $x = 1$, $y = 2$

43. The exercises all have 1x + 2y = 5 as one equation, and ax + 4y = 6 as the other. Changing *a* changes the slope of the graph. The y-intercept remains constant.

45. 5x = 28.2 + 2y, 5x - 2y = 28.2

$\begin{bmatrix} 5 & -2 \\ 3 & 5 \end{bmatrix}\begin{bmatrix} x \\ y \end{bmatrix} = \begin{bmatrix} 28.2 \\ -5.4 \end{bmatrix}$, $x = 4.2$, $y = -3.6$

47. 3y + 6 = 2x, 2x - 3y = 6, -2x + 3y = 12

$\begin{bmatrix} 2 & -3 \\ -2 & 3 \end{bmatrix}\begin{bmatrix} x \\ y \end{bmatrix} = \begin{bmatrix} 6 \\ 12 \end{bmatrix}$, det[A] = 0;

parallel lines

49. 2x - 3y = 7

$\begin{bmatrix} 1 & -1.5 \\ 2 & -3 \end{bmatrix}\begin{bmatrix} x \\ y \end{bmatrix} = \begin{bmatrix} 3.5 \\ 7 \end{bmatrix}$, det[A] = 0;

x - 1.5y = 3.5, x = 1.5y + 3.5

2x - 7 = 3y, 2x = 3y + 7, x = 1.5y + 3.5

Coincident lines

51. Let x = kg of Honduran coffee and

y = kg of Indonesian coffee

x + y = 200,

18.70x + 23.65y = 19.80(200)

$\begin{bmatrix} 1 & 1 \\ 18.70 & 23.65 \end{bmatrix}\begin{bmatrix} x \\ y \end{bmatrix} = \begin{bmatrix} 200 \\ 3960 \end{bmatrix}$

x ≈ 156, y ≈ 44

156 kg Honduran, 44 kg Indonesian

53. Let x = liters of 3% solution and

y = liters of 20% solution

x + y = 1000

0.03x + 0.20y = 0.064(1000)

$\begin{bmatrix} 1 & 1 \\ 0.03 & 0.20 \end{bmatrix}\begin{bmatrix} x \\ y \end{bmatrix} = \begin{bmatrix} 1000 \\ 64 \end{bmatrix}$,

x = 800, y = 200

800 L 3% solution, 200 L 20% solution

Section 8.2 (con't)

55. Convert to % water, 20% hydrogen peroxide is 80% water, 3% hydrogen peroxide is 97% water.

Let x = gallons of water and

y = gallons of 80% water solution

$x + y = 1000, \; x = 1000 - y$

$x + 0.80y = 0.97(1000)$

$1000 - y + 0.80y = 970, \; -0.2y = -30,$

$y = 150, \; x = 1000 - 150, x = 850$

850 gallons of water, 150 gallons of 20% hydrogen peroxide solution

57. Let d = weight of dimes, q = weight of quarters.

$9d + 13q = 3.5,$

$12d + 9q = 3$

$$\begin{bmatrix} 9 & 13 \\ 12 & 9 \end{bmatrix}\begin{bmatrix} d \\ q \end{bmatrix} = \begin{bmatrix} 3.5 \\ 3 \end{bmatrix}$$

d = 0.1 oz, q = 0.2 oz

59. Let x = amount invested at 4.5% and

y = amount invested at 8.5%

$x + y = 23{,}600, \; 0.045x + 0.085y = 1782$

$$\begin{bmatrix} 1 & 1 \\ 0.045 & 0.085 \end{bmatrix}\begin{bmatrix} x \\ y \end{bmatrix} = \begin{bmatrix} 23{,}600 \\ 1782 \end{bmatrix}$$

x = \$5600, y = \$18,000

61. a. $a + n = a, \; n = 0$

b. $a \cdot n = a, \; n = 1$

c. The product of a number and its multiplicative inverse is, the identity, 1.

d. The identity function is y = x or f(x) = x

e. The matrix [I] is called the identity matrix because when we multiply a matrix by [I] we get the original matrix.

63. $4 \cdot 5 = 5 \cdot 4$ is the commutative property of multiplication. There is no similar property for multiplication of matrices.

Mid-Chapter 8 Test

1. $5x - 2y = 4$, $-2y = -5x + 4$, $y = \frac{5}{2}x - 2$

2. $2x - \frac{1}{2}y = 5$, $-\frac{1}{2}y = -2x + 5$, $y = 4x - 10$

3. $2(1.5x - 3) - 3x = 8$, $3x - 6 - 3x = 8$,

$-6 = 8$, no real number solution

4. $3(-\frac{5}{6}x + 1) = 2x - 6$, $-\frac{5}{2}x + 3 = 2x - 6$,

$9 = \frac{5}{2}x + \frac{4}{2}x$, $9 = \frac{9}{2}x$, $x = 2$,

$y = -\frac{5}{6}(2) + 1$, $y = -\frac{5}{3} + 1$, $y = -\frac{2}{3}$

5. $0.8x + y = -1.8$, $1.2x + y = 3$

$\begin{bmatrix} 0.8 & 1 \\ 1.2 & 1 \end{bmatrix}\begin{bmatrix} x \\ y \end{bmatrix} = \begin{bmatrix} -1.8 \\ 3 \end{bmatrix}$, det[A] = -0.4

$x = 12$, $y = -11.4$

6. $-2.5x + y = 4$, $-5x + 2y = 8$

$\begin{bmatrix} -2.5 & 1 \\ -5 & 2 \end{bmatrix}\begin{bmatrix} x \\ y \end{bmatrix} = \begin{bmatrix} 4 \\ 8 \end{bmatrix}$, det[A] = 0,

Solving for y, second equation is the same as the first, coincident lines.

7. Equations will vary, when equations are in the form $y = mx + b$ both will have the same value for m, det[A] = 0

8. Let x = Wilt's points, y = Elgin's points

$x + y = 54{,}568$, $x = y + 8270$;

$y + 8270 + y = 54{,}568$, $2y = 44{,}298$

$y = 22{,}149$, $x = 22{,}149 + 8270$,

$x = 30{,}419$

Wilt Chamberlain scored 30,419 points

Elgin Baylor scored 22,149 points

9. $-0.05 = 2a(0) + b$, $b = -0.05$,

$0.03 = 2a(2000) - 0.05$, $0.08 = 4000a$,

$a = 0.00002$ or $a = \frac{0.08}{4000} = \frac{8}{400{,}000} = \frac{1}{50{,}000}$

$600 = 0.00002(0)^2 - 0.05(0) + c$, $c = 600$

$a = 0.00002$, $b = -0.05$, $c = 600$

10. Convert to % water. Let x = pts water, y = pts of 3% hydrogen peroxide

Q, pts	**V**	**Q · V**
x	1	x
y	0.97	0.97y
1	0.98	1(0.98)

$x + y = 1$, $x + 0.97y = 0.98$;

$1 - y + 0.97y = 0.98$, $-0.03y = -0.02$,

$y = \frac{2}{3}$ pint, $x = 1 - \frac{2}{3}$, $x = \frac{1}{3}$ pint

$\frac{1}{3}$ pint water, $\frac{2}{3}$ pint 3% hydrogen peroxide solution

Section 8.3

1. **a.** Planes 1 and 2 are coincident, they have a line of intersection with plane 3, the equations are dependent.

b. The planes are coincident, the intersection is a plane, equations are dependent.

c. There is no common point of intersection, equations are inconsistent.

d. There is no common point of intersection, equations are inconsistent.

3. $$\begin{bmatrix} 1 & 1 & -1 \\ 1 & -1 & 1 \\ -1 & 1 & 1 \end{bmatrix}\begin{bmatrix} x \\ y \\ z \end{bmatrix} = \begin{bmatrix} 2 \\ 3 \\ 4 \end{bmatrix}, \begin{bmatrix} x \\ y \\ z \end{bmatrix} = \begin{bmatrix} 2.5 \\ 3 \\ 3.5 \end{bmatrix}$$

5. $$\begin{bmatrix} 0 & 3 & 5 \\ 3 & 0 & 7 \\ 1 & 1 & 4 \end{bmatrix}\begin{bmatrix} x \\ y \\ z \end{bmatrix} = \begin{bmatrix} 0 \\ 0 \\ 3 \end{bmatrix}, \det[A] = 0$$

$3y + 5z = 0,\ 3y = -5z,\ y = -\frac{5}{3}z;$

$3x + 7z = 0,\ 3x = -7z,\ x = -\frac{7}{3}z;$

$-\frac{7}{3}z - \frac{5}{3}z + 4z = 3,\ -4z + 4z = 3,$

$0 = 3$, false result, equations are inconsistent.

7. $$\begin{bmatrix} 2 & -1 & 0 \\ 1 & 0 & 2 \\ -2 & 1 & 1 \end{bmatrix}\begin{bmatrix} x \\ y \\ z \end{bmatrix} = \begin{bmatrix} 3 \\ 4 \\ 1 \end{bmatrix}, \begin{bmatrix} x \\ y \\ z \end{bmatrix} = \begin{bmatrix} -4 \\ -11 \\ 4 \end{bmatrix}$$

9. $$\begin{bmatrix} -2 & 0 & 3 \\ -1 & 1 & 3 \\ 1 & 1 & 0 \end{bmatrix}\begin{bmatrix} x \\ y \\ z \end{bmatrix} = \begin{bmatrix} 0 \\ -1 \\ -1 \end{bmatrix} \det[A] = 0$$

$3z = 2x,\ z = \frac{2}{3}x,\ y = -x - 1,$

$-x + (-x - 1) + 3(\frac{2}{3}x) = -1,$

$-2x - 1 + 2x = -1,\ -1 = -1$, true result, equations are dependent

11. $$\begin{bmatrix} 1 & 2 & 3 \\ 4 & 5 & 6 \\ 1 & 1 & 1 \end{bmatrix}\begin{bmatrix} a \\ b \\ c \end{bmatrix} = \begin{bmatrix} 2 \\ 3 \\ 1 \end{bmatrix}, \det[A] = 0$$

subtract equation 3 from equation 1;

$(a - a) + (2b - b) + (3c - c) = (2 - 1),$

$b + 2c = 1,\ b = 1 - 2c;$

$a + 2(1 - 2c) + 3c = 2,\ a + 2 - 4c + 3c = 2$

$a - c = 0,\ a = c$

$4c + 5(1 - 2c) + 6c = 3,$

$4c + 5 - 10c + 6c = 3,\ 5 = 3$, false result, equations are inconsistent.

13. $$\begin{bmatrix} 1 & 1 & 1 \\ 1 & -1 & -1 \\ -1 & 1 & -1 \end{bmatrix}\begin{bmatrix} a \\ b \\ c \end{bmatrix} = \begin{bmatrix} 4 \\ 0 \\ 2 \end{bmatrix}, \begin{bmatrix} a \\ b \\ c \end{bmatrix} = \begin{bmatrix} 2 \\ 3 \\ -1 \end{bmatrix}$$

Section 8.3 (con't)

15. p + f + c= 32,

c = f + 12; -f + c = 12

4p + 9f + 4c = 148

$$\begin{bmatrix} 1 & 1 & 1 \\ 0 & -1 & 1 \\ 4 & 9 & 4 \end{bmatrix} \begin{bmatrix} p \\ f \\ c \end{bmatrix} = \begin{bmatrix} 32 \\ 12 \\ 148 \end{bmatrix}, \begin{bmatrix} p \\ f \\ c \end{bmatrix} = \begin{bmatrix} 12 \\ 4 \\ 16 \end{bmatrix}$$

17. 0.05n + 0.10d + 0.25q = 5.80

n + d + q = 44

q = d + 3; -d + q = 3

$$\begin{bmatrix} 0.05 & 0.10 & 0.25 \\ 1 & 1 & 1 \\ 0 & -1 & 1 \end{bmatrix} \begin{bmatrix} n \\ d \\ q \end{bmatrix} = \begin{bmatrix} 5.80 \\ 44 \\ 3 \end{bmatrix},$$

$$\begin{bmatrix} n \\ d \\ q \end{bmatrix} = \begin{bmatrix} 17 \\ 12 \\ 15 \end{bmatrix}$$

19. AO + CS - CP = 412,545

-AO + CS = 450,907

AO + CP = 1,445,464

$$\begin{bmatrix} 1 & 1 & -1 \\ -1 & 1 & 0 \\ 1 & 0 & 1 \end{bmatrix} \begin{bmatrix} AO \\ CS \\ CP \end{bmatrix} = \begin{bmatrix} 412,545 \\ 450,907 \\ 1,445,464 \end{bmatrix},$$

$$\begin{bmatrix} AO \\ CS \\ CP \end{bmatrix} = \begin{bmatrix} 469,034 \\ 919,941 \\ 976,430 \end{bmatrix}$$

21. E + F + G + C = 1000

C = 2G; 2G - C = 0

F = 4E; 4E - F = 0

10.95E + 8.95F + 8.50G + 8.25C

= 8.59(1000)

$$\begin{bmatrix} 1 & 1 & 1 & 1 \\ 0 & 0 & 2 & -1 \\ 4 & -1 & 0 & 0 \\ 10.95 & 8.95 & 8.50 & 8.25 \end{bmatrix} \begin{bmatrix} E \\ F \\ G \\ C \end{bmatrix}$$

$$= \begin{bmatrix} 1000 \\ 0 \\ 0 \\ 8590 \end{bmatrix}$$, rounding to nearest 50,

$$\begin{bmatrix} E \\ F \\ G \\ C \end{bmatrix} = \begin{bmatrix} 50 \\ 200 \\ 250 \\ 500 \end{bmatrix}$$

23. Substitute x & y into the equation;

$y = ax^4 + bx^3 + cx^2 + dx + e$;

a(16) + b(-8) + c(4) + d(-2) + e = -7

a(1) + b(-1) + c(1) + d(-1) + e = -9

a(0) + b(0) + c(0) + d(0) + e = -5

a(1) + b(1) + c(1) + d(1) + e = -1

a(16) + b(8) + c(4) + d(2) + e = 45

Now set up a matrix;

Section 8.3 (con't)

23. (continued)

$$\begin{bmatrix} 16 & -8 & 4 & -2 & 1 \\ 1 & -1 & 1 & -1 & 1 \\ 0 & 0 & 0 & 0 & 1 \\ 1 & 1 & 1 & 1 & 1 \\ 16 & 8 & 4 & 2 & 1 \end{bmatrix} \begin{bmatrix} a \\ b \\ c \\ d \\ e \end{bmatrix} = \begin{bmatrix} -7 \\ -9 \\ -5 \\ -1 \\ 45 \end{bmatrix}$$

$$\begin{bmatrix} a \\ b \\ c \\ d \\ e \end{bmatrix} = \begin{bmatrix} 2 \\ 3 \\ -2 \\ 1 \\ -5 \end{bmatrix}$$

$y = 2x^4 + 3x^3 - 2x^2 + x - 5$

25. $s + m + l = 620$

$s - m = 0$

$l = 20$

$$\begin{bmatrix} 1 & 1 & 1 \\ 1 & -1 & 0 \\ 0 & 0 & 1 \end{bmatrix} \begin{bmatrix} s \\ m \\ l \end{bmatrix} = \begin{bmatrix} 620 \\ 0 \\ 20 \end{bmatrix}$$

27. a. [A] is a 1 x 3, [B] is a 3 x 4; the product will be a 1 x 4 matrix.

b. $[C] = \begin{bmatrix} \$4800 \\ \$5100 \\ \$5400 \\ \$6000 \end{bmatrix}$

29. a. small = $\frac{4.5}{16}$ lb, medium = $\frac{3}{4}$ lb, large = 3 lb

b. Create a new matrix [W] for weights then multiply by [B]

$$\begin{bmatrix} \frac{4.5}{16} & \frac{3}{4} & 3 \end{bmatrix} \begin{bmatrix} 500 & 450 & 400 & 300 \\ 100 & 150 & 200 & 300 \\ 20 & 20 & 20 & 2 \end{bmatrix}$$

$$= \begin{bmatrix} 275.6 \\ 299.1 \\ 322.5 \\ 369.4 \end{bmatrix}$$

c. $\$3.50 \begin{bmatrix} 275.6 \\ 299.1 \\ 322.5 \\ 369.4 \end{bmatrix} = \begin{bmatrix} \$965 \\ \$1047 \\ \$1129 \\ \$1293 \end{bmatrix}$

31. $\begin{bmatrix} \$4800 \\ \$5100 \\ \$5400 \\ \$6000 \end{bmatrix} - \begin{bmatrix} \$522 \\ \$507 \\ \$492 \\ \$462 \end{bmatrix} - \begin{bmatrix} \$965 \\ \$1047 \\ \$1129 \\ \$1293 \end{bmatrix}$

$$= \begin{bmatrix} \$3313 \\ \$3546 \\ \$3779 \\ \$4245 \end{bmatrix}$$

Section 8.4

1. The easiest points will be 3 units left and right of the origin on the x-axis and 3 units above and below the origin on the y-axis, $(\pm 3, 0)$, $(0, \pm 3)$

For exercises 3 to 19 refer to text for standard form of equations.

3. $x^2 + y^2 = 5^2$, $x^2 + y^2 = 25$

5. $\frac{x^2}{5^2} + \frac{y^2}{2^2} = 1$, $\frac{x^2}{25} + \frac{y^2}{4} = 1$

7. $\frac{y^2}{4^2} - \frac{x^2}{3^2} = 1$, $\frac{y^2}{16} - \frac{x^2}{9} = 1$

9. $3 = a(-1)^2$, $a = 3$, $x = 3y^2$

11. $-1 = a(3^2)$, $a = -\frac{1}{9}$, $y = -\frac{1}{9}x^2$

13. $x^2 + y^2 = 4^2$, $x^2 + y^2 = 16$

15. $\frac{x^2}{1^2} + \frac{y^2}{5^2} = 1$, $x^2 + \frac{y^2}{25} = 1$

17. $\frac{y^2}{3^2} - \frac{x^2}{5^2} = 1$, $\frac{y^2}{9} - \frac{x^2}{25} = 1$

19. $1 = a(-2)^2$, $a = \frac{1}{4}$, $y = \frac{1}{4}x^2$

21. $r < 0$ has no meaning; $r = 0$ describes a point.

23. $y = x^2 + 2$ is a parabola, no x-intercept, y-intercept is $(0, 2)$

25. $x^2 + y^2 = 9$ is a circle with radius of 3, x-intercepts $(\pm 3, 0)$, y-intercepts $(0, \pm 3)$

For graphing on a calculator, solve for y,

$y^2 = 9 - x^2$, $y = \pm\sqrt{9 - x^2}$, $-3 \le x \le 3$

27. $x^2 - y^2 = 4$, $\frac{x^2}{4} - \frac{y^2}{4} = 1$; hyperbola with x-intercepts $(\pm 2, 0)$ and no y-intercepts,

$y^2 = x^2 - 4$, $y = \pm\sqrt{x^2 - 4}$, $x \le -2$ or $x \ge 2$

29. $\frac{x^2}{4} + \frac{y^2}{1} = 1$, is an ellipse;

x-intercepts $(\pm 2, 0)$, y-intercepts $(0, \pm 1)$

$y^2 = 1 - \frac{x^2}{4}$, $y = \pm\sqrt{1 - \frac{x^2}{4}}$, $-2 \le x \le 2$

31. $4x^2 + y^2 = 100$, $\frac{x^2}{25} + \frac{y^2}{100} = 1$, is an ellipse, x-intercepts $(\pm 5, 0)$, y-intercepts $(0, \pm 10)$;

$y^2 = 100 - 4x^2$, $y = \pm\sqrt{100 - 4x^2}$ or

$y = \pm 2\sqrt{25 - x^2}$, $-5 \le x \le 5$

33. $2x^2 + 2y^2 = 8$, $x^2 + y^2 = 4$, circle;

x-intercepts $(\pm 2, 0)$, y-intercepts $(0, \pm 2)$

$y^2 = 4 - x^2$, $y = \pm\sqrt{4 - x^2}$, $-2 \le x \le 2$

35. $y = x$ is a straight line, intercepts $(0, 0)$

Section 8.4 (con't)

37. $x^2 - 4y^2 = 1$ is a hyperbola,

x-intercepts $(\pm 1, 0)$, no y-intercepts

$4y^2 = x^2 - 1,\ y = \pm \frac{1}{2}\sqrt{x^2 - 1}$

39. $x = 4y^2$ is a parabola, intercepts (0, 0)

$y = \pm \frac{1}{2}\sqrt{x},\ x \ge 0$

41. $\frac{y^2}{1} - \frac{x^2}{4} = 1$, hyperbola, no x-intercepts,

y-intercepts $(0, \pm 1)$, $y^2 = \frac{x^2}{4} + 1$,

$y = \pm\sqrt{\frac{x^2}{4} + 1}$

43. $x = -y^2$, is a parabola; intercepts (0, 0),

$y^2 = -x,\ y = \pm\sqrt{-x},\ x \le 0$

45.

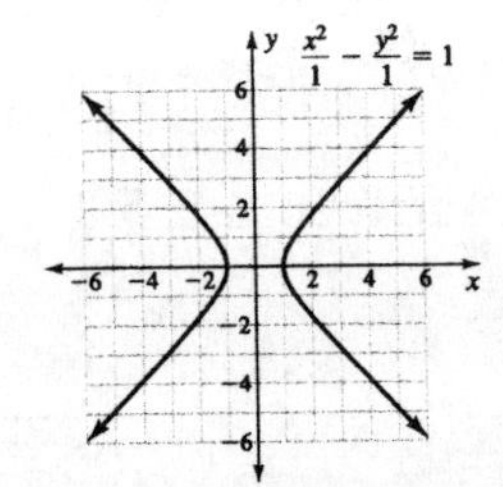

Branches become steeper

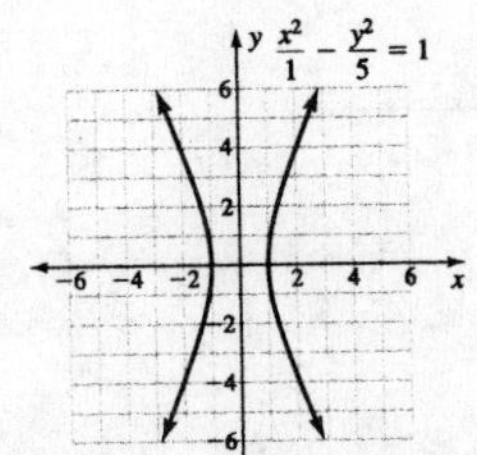

47. a.

x	y
-9	$-9 = -(\pm 3)^2$
-4	$-4 = -(\pm 2)^2$
-1	$-1 = -(\pm 1)^2$
0	$0 = -(0)^2$
1	not possible
4	not possible
9	not possible

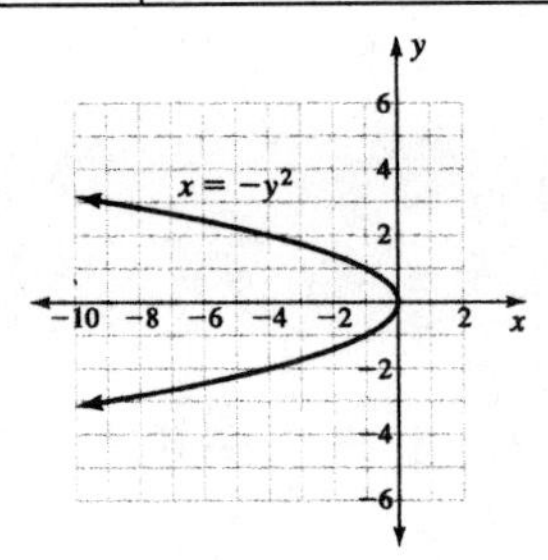

b. The opposite of x makes the radicand appear negative.

c. The equation is meaningful when

$x \le 0$

Section 8.4 (con't)

49. $144 + 25 + D(-12) + E(5) + F = 0$

$144 + 25 + D(12) + E(5) + F = 0$

$25 + 144 + D(5) + E(-12) + F = 0$

$-12D + 5E + F = -169$

$12D + 5E + F = -169$

$5D - 12E + F = -169$

$$\begin{bmatrix} -12 & 5 & 1 \\ 12 & 5 & 1 \\ 5 & -12 & 1 \end{bmatrix} \begin{bmatrix} D \\ E \\ F \end{bmatrix} = \begin{bmatrix} -169 \\ -169 \\ -169 \end{bmatrix}$$

$$\begin{bmatrix} D \\ E \\ F \end{bmatrix} = \begin{bmatrix} 0 \\ 0 \\ -169 \end{bmatrix}, \ x^2 + y^2 - 169 = 0,$$

$x^2 + y^2 = 169$

51. $16 + 1 + D(-4) + E(-1) + F = 0$

$25 + 36 + D(-5) + E(6) + F = 0$

$9 + 36 + D(3) + E(6) + F = 0$

$-4D - E + F = -17$

$-5D + 6E + F = -61$

$3D + 6E + F = -45$

$$\begin{bmatrix} -4 & -1 & 1 \\ -5 & 6 & 1 \\ 3 & 6 & 1 \end{bmatrix} \begin{bmatrix} D \\ E \\ F \end{bmatrix} = \begin{bmatrix} -17 \\ -61 \\ -45 \end{bmatrix}$$

$$\begin{bmatrix} D \\ E \\ F \end{bmatrix} = \begin{bmatrix} 2 \\ -6 \\ -15 \end{bmatrix}, \ x^2 + y^2 + 2x - 6y - 15 = 0$$

Section 8.5

1. The equations represent a parabola and a line. Set equal to each other and solve.

$x^2 - 3x + 2 = -x + 5,\ x^2 - 2x - 3 = 0,$

$(x - 3)(x + 1) = 0,\ x - 3 = 0, x = 3,$

$x + 1 = 0, x = -1$

$y = -(3) + 5, y = 2, y = -(-1) + 5, y = 6$

$(-1, 6), (3, 2)$

3. Equations represent a hyperbola and a parabola. Solve the parabola equation for x^2 and substitute into the hyperbola equation.

$y = x^2 - 4,\ x^2 = y + 4;$

$\frac{y+4}{4} - \frac{y^2}{9} = 1,\ 36\left(\frac{y+4}{4} - \frac{y^2}{9}\right) = 36,$

$9y + 36 - 4y^2 = 36, -4y^2 + 9y = 0,$

$y(-4y + 9) = 0,\ y = 0, \text{ or } -4y + 9 = 0,$

$-4y = -9,\ y = \frac{9}{4};$

$x^2 = 0 + 4, x = \pm 2;$

$x^2 = \frac{9}{4} + 4,\ x^2 = \frac{25}{4},\ x = \pm \frac{5}{2}$

$(\pm\frac{5}{2}, \frac{9}{4}), (\pm 2, 0)$

5. Both equations are parabolas, set equal to each other and solve.

$5x^2 + 2x = -x^2 + x + 2,\ 6x^2 + x - 2 = 0,$

$(3x + 2)(2x - 1) = 0,\ 3x + 2 = 0, 3x = -2,$

$x = -\frac{2}{3};\ 2x - 1 = 0, 2x = 1, x = \frac{1}{2};$

$y = 5(-\frac{2}{3})^2 + 2(-\frac{2}{3}),\ y = \frac{20}{9} - \frac{4}{3}, y = \frac{8}{9}$

$y = 5(\frac{1}{2})^2 + 2(\frac{1}{2}), y = \frac{5}{4} + 1,\ y = \frac{9}{4};$

$(-\frac{2}{3}, \frac{8}{9}), (\frac{1}{2}, \frac{9}{4})$

7. Equations are an ellipse and a parabola. Use substitution.

$y = x^2 - 1,\ x^2 = y + 1;\ y + 1 + \frac{y^2}{4} = 1,$

$4y + 4 + y^2 = 4, y^2 + 4y = 0, y(y + 4) = 0$

$y = 0 \text{ or } y + 4 = 0, y = -4;$

$0 = x^2 - 1,\ x^2 = 1, x = \pm 1;$

$-4 = x^2 - 1, x^2 = -3$ not possible, discard extraneous answer. $(\pm 1, 0)$

9. Equations are a circle and a parabola.

$y = x^2 - 4,\ x^2 = y + 4; y + 4 + y^2 = 16,$

$y^2 + y - 12 = 0, (y + 4)(y - 3) = 0;$

$y + 4 = 0, y = -4; y - 3 = 0,\ y = 3;$

$-4 = x^2 - 4, x^2 = 0,\ x = 0;$

$3 = x^2 - 4, x^2 = 7, x = \pm\sqrt{7},$

$(0, -4), (\pm\sqrt{7}, 3)$

Section 8.5 (con't)

11. Equations are both parabolas.

$x^2 - 6 = -x^2 + 2,\ 2x^2 = 8,\ x^2 = 4,$

$x = \pm 2;\ y = 4 - 6,\ y = -2;$

$(\pm 2, -2)$

13. Equations are an ellipse and a parabola.

$y = x^2 - 1,\ x^2 = y + 1;\ \frac{y+1}{4} + y^2 = 1,$

$y + 1 + 4y^2 = 4,\ 4y^2 + y - 3 = 0,$

$(4y - 3)(y + 1) = 0,\ y + 1 = 0,\ y = -1,$

$4y - 3 = 0,\ 4y = 3,\ y = 0.75;$

$-1 = x^2 - 1,\ x^2 = 0,\ x = 0,$

$0.75 = x^2 - 1,\ x^2 = 1.75,\ x = \pm\sqrt{1.75}$

$(0, -1), (\pm\sqrt{1.75}, 0.75)$

15. Equations are a circle and a hyperbola.

Use elimination, add equations.

$(x^2 + x^2) + (y^2 - y^2) = 4 + 4,\ 2x^2 = 8,$

$x^2 = 4,\ x = \pm 2;\ 4 + y^2 = 4,\ y^2 = 0,\ y = 0,$

$(\pm 2, 0)$

17. Equations are a hyperbola and a line.

$x + y = 4,\ y = 4 - x;\ 4 - x = \frac{1}{x},$

$4x - x^2 = 1,\ x^2 - 4x + 1 = 0,$

$$x = \frac{-(-4) \pm \sqrt{(-4)^2 - 4(1)(1)}}{2(1)} = \frac{4 \pm \sqrt{12}}{2},$$

$$x = \frac{4 \pm 2\sqrt{3}}{2},\quad x = 2 \pm \sqrt{3};$$

$y = 4 - (2 + \sqrt{3}),\ y = 2 - \sqrt{3},$

$y = 4 - (2 - \sqrt{3}),\ y = 2 + \sqrt{3};$

$(2 \pm \sqrt{3},\ 2 \pm \sqrt{3})$

19. Equations are a circle and an ellipse.

$x^2 = 9 - y^2,\ \frac{9 - y^2}{16} + \frac{y^2}{4} = 1,$

$9 - y^2 + 4y^2 = 16,\ 3y^2 = 7,\ y^2 = \frac{7}{3},$

$$y = \pm\sqrt{\tfrac{7}{3}},\ y = \pm\frac{\sqrt{7}}{\sqrt{3}},\quad y = \pm\frac{\sqrt{21}}{3};$$

$x^2 = 9 - \frac{7}{3},\ x^2 = \frac{20}{3},\ x = \pm\sqrt{\frac{20}{3}},$

$$x = \pm\frac{2\sqrt{5}}{\sqrt{3}},\quad x = \pm\frac{2\sqrt{15}}{3};$$

$$\left(\pm\frac{2\sqrt{15}}{3},\ \pm\frac{\sqrt{21}}{3}\right)$$

21. A possible window is Xmin = -1.1, Xmax = 0.1, Ymin = -1, Ymax = 0.1

Section 8.5 (con't)

23. $0 = -2(-1)^2 + 2,\ 0 = -2 + 2,\ 0 = 0$

$(-1)^2 + (0 - 2)^2 = 4,\ 1 + 4 = 4,\ 5 \neq 4$

$(1)^2 + (0 - 2)^2 = 4,\ 1 + 4 = 4,\ 5 \neq 4$

The points are not solutions, they do not satisfy both equations.

25. $3^x = 2 - 3^x,\ 2 \cdot 3^x = 2,\ 3^x = 1,\ x = 0;$

$y = 3^0,\ y = 1;\ (0, 1)$

27. $2^{x+2} = 4 + 2^x,\ 2^2 \cdot 2^x = 4 + 2^x,$

$4 \cdot 2^x - 2^x = 4,\ 3 \cdot 2^x = 4,\ 2^x = \frac{4}{3},$

$\log 2^x = \log \frac{4}{3},\ x\log 2 = \log \frac{4}{3},$

$x = \frac{\log \frac{4}{3}}{\log 2},\ x \approx 0.415;\ y \approx 4 + 2^{0.415}$

$y \approx 5.333;\quad (0.415, 5.333)$

29. $\log x = 3 - \log x,\ 2\log x = 3,$

$\log x = 1.5,\ 10^{1.5} = x,\ x \approx 31.6;$

$y = \log 10^{1.5},\ y = 1.5\quad (31.6, 1.5)$

31. $\log_4 x = 2 - \log_4 x,\ 2\log_4 x = 2,$

$\log_4 x = 1,\ 4^1 = x,\ x = 4;\ y = \log_4 4,$

$y = 1;\quad (4, 1)$

33. $2^{x+2} = 2^2 \cdot 2^x = 4 \cdot 2^x\quad 4 \cdot \frac{1}{3} = \frac{4}{3}$

35. $3^{x+1} = 3^1 \cdot 3^x = 3 \cdot 3^x\quad 3 \cdot 2 = 6$

37. $2^{x-1} = 2^{-1} \cdot 2^x = \frac{1}{2} \cdot 2^x\quad \frac{1}{2} \cdot 5 = \frac{5}{2}$

39. $4^{x-2} = 4^{-2} \cdot 4^x = \frac{1}{16} \cdot 4^x\quad \frac{1}{16} \cdot 3 = \frac{3}{16}$

41. $\log_2 x^2 = 1,\ 2^1 = x^2,\ x = \pm\sqrt{2}$, The positive root is the same as the original solution, the negative root must be discarded.

Section 8.6

1. x < 0 and y > 0 is in quadrant 2

3. x < 0 and y < 0 is in quadrant 3

5. y < 0 is in quadrants 3 and 4

7. x < 0 is in quadrants 2 and 3

9. In quadrant 2 x < 0 and y > 0

11. In quadrants 2 and 4 x < 0 and y > 0 or

x > 0 and y < 0

13. 300 ≤ y ≤ 2800 and x ≥ 0

15. y ≤ x and x ≤ 0

17.

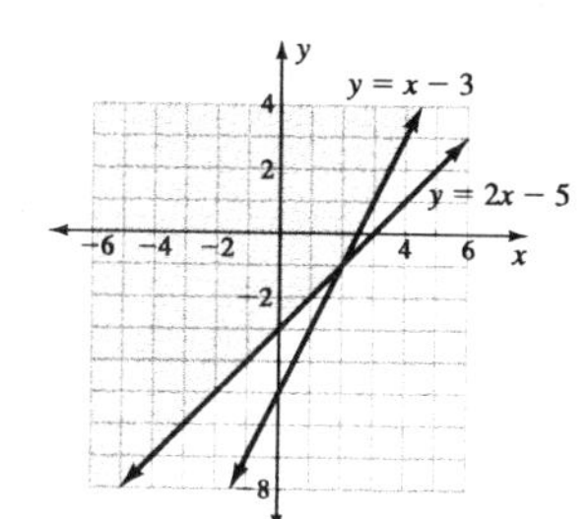

19.

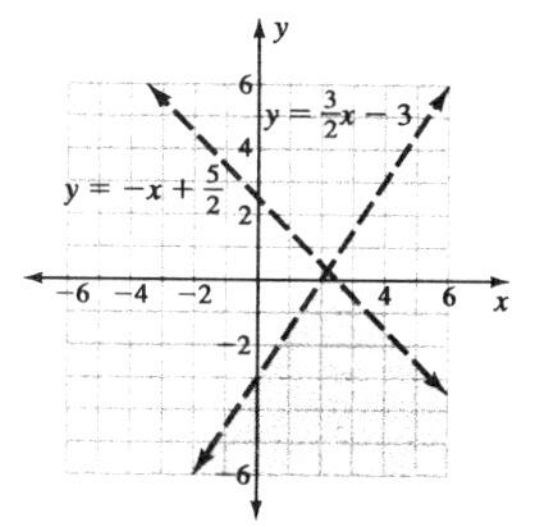

21.

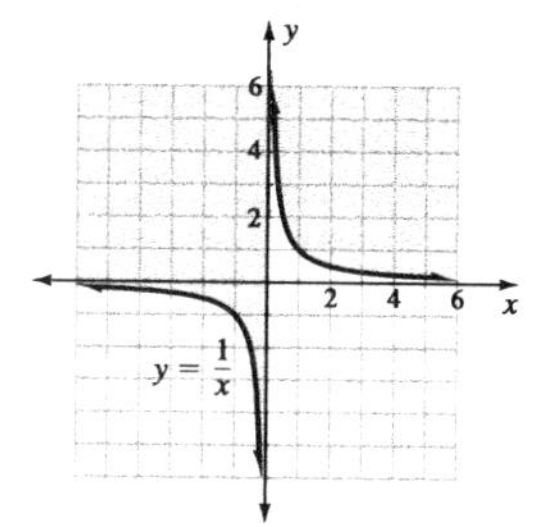

23.

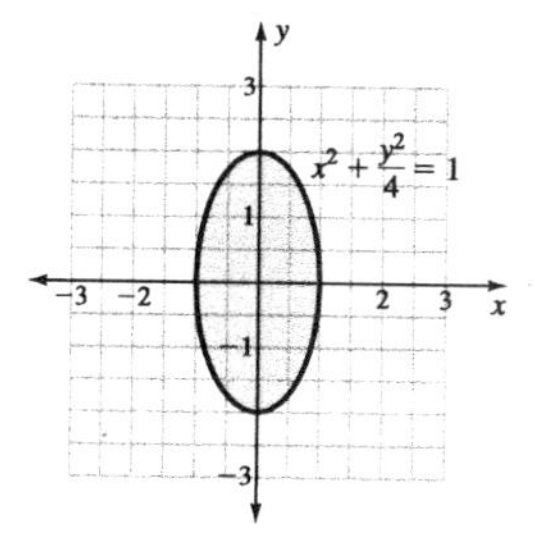

25.

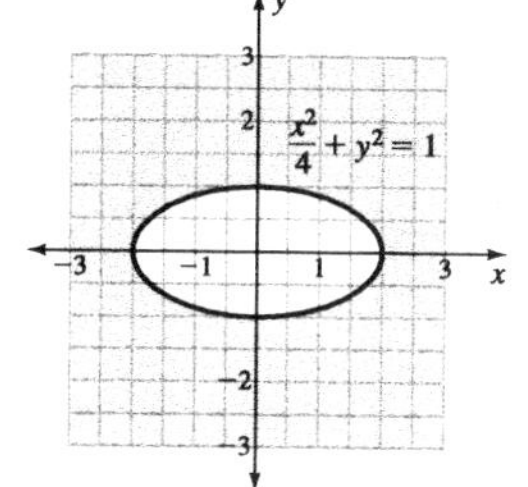

27.

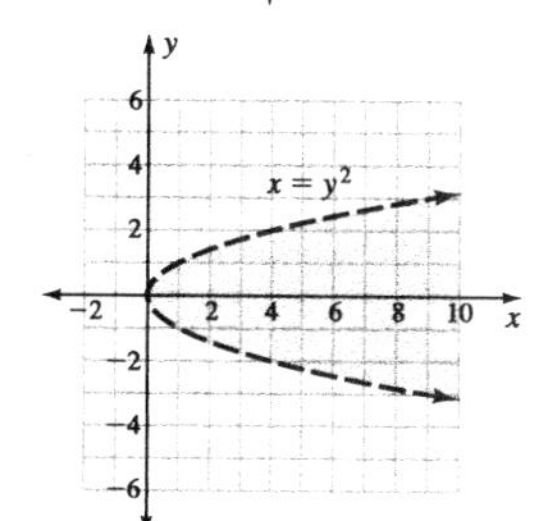

29.

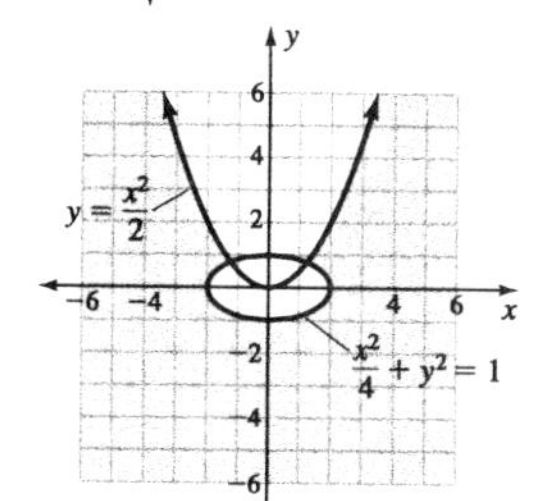

31.

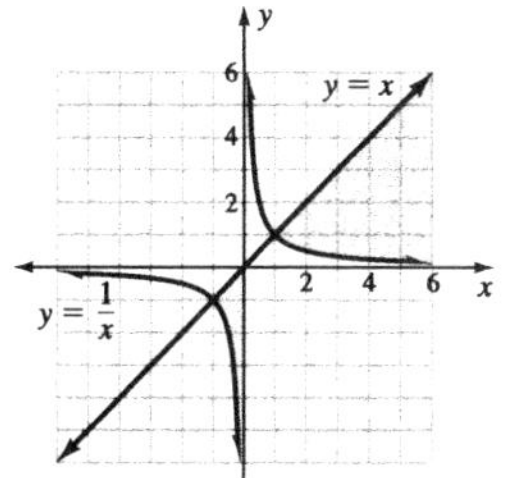

33.

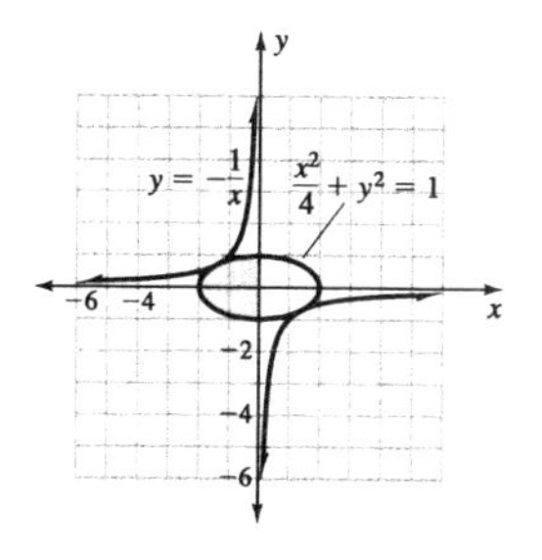

Chapter 8 Review

1. $x + y + z = 12$, $xy = 15$, $yz = 12$;

$x = \frac{15}{y}$, $z = \frac{12}{y}$, $\frac{15}{y} + y + \frac{12}{y} = 12$;

$15 + y^2 + 12 = 12y$, $y^2 - 12y + 27 = 0$,

$(y - 3)(y - 9) = 0$, $y - 3 = 0$, $y = 3$;

$y - 9 = 0$, $y = 9$;

$x = \frac{15}{9}$, $x = 1\frac{2}{3}$; $x = \frac{15}{3}$, $x = 5$;

$z = \frac{12}{9}$, $z = 1\frac{1}{3}$; $z = \frac{12}{3}$, $z = 4$;

{5 in, 4 in, 3 in} or { $1\frac{2}{3}$ in, $1\frac{1}{3}$ in, 9 in }

3. Let x = length, y = width and z = height

$x = y + z$, $y = z + 2$, $xyz = 240$;

$x = (z + 2) + z$, $x = 2z + 2$

$(2z + 2)(z + 2)(z) = 240$;

$(2z^2 + 6z + 4)(z) = 240$;

$2z^3 + 6z^2 + 4z = 240$,

$2z^3 + 6z^2 + 4z - 240 = 0$,

$z^3 + 3z^2 + 2z - 120 = 0$

Guess and check or graph the equation and find the intercept.

$z = 4$, $y = 4 + 2$, $y = 6$, $x = 6 + 4 = 10$

length = 10 in, width = 6 in, height = 4 in

5. $4(3x + 5y) = 4(5)$; $12x + 20y = 20$

$5(2x - 4y) = 5(18)$, $10x - 20y = 90$

$(12x + 10x) + (20y - 20y) = (20 + 90)$

$22x = 110$, $x = 5$;

$3(5) + 5y = 5$, $15 + 5y = 5$, $5y = -10$,

$y = -2$; $(5, -2)$

$$\begin{bmatrix} 3 & 5 \\ 2 & -4 \end{bmatrix}\begin{bmatrix} x \\ y \end{bmatrix} = \begin{bmatrix} 5 \\ 18 \end{bmatrix}, \begin{bmatrix} x \\ y \end{bmatrix} = \begin{bmatrix} 5 \\ -2 \end{bmatrix}$$

7. $1.5x - 3.5y = 19$, $6(1.5x - 3.5y) = 6(19)$,

$9x - 21y = 114$

$14(5.5x + 1.5y) = 14(41)$,

$77x + 21y = 574$

$(9x + 77x) + (-21y + 21y) = (114 + 574)$,

$86x = 688$, $x = 8$;

$1.5(8) = 3.5y + 19$, $12 - 19 = 3.5y$,

$3.5y = -7$, $y = -2$; $(8, -2)$

$$\begin{bmatrix} 1.5 & -3.5 \\ 5.5 & 1.5 \end{bmatrix}\begin{bmatrix} x \\ y \end{bmatrix} = \begin{bmatrix} 19 \\ 41 \end{bmatrix}, \begin{bmatrix} x \\ y \end{bmatrix} = \begin{bmatrix} 8 \\ -2 \end{bmatrix}$$

9. $-3x + 4(\frac{3}{4}x + 2) = 5$, $-3x + 3x + 8 = 5$,

$8 = 5$, false result, no solution

$$\begin{bmatrix} -3 & 4 \\ -\frac{3}{4} & 1 \end{bmatrix}\begin{bmatrix} x \\ y \end{bmatrix} = \begin{bmatrix} 5 \\ 2 \end{bmatrix}, \ \det[A] = 0$$

Chapter 8 Review (con't)

11. x = z - 2, y = 4z - 10;

3(z - 2) - (4z - 10) + z = 4;

3z - 6 - 4z + 10 + z = 4, 4 = 4;

true result, infinite number of solutions

$$\begin{bmatrix} 3 & -1 & 1 \\ -1 & 0 & 1 \\ 0 & -1 & 4 \end{bmatrix}\begin{bmatrix} x \\ y \\ z \end{bmatrix} = \begin{bmatrix} 4 \\ 2 \\ 10 \end{bmatrix}, \quad \det[A] = 0$$

13. Add equations 2 and 3, subtract result from equation 1.

(x + x) + (y - y) + (2z - 3z) = 4 + 6,

2x - z = 10;

(2x - 2x) + y + [- z -(- z)] = 3 - 10

y = -7;

x - 7 + 2z = 4, x + 2z = 11

x - (-7) - 3z = 6, x - 3z = -1

(x - x) + [2z -(-3z)] = 11 - (-1)

5z = 12, z = 2.4;

x - 7 + 2(2.4) = 4, x = 6.2;

(6.2, -7, 2.4)

$$\begin{bmatrix} 2 & 1 & -1 \\ 1 & 1 & 2 \\ 1 & -1 & -3 \end{bmatrix}\begin{bmatrix} x \\ y \\ z \end{bmatrix} = \begin{bmatrix} 3 \\ 4 \\ 6 \end{bmatrix}, \begin{bmatrix} x \\ y \\ z \end{bmatrix} = \begin{bmatrix} 6.2 \\ -7 \\ 2.4 \end{bmatrix}$$

15. $$\begin{bmatrix} 2 & 4 \\ 1 & 3 \end{bmatrix} + \begin{bmatrix} 1 & 2 \\ 4 & 3 \end{bmatrix} = \begin{bmatrix} 2+1 & 4+2 \\ 1+4 & 3+3 \end{bmatrix}$$

$$= \begin{bmatrix} 3 & 6 \\ 5 & 6 \end{bmatrix}$$

17. $$\begin{bmatrix} 2 & 4 \\ 1 & 3 \end{bmatrix}\begin{bmatrix} 1 & 2 \\ 4 & 3 \end{bmatrix} = \begin{bmatrix} 2(1)+4(4) & 2(2)+4(3) \\ 1(1)+3(4) & 1(2)+3(3) \end{bmatrix}$$

$$= \begin{bmatrix} 18 & 16 \\ 13 & 11 \end{bmatrix}$$

19. $$\begin{bmatrix} 1.5 & -2 \\ -0.5 & 1 \end{bmatrix}\begin{bmatrix} 2 & 4 \\ 1 & 3 \end{bmatrix} = \begin{bmatrix} 3-2 & 6-6 \\ -1+1 & -2+3 \end{bmatrix}$$

$$= \begin{bmatrix} 1 & 0 \\ 0 & 1 \end{bmatrix}$$

21. Equation will have the form $x = ay^2$,

$-3 = a(1)^2$, a = -3, $x = -3y^2$

23. Equation will have the form $\frac{x^2}{a^2} + \frac{y^2}{b^2} = 1$,

$$\frac{x^2}{3^2} + \frac{y^2}{2^2} = 1, \ \frac{x^2}{9} + \frac{y^2}{4} = 1$$

25. Equations are a parabola and a line.

$x^2 + x - 2 = -x - 3$, $x^2 + 2x + 1 = 0$,

(x + 1)(x + 1) = 0; x + 1 = 0, x = -1

y = -(-1) - 3, y = -2; (-1, -2)

Chapter 8 Review (con't)

27. Equations are a parabola and a line.

$x^2 - 4x - 1 = -x + 3,\ x^2 - 3x - 4 = 0,$

$(x - 4)(x + 1) = 0,\ x - 4 = 0, x = 4,$

$x + 1 -= 0, x = -1;$

$y = -(4) + 3,\ y = -1,$

$y = -(-1) + 3,\ y = 4$

(4, -1), (-1, 4)

29. Equations are a circle and a parabola.

$x^2 = y - 1,\ y - 1 + y^2 = 16,$

$y^2 + y - 17 = 0;$

$$y = \frac{-1 \pm \sqrt{1^2 - 4(1)(-17)}}{2(1)},\ y = \frac{-1 \pm \sqrt{69}}{2},$$

$y \approx 3.653$ or $y \approx -4.653$

$3.653 \approx x^2 + 1, 2.653 \approx x^2,$

$x \approx \pm\sqrt{2.653},\ x \approx \pm 1.629$

$-4.653 \approx x^2 + 1, -3.653 = x^2$ not possible

(± 1.629, 3.653)

31. Equations are both parabolas.

$x^2 - x = -x^2 + 3,\ 2x^2 - x - 3 = 0,$

$(2x - 3)(x + 1) = 0, x + 1 = 0, x = -1;$

$2x - 3 = 0, 2x = 3, x = 1.5;$

$y = -(-1)^2 + 3,\ y = 2;$

$y = -(1.5)^2 + 3, y = 0.75$

(-1, 2), (1.5, 0.75)

33. Equations are both parabolas.

$x^2 + 1 = x^2 - 5,\ 1 = -5$ false result,

no point of intersection

35. $2^x = 4 - 2^x, 2 \cdot 2^x = 4,\ 2^x = 2, x = 1$

$y = 2^1, y = 2$ (1, 2)

37. $y = 3 - 2y,\ 3y = 3, y = 1;$

$1 = \log x,\ 10^1 = x, x = 10$

39. JN + TJ = 40, JN = TJ + 2;

TJ + 2 + TJ = 40, 2TJ = 38, TJ = 19,

JN = 19 + 2, JN = 21

Jeanne Maiden-Naccarato = 21

Tish Johnson = 19

Chapter 8 Review (con't)

41. c + p + i = 215, p = 2c - 1, p = 47i;

-2c + p = -1; p - 47i = 0;

$$\begin{bmatrix} 1 & 1 & 1 \\ -2 & 1 & 0 \\ 0 & 1 & -47 \end{bmatrix}\begin{bmatrix} c \\ p \\ i \end{bmatrix} = \begin{bmatrix} 215 \\ -1 \\ 0 \end{bmatrix}, \begin{bmatrix} c \\ p \\ i \end{bmatrix} = \begin{bmatrix} 71 \\ 141 \\ 3 \end{bmatrix}$$

71 mg calcium, 141 mg phosphorus and 3 mg iron

43. 4p + 9f + 4c = 184, p + f + c = 26,

f = p + 9, -p + f = 9

$$\begin{bmatrix} 4 & 9 & 4 \\ 1 & 1 & 1 \\ -1 & 1 & 0 \end{bmatrix}\begin{bmatrix} p \\ f \\ c \end{bmatrix} = \begin{bmatrix} 184 \\ 26 \\ 9 \end{bmatrix}, \begin{bmatrix} p \\ f \\ c \end{bmatrix} = \begin{bmatrix} 7 \\ 16 \\ 3 \end{bmatrix}$$

7 g protein, 16 g fat, 29 g carbohydrates

45. t + f + s = 18, 10t + 5f + s = 86,

f = t + s - 2, -t + f - s = -2

$$\begin{bmatrix} 1 & 1 & 1 \\ 10 & 5 & 1 \\ -1 & 1 & -1 \end{bmatrix}\begin{bmatrix} t \\ f \\ s \end{bmatrix} = \begin{bmatrix} 18 \\ 86 \\ -2 \end{bmatrix}, \begin{bmatrix} t \\ f \\ s \end{bmatrix} = \begin{bmatrix} 4 \\ 8 \\ 6 \end{bmatrix}$$

4 \$10 bills, 8 \$5 bills, 6 \$1 bills

47. At the point 2800 feet to the left, x = 0

-0.03 = 2a(0) + b, b = -0.03,

0.04 = 2a(2800) - 0.03, 0.07 = 5600a,

$a = 1.25 \times 10^{-5}$ or $a = \frac{1}{80{,}000}$

$300 = \frac{1}{80{,}000}(2800)^2 - 0.03(2800) + c$,

300 = 98 - 84 + c, c = 286

$a = \frac{1}{80{,}000}$, b = -0.03, c = 286

49.

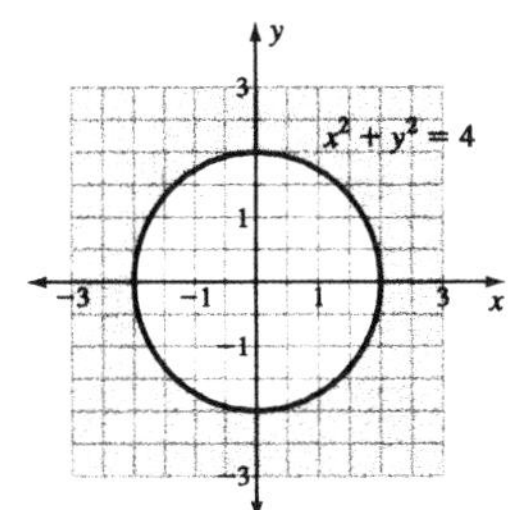

51.

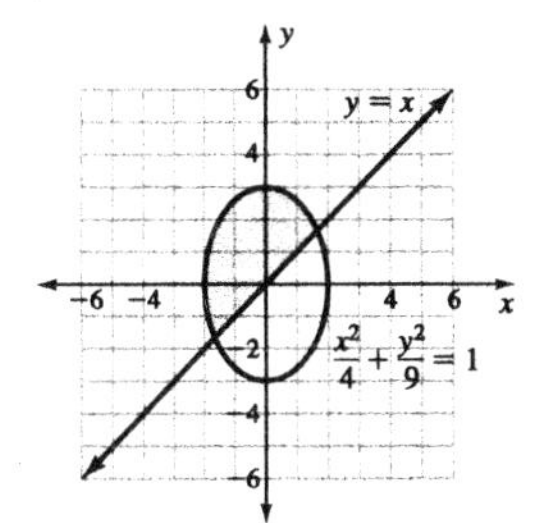

53.

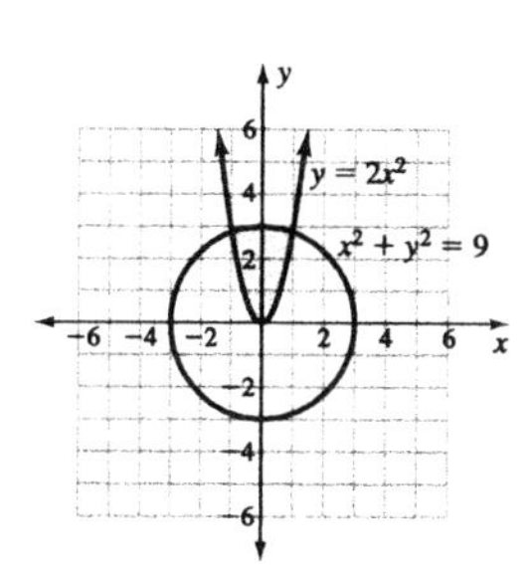

Chapter 8 Test

1. $l + w + h = 24$, $lwh = 240$, $lw = 20$,

$20h = 240$, $h = 12$, $l + w + 12 = 24$,

$l + w = 12$, and $lw = 20$, $l = 10$, $w = 2$;

dimensions are 2 in. by 10 in. by 12 in.

2. $\begin{bmatrix} 2 & -3 \\ 4 & 5 \end{bmatrix}\begin{bmatrix} x \\ y \end{bmatrix} = \begin{bmatrix} 6 \\ -32 \end{bmatrix}, \begin{bmatrix} x \\ y \end{bmatrix} = \begin{bmatrix} -3 \\ -4 \end{bmatrix}$

(-3, -4)

3. $x = -2y - 8$,

$0.5x = -y - 4$, $x = -2y - 8$

Coincident lines, infinite number of solutions.

4. $x + 3z = 4$, $3z = 4 - x$, $z = \frac{4}{3} - \frac{x}{3}$,

$3y = -4x$, $y = -\frac{4}{3}x$

$2x + (-\frac{4}{3}x) + 2(\frac{4}{3} - \frac{x}{3}) = 5$,

$2x - \frac{4}{3}x - \frac{2}{3}x + \frac{8}{3} = 5$, $\frac{8}{3} = 5$, false result, no point of intersection

5. $\begin{bmatrix} 3 & 1 & 1 \\ 2 & -1 & -2 \\ -1 & -1 & 3 \end{bmatrix}\begin{bmatrix} x \\ y \\ z \end{bmatrix} = \begin{bmatrix} 8 \\ 17 \\ -9 \end{bmatrix}, \begin{bmatrix} x \\ y \\ z \end{bmatrix} = \begin{bmatrix} 4.5 \\ -3 \\ -2.5 \end{bmatrix}$

(4.5, -3, -2.5)

6. Equations are a parabola and a line.

$3x^2 - 4x - 5 = x - 3$, $3x^2 - 5x - 2 = 0$,

$(3x + 1)(x - 2) = 0$, $x - 2 = 0$, $x = 2$,

$3x + 1 = 0$, $x = \frac{1}{3}$;

$y = 2 - 3$, $y = -1$; $y = \frac{1}{3} - 3$, $y = -\frac{10}{3}$

7. Equations are a parabola and a line.

$x^2 - 3x + 2 = 0.5x - 2$, $x^2 - 3.5x + 4 = 0$,

$x = \frac{-(-3.5) \pm \sqrt{(-3.5)^2 - 4(1)(4)}}{2(1)}$,

$x = \frac{3.5 \pm \sqrt{-3.75}}{2}$, not a real number, no point of intersection

8. Equations are a circle and a hyperbola.

$y^2 = x^2 + 4$, $x^2 + (x^2 + 4) = 4$, $2x^2 = 0$,

$x = 0$; $y^2 = 4$, $y = \pm 2$ $(0, \pm 2)$

9. Equations are a parabola and a circle.

$x^2 = y - 1$, $(y - 1) + y^2 = 4$, $y^2 + y - 5 = 0$

$y = \frac{-1 \pm \sqrt{1^2 - 4(1)(-5)}}{2(1)}$,

$y = \frac{-1 \pm \sqrt{21}}{2}$, $y \approx 1.791$ or $y \approx -2.791$

$1.791 \approx x^2 + 1$, $x^2 \approx 0.791$, $x \approx \pm 0.889$,

$-2.791 \approx x^2 + 1$, $-1.791 - x^2$, not possible

$(\pm 0.889, 1.791)$

Chapter 8 Test (con't)

10. $3^x = 5 - 3^x$, $2 \cdot 3^x = 5$, $3^x = 2.5$,

$\log 3^x = \log 2.5$, $x\log 3 = \log 2.5$,

$x = \dfrac{\log 2.5}{\log 3}$, $x \approx 0.834$,

$y = 3^x$, $y = 2.5$ (0.834, 2.5)

11. $y = 4 - y$, $2y = 4$, $y = 2$;

$2 = \log_3 x$, $3^2 = x$, $x = 9$ (9, 2)

12. $C + 10 = P$, $C + P = 3922$,

$C + C + 10 = 3922$, $2C = 3912$,

$C = 1956$, $P = 1956 + 10$, $P = 1966$

Carol Heiss 1956, Peggy Fleming 1966

13. $p + f + c = 26$, $4p + 9f + 4c = 199$,

$p = c - 3$, $p - c = -3$;

$$\begin{bmatrix} 1 & 1 & 1 \\ 4 & 9 & 4 \\ 1 & 0 & -1 \end{bmatrix}\begin{bmatrix} p \\ f \\ c \end{bmatrix} = \begin{bmatrix} 26 \\ 199 \\ -3 \end{bmatrix}, \begin{bmatrix} p \\ f \\ c \end{bmatrix} = \begin{bmatrix} 2 \\ 19 \\ 5 \end{bmatrix}$$

2 g protein, 19 g fat, 5 g carbohydrates

14.

Q	V	QV
w	1	w
a	0.84	0.84a
400	0.986	400(0.986)

$w + a = 400$, $w + 0.84a = 394.4$,

$w = 400 - a$, $400 - a + 0.84a = 394.4$,

$-0.16a = -5.6$, $a = 35$, $w = 400 - 35$,

$w = 365$,

365 gallons of water, 35 gallons of polyvinyl solution

15. $p + n + q = 59$, $q = n + 10$, $-n + q - 10$

$0.01p + 0.05n + 0.24q = 6.35$,

$$\begin{bmatrix} 1 & 1 & 1 \\ 0 & -1 & 1 \\ 0.01 & 0.05 & 0.25 \end{bmatrix}\begin{bmatrix} p \\ n \\ q \end{bmatrix} = \begin{bmatrix} 59 \\ 10 \\ 6.35 \end{bmatrix}, \begin{bmatrix} p \\ n \\ q \end{bmatrix} = \begin{bmatrix} 25 \\ 12 \\ 22 \end{bmatrix}$$

25 pennies, 12 nickels, 22 quarters

16. $j + a + c = 11$, $17j + 8a + 16c = 158$,

$0.89j + 0.59a + 3.98c = 15.07$

$$\begin{bmatrix} 1 & 1 & 1 \\ 17 & 8 & 16 \\ 0.89 & 0.59 & 3.98 \end{bmatrix}\begin{bmatrix} j \\ a \\ c \end{bmatrix} = \begin{bmatrix} 11 \\ 158 \\ 15.07 \end{bmatrix}, \begin{bmatrix} j \\ a \\ c \end{bmatrix} = \begin{bmatrix} 6 \\ 3 \\ 2 \end{bmatrix}$$

6 juice, 3 apples, 2 packages of cookies

Chapter 8 Test (con't)

17. -0.03 = 2a(0) + b, b = -0.03

0.06 = 2a(3200) - 0.03, 0.09 = 6400a,

a = $\frac{9}{640{,}000}$,

1500 = a(0) + b(0) + c, c = 1500

a = $\frac{9}{640{,}000}$, b = -0.03, c = 1500

18. a. $\begin{bmatrix} 3 & -5 \\ -1 & 2 \end{bmatrix}\begin{bmatrix} 2 & 5 \\ 1 & 3 \end{bmatrix} = \begin{bmatrix} 6-5 & 15-15 \\ -2+2 & -5+6 \end{bmatrix}$

$= \begin{bmatrix} 1 & 0 \\ 0 & 1 \end{bmatrix}$

b. $[A]^{-1} = \begin{bmatrix} 2 & 5 \\ 1 & 3 \end{bmatrix}$

c. det[B] = 1

19. $\begin{bmatrix} 2 & -3 \\ 4 & 5 \end{bmatrix}\begin{bmatrix} x \\ y \end{bmatrix} = \begin{bmatrix} 6 \\ -32 \end{bmatrix}, \begin{bmatrix} x \\ y \end{bmatrix} = \begin{bmatrix} -3 \\ -4 \end{bmatrix}$

20.

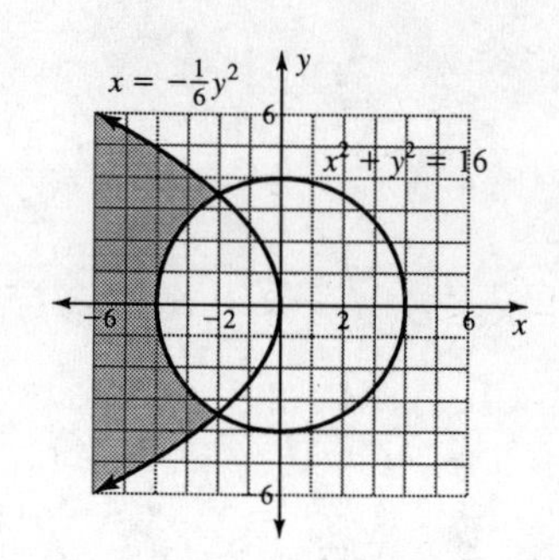

Final Exam Review Part 1

1. $-(-2)^2 = -4$

3. $\dfrac{xy^2}{x^2y} = \dfrac{y}{x}$

5. x - 6 + 6 - 3x = 8, -2x = 8, x = -4

7. $3 - x \le 8,\ 3 \le 8 + x,\ -5 \le x$

9. The -4 in -4xyz is the *coefficient* of xyz.

11. A line parallel to the x-axis may be described by *a constant* function.

13. y = -2x + 3 is a decreasing function.

15. The range of the absolute value function is $f(x) \ge 0$.

17. Using (-40, -40) and (100, 212);

$$m = \frac{212-(-40)}{100-(-40)},\quad m = \frac{9}{5},$$

$b = 32,\ y = \frac{9}{5}x + 32$

19. $f(a + b) = 2(a + b)^2 = 2(a^2 + 2ab + b^2)$

$= 2a^2 + 4ab + 2b^2$

21. *Parabola* is the name given to the graph of a quadratic equation, $y = ax^2 + bx + c$.

23. The Pythagorean theorem formula is $a^2 + b^2 = c^2$.

25. $\left(x+\frac{5}{2}\right)^2 = x^2 + 5x + \frac{25}{4}$

27. Some possible answers are π, e, $\sqrt{2}$, $\sqrt{3}$

29. $(2-i)(3+i) = 6 + 2i - 3i - i^2$

$= 6 - i - (-1) = 7 - i$

31. $\sqrt{-16} = \sqrt{16}\sqrt{-1} = 4i$

33. $x = \dfrac{-2 \pm \sqrt{2^2 - 4(1)(5)}}{2(1)},\ x = \dfrac{-2 \pm \sqrt{-16}}{2}$

$x = \dfrac{-2 \pm 4i}{2},\ x = -1 \pm 2i$

35. Largest exponent on x is 4, there are a maximum of 4 possible solutions.

37. $x^3 - x = x(x^2 - 1) = x(x + 1)(x - 1)$

$x^2 + x = x(x + 1)$; the missing expression is x - 1. Others combinations are possible.

39. $\dfrac{25}{11} = \dfrac{27}{x},\ 25x = 297,\ x \approx 11.9$

$\dfrac{25}{11} = \dfrac{y}{9},\ 11y = 225,\ y \approx 20.5$

41. $I = \dfrac{P}{r^2},\ r^2 = \dfrac{P}{I},\ r = \sqrt{\dfrac{P}{I}}$

43. $\dfrac{12x}{12} + \dfrac{24x}{3x} = x + 8$

45. $\dfrac{a(a+2)}{2(a-3)} \cdot \dfrac{(a+4)(a-3)}{(a+2)} = \dfrac{a(a+4)}{2}$

Final Exam Review Part 1 (con't)

47. $\frac{x-2}{(x-2)(x-1)} - \frac{2x(x-1)}{(x-2)(x-1)}$

$= \frac{x-2-2x^2+2x}{(x-2)(x-1)} = \frac{-2x^2+3x-2}{(x-2)(x-1)}$

$= -\frac{2x^2-3x+2}{(x-2)(x-1)}$

49. $x^2 + 2x^2 - 3x + 4 \div x + 1$

$$\begin{array}{r} x^2 + x - 4 \\ x+1\overline{)x^3 + 2x^2 - 3x + 4} \\ \underline{-(x^3 + x^2)} \\ x^2 - 3x \\ \underline{-(x^2 + x)} \\ -4x + 4 \\ \underline{-(-4x-4)} \\ 8 \end{array}$$

$= x^2 + x - 4 + \frac{8}{x+1}$

51. Defined for all real numbers,

$12(x - 5) = 10(x + 3)$,

$12x - 60 = 10x + 30$, $2x = 90$, $x = 45$

53. Defined for all real numbers $x \neq 2$,

$1(x + 3) = 2(x - 2)$, $x + 3 = 2x - 4$,

$x = 7$

55. Defined for all real numbers $x \neq 0, 4$,

$3x(x-4)\left(\frac{2}{x-4} - \frac{1}{x}\right) = \left(\frac{x+4}{3x}\right)3x(x-4)$,

$6x - 3(x - 4) = (x + 4)(x - 4)$,

$6x - 3x + 12 = x^2 - 16$, $x^2 - 3x - 28 = 0$,

$(x - 7)(x + 4) = 0$, $x - 7 = 0$, $x = 7$,

$x + 4 = 0$, $x = -4$ $\{-4, 7\}$

57. $\sqrt[3]{8x^6y^{12}} = \sqrt[3]{2^3(x^2)^3(y^4)^3} = 2x^2y^4$

59. 5.67×10^{-7}

61. $d = \sqrt{(x_2 - x_1)^2 + (y_2 - y_1)^2}$

63. If $3^y = 1$ then $y = 0$.

65. $(3x + 2)^2 = (3x)^2 + 2(2)(3x) + 2^2$

$= 9x^2 + 12x + 4$

67. $\left(\sqrt{3}+2\right)^2 = (\sqrt{3})^2 + 2(2)(\sqrt{3}) + 2^2$

$= 3 + 4\sqrt{3} + 4 = 7 + \sqrt{3}$

69. a. $x^{1.5} = x^{\frac{3}{2}} = \sqrt{x^3} = (\sqrt{x})^3$

b. $x^{\frac{n}{m}} = \sqrt[m]{x^n} = (\sqrt[m]{x})^n$

c. $x^{\frac{1}{2}}(y^2)^{\frac{1}{4}} = x^{\frac{1}{2}}y^{\frac{1}{2}} = \sqrt{x}\sqrt{y} = \sqrt{xy}$

71. a. $\sqrt[3]{a^3b^4c^5} = abc\sqrt[3]{bc^2}$

b. $\sqrt[4]{w^4x^5y^8z^{12}} = |wz^3|xy^2\sqrt[4]{x},\ x \geq 0$

Final Exam Review Part 1 (con't)

73. If $a^x = a^y$ and $a > 0$, $a \neq 1$, then $x = y$.

75. $16^x = 8$, $(2^4)^x = 2^3$, $4x = 3$, $x = \frac{3}{4}$

77. $0.01^x = 100$, $(10^{-2})^x = 10^2$, $-2x = 2$, $x = -1$

79. $9^{x+1} = 27^{-1}$, $(3^2)^{x+1} = (3^3)^{-1}$, $2x + 2 = -3$,

$2x = -5$, $x = -\frac{5}{2}$

81. $10^y = x$

83. $b^c = a$

85. $y = \dfrac{\log_{10} a}{\log_{10} b}$

87. The base in $2^6 = 64$ is 2.

89. When we write a logarithm without a base, we mean log base 10.

91. Calculating y = antilog 2.3 means raising 10 to the power of 2.3.

93. $\log 2^{x+1} = \log 19$, $(x + 1)\log 2 = \log 19$,

$x + 1 = \dfrac{\log 19}{\log 2}$, $x = \dfrac{\log 19}{\log 2} - 1$, $x \approx 3.248$

95. $\log 17 = x$, $x \approx 1.230$

97. $2^1 = x$, $x = 2$

99. $\log 2 = x$, $x \approx 0.301$

101. $\log 1000 = x$, $x = 3$

103. $\log_9 27 = x$, $9^x = 27$, $3^{2x} = 3^3$, $2x = 3$,

$x = 1.5$

105. $\ln x = 2$, $e^2 = x$, $x \approx 7.389$

107. $2500 = 1000(1 + \frac{0.094}{2})^{2t}$, $2.5 = (1.047)^{2t}$,

$\log 2.5 = \log 1.047^{2t}$,

$\log 2.5 = 2t\log 1.047$, $t = \dfrac{\log 2.5}{2\log 1.047}$,

$t \approx 9.98$ years

109. $2x - 3(79 - x) = 28$, $2x - 237 + 3x = 28$,

$5x = 265$, $x = 53$; $y = 79 - 53$, $y = 26$

(53, 26)

111.

$$\begin{bmatrix} 2 & 3 & -1 \\ 3 & -4 & 1 \\ 4 & -1 & 2 \end{bmatrix} \begin{bmatrix} x \\ y \\ z \end{bmatrix} = \begin{bmatrix} 10.7 \\ 4.2 \\ 19.6 \end{bmatrix}, \begin{bmatrix} x \\ y \\ z \end{bmatrix} = \begin{bmatrix} 3.5 \\ 2.6 \\ 4.1 \end{bmatrix}$$

113. A system of two equations that has an infinite number of solutions might contain *coincident lines.*

115. The equations are a circle and a hyperbola. $x^2 = y^2 + 7$,

$y^2 + 7 + y^2 = 11$, $2y^2 = 4$, $y^2 = 2$,

$y = \pm\sqrt{2}$; $x^2 = 2 + 7$, $x = \pm 3$

$(\pm 3, \pm\sqrt{2})$

117.

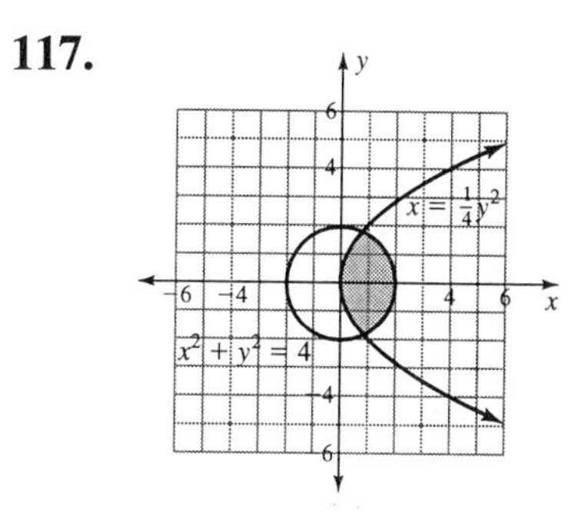

Final Exam Review Part 2

1. a. $h = 0.8(200 - 100)$, $h = 80$;

$h = 0.8(200 - 150)$, $h = 40$;

$h = 0.8(200 - 200)$, $h = 0$

$h = 0.8(200 - 250)$, $h = -40$, not possible

b. $y = a + 0.8(200 - a)$ for $a < 200$, $y = a$

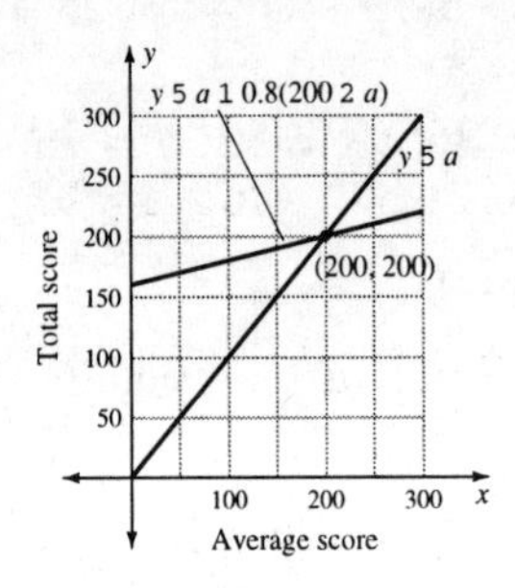

3. Start by finding the vertex.

$x = \dfrac{-2}{2(-4)}$, $x = 0.25$,

$y = -4(0.25)^2 + 2(0.25) - 5$, $y = -4.75$,

a is negative so graph turns downward.

Domain is all real numbers;

Range is $y \leq -4.75$;

Axis of symmetry is $x = 0.25$;

Vertex is at $(0.25, -4.75)$.

y coordinate of vertex determines highest or lowest point in the range, x coordinate of the vertex is on the line of symmetry.

5. $5x^2 + 8x - 4 = 0$, $(5x - 2)(x + 2) = 0$,

$5x - 2 = 0$, $5x = 2$, $x = 0.4$;

$x + 2 = 0$, $x = -2$; $\{-2, 0.4\}$

$x = \dfrac{-8 \pm \sqrt{8^2 - 4(5)(-4)}}{2(5)}$, $x = \dfrac{-8 \pm \sqrt{144}}{10}$,

$x = \dfrac{-8 \pm 12}{10}$, $x = \dfrac{-4 \pm 6}{5}$, $x = 0.4$, $x = -2$

$\{-2, 0.4\}$

Third method is to graph the equation and locate the x-intercept points.

7. $a = 1$, $b = 2$, $60 = -4(1)(c)$, $c = -15$

$y = x^2 + 2x - 15$; $x = -5$ or $x = 3$;

solutions indicate there are 2 intercepts on the horizontal axis.

9. Using quadratic regression:

$y = x^2 + 2x - 5$

11. Equations will have the form:

$a(x + 2)(x - 3) = 0$ where a is any real number.

Final Exam Review Part 2 (con't)

13. $x^2 - 3x - 10$ is an expression;

$y = x^2 - 3x - 10$ is an equations with a parabolic graph, an infinite number of ordered pairs will make it true.

$0 = x^2 - 3x - 10$ is an equation with 2 solutions; $x = -2$ and $x = 5$; these solutions are the x-intercepts for the previous equation.

15. *a* determines whether the graph turns up or down and contributes to the position of the vertex.

17. $(-3 - 2i)^2 + 6(-3 - 2i) + 13 = 0$,

$9 + 12i - 4 - 18 - 12i + 13 = 0$,

$22 - 22 = 0, 0 = 0$, $-3 - 2i$ is a solution; the conjugate $-3 + 2i$ will also be a solution.

19. $$\frac{y+\frac{1}{y}}{y-\frac{1}{y}} = \frac{\frac{y^2+1}{y}}{\frac{y^2-1}{y}} = \frac{y^2+1}{y}\cdot\frac{y}{y^2-1} = \frac{y^2+1}{y^2-1}$$

21. $$\left(\frac{x}{2}+\frac{3}{x}\right)^{-1} = \left(\frac{x^2}{2x}+\frac{6}{2x}\right)^{-1} = \left(\frac{x^2+6}{2x}\right)^{-1} = \frac{2x}{x^2+6}$$

23. All answers are correct, the answer in part d contains no fractions in the numerator or denominator and is a single term, this is the preferred solution.

25. $\left(\sqrt{3x-11}\right)^2 = 1^2$, $3x-11=1$, $3x=12$,

$x = 4$

27. $\left(\sqrt{x-9}\right)^2 = \left(9-\sqrt{x}\right)^2$,

$x-9 = 81-18\sqrt{x}+x$, $18\sqrt{x} = 90$, $\sqrt{x} = 5$,

$x = 25$

29. $2 = (1 + r)^{10}$, $\log 2 = \log (1 + r)^{10}$,

$\log 2 = 10\log (1 + r)$, $\frac{\log 2}{10} = \log(1 + r)$,

$\log(1 + r) \approx 0.0301$, $10^{0.0301} \approx 1 + r$,

$r \approx 10^{0.0301} - 1$, $r \approx 0.072$ or 7.2%

Using the same method for 8 years;

$\frac{\log 2}{8} = \log(1 + r)$, $\log(1 + r) \approx 0.0376$,

$10^{0.0376} - 1 \approx r$, $r \approx 0.09$ or 9 %;

And for 5.375 years:

$\frac{\log 2}{5.375} = \log(1 + r)$, $\log(1 + r) \approx 0.056$,

$10^{0.056} - 1 \approx r$, $r \approx 0.138$ or 13.8%

Final Exam Review Part 2 (con't)

31. Common ratio $\frac{1}{3}$, geometric sequence;

$a_n = 27(\frac{1}{3})^{n-1}$; $y = 81(\frac{1}{3})^x$;

$27(\frac{1}{3})^{-1}(\frac{1}{3})^n = 27(3)(\frac{1}{3})^n = 81(\frac{1}{3})^n$

$S_{10} = \dfrac{27(1-(\frac{1}{3})^{10})}{1-\frac{1}{3}}$, $S_{10} \approx 40.4993$

$S_\infty = \dfrac{27}{1-\frac{1}{3}}$, $S_\infty = 40.5$

33. First differences; 5, 9, 13, 17

Second differences; 4, 4, 4

Quadratic sequence, $y = 2x^2 - x$

35. $y \approx 175{,}200\,(1.206)^x$, where x is years since 1945. Annual growth rate ≈ 20.6%

37. J + F = 3924, F = J + 52;

J + J + 52 = 3924, 2J = 3872, J = 1936;

F = 1936 + 52, F = 1988

Jesse Owen 1936,

Florence Griffith Joyner 1988

39. $-0.05 = 2a(0) + b$, $b = -0.05$

$0.04 = 2a(3200) - 0.05$, $0.09 = 6400a$,

$a = \frac{9}{640{,}000}$;

$1000 = \frac{9}{640{,}000}(3200)^2 - 0.05(3200) + c$,

$1000 = 144 - 160 + c$, $c = 1016$;

$a = \frac{9}{640{,}000}$, $b = -0.05$, $c = 1016$

Appendix 2 Arithmetic Sequences

1. First find a_{20}, $a_n = 18 + (n - 1)(-2)$,

$a_n = 20 - 2n$; $a_{20} = 20 - 2(20)$, $a_{20} = -20$;

$S_{20} = \frac{20}{2}[18 + (-20)]$, $S_{20} = -20$

3. $a_{20} = 10 + (20 - 1)8$, $a_{20} = 162$;

$S_{20} = \frac{20}{2}(10 + 162)$, $S_{20} = 1720$

5. This is not an arithmetic sequence

7. The sequence is not arithmetic, pairs have different sums.

9. $113 = 47 + (n - 1)$, $113 = 46 + n$, $n = 67$

Dominique has 67 seats.

11. $2000 = 1896 + (n - 1)4$,

$2000 = 1896 + 4n - 4$; $4n = 108$, $n = 27$

There should have been 27 games.

Games were canceled in 1916 for W.W.I and in 1940 and 1944 for W.W.II

13. $145 = 1 + (n - 1)3$, $145 = 1 + 3n - 3$,

$3n = 147$, $n = 49$ tests

Appendix 2 Geometric Sequences

17. $r = 3,\ S_{10} = \dfrac{9(1-3^{10})}{1-3},\ S_{10} = 265{,}716$

ratio is not between 0 and 1, infinite sum is not appropriate.

19. $r = 0.5,\ S_{10} = \dfrac{64(1-0.5^{10})}{1-0.5},$

$S_{10} = 127.875;$

$S_{\infty} = \dfrac{64}{1-0.5},\ S_{\infty} = 128$

21. $r = 3,\ S_{10} = \dfrac{3(1-3^{10})}{1-3},\ S_{10} = 88{,}572,$

ratio is not between 0 and 1, infinite sum is not appropriate.

23. $S_{20} = \dfrac{108[1-(\frac{2}{3})^{20}]}{1-\frac{2}{3}},\ S_{20} \approx 323.9$ in

25. $S_{20} = \dfrac{135[1-(\frac{2}{3})^{20}]}{1-\frac{2}{3}},\ S_{20} \approx 404.88$ in

27. $r = 0.5,\ S_{\infty} = \dfrac{1}{1-0.5},\ S_{\infty} = 2$

29. $S_{\infty} = \dfrac{18}{1-\frac{5}{6}},\ S_{\infty} = 108$ ft

Appendix 3

1. $(x + 1)^4 = (x + 1)^2(x + 1)^2$

$= (x^2 + 2x + 1)(x^2 + 2x + 1)$

	x^2	$+2x$	$+1$
x^2	x^4	$+2x^3$	$+x^2$
$+2x$	$+2x^3$	$+4x^2$	$+2x$
$+1$	$+x^2$	$+2x$	$+1$

$= x^4 + 4x^3 + 6x^2 + 4x + 1$

3. $(1 - 3a)^4 = (1 - 3a)^2(1 - 3a)^2$

$= (1 - 6a + 9a^2)(1 - 6a + 9a^2)$

	1	$-6a$	$+9a^2$
1	1	$-6a$	$+9a^2$
$-6a$	$-6a$	$+36a^2$	$-54a^3$
$+9a^2$	$+9a^2$	$-54a^3$	$+81a^4$

$= 1 - 12a + 54a^2 - 108a^3 + 81a^4$

5. Continuing from where the text leaves off;

1, 8, 28, 56, 70, 56, 28, 8, 1

1, 9, 36, 84, 126, 126, 84, 36, 9, 1

1, 10, 45, 120, 210, 252, 210, 120, 45, 10, 1

7. $126a^5b^4 + 126a^4b^5 + 84a^3b^6 + 36a^2b^7 + 9ab^8 + b^9$

9. $70a^4b^4 - 56a^3b^5 + 28a^2b^6 - 8ab^7 + b^8$

11. $(a + b)^2 = a^2 + 2ab + b^2$ from the 2nd row of Pascal's triangle. The table contains 4 terms.

13. $(x - y)^3 = x^3 - 3x^2y + 3xy^2 - y^3$

15. $(x + y)^6 = x^6 + 6x^5y + 15x^4y^2 + 20x^3y^3 + 15x^2y^4 + 6xy^5 + y^6$

17. $(b - 1)^4 = b^4 - 4b^3 + 6b^2 - 4b + 1$

19. $(1 + z)^5 = 1 + 5z + 10z^2 + 10z^3 + 5z^4 + z^5$

21. $(x + 3)^4 = x^4 + 4x^3 3 + 6x^2 3^2 + 4x3^3 + 3^4$

$= x^4 + 12x^3 + 54x^2 + 108x + 81$

23. $(x - 3y)^3 = x^3 - 3x^2 3y + 3x3^2y^2 - 3^3y^3$

$= x^3 - 9x^2y + 27xy^2 - 27y^3$

25. $2^4 = 16, 2^5 = 32, 2^6 = 64, 2^7 = 128,$

$2^8 = 256, 2^9 = 512, 2^{10} = 1024$

The powers of 2 correspond to the sums of the rows in Pascal's triangle.

27. $11^0 = 1, 11^1 = 11, 11^2 = 121, 11^3 = 1331,$

$11^4 = 14641, 11^5 = 161051, 11^6 = 1771561$; The digits in 11^0 through 11^4 match the rows in Pascal's triangle.